A New Idea Each Morning

How food and agriculture came together in one international organisation

A New Idea Each Morning

How food and agriculture came together in one international organisation

Wendy Way

E PRESS

Published by ANU E Press
The Australian National University
Canberra ACT 0200, Australia
Email: anuepress@anu.edu.au
This title is also available online at http://epress.anu.edu.au

---

National Library of Australia Cataloguing-in-Publication entry

Author:	Way, Wendy.
Title:	A new idea each morning : how food and agriculture came together in one international organisation / Wendy Way.
ISBN:	9781922144102 (pbk.) 9781922144119 (ebook)
Notes:	Includes bibliographical references and index.
Subjects:	International organization. Intellectual cooperation. Food-supply--International cooperation. Agriculture--International cooperation.
Dewey Number:	327.17

All rights reserved. No part of this publication may be reproduced, stored in a retrieval system or transmitted in any form or by any means, electronic, mechanical, photocopying or otherwise, without the prior permission of the publisher.

Cover design and layout by ANU E Press

This edition © 2013 ANU E Press

Contents

Illustration vii
Preface and Acknowledgments xi
List of Figures xv
List of Tables xvii
Imperial Currency and Weights xix
Abbreviations xxi
Introduction 1

Part A: Visions of Empire

Prologue 7
1. Renmark 15
2. Working with Bruce 37
3. A Vision of Empire 61
4. The Vision Ends 93

Part B: The League of Nations

Prologue 123
5. The Wheat Crisis of the 1930s 129
6. 'The Marriage of Health and Agriculture' 153
7. A Vision for the League of Nations 175

Part C: A New Deal for the World

Prologue 211
8. The War of Ideas 217
9. 'A Keen Outsider' 243
10. 'A New Era in the World's History' 271
Afterword 303
Appendix 1 307
Appendix 2 309
Bibliography 315
Index 333

Figure 1 F. L. McDougall in Vienna, 1930s.

Source: E. McDougall.

'McDougall brings me a new idea every morning.'

— S. M. Bruce

For John

Preface and Acknowledgments

It is now some 30 years since I joined the Historical Documents Section of the Department of Foreign Affairs, to find that my first major task would be to edit a volume of letters written throughout the 1920s by Frank Lidgett McDougall to Australian Prime Minister Stanley Melbourne Bruce.[1] I knew a little of Bruce, having read some of his papers concerning his wartime work as High Commissioner in London. The name McDougall had cropped up occasionally in those papers, attached to cables about wheat, but I knew nothing more of him.

W. J. (Bill) Hudson, Editor of Historical Documents in the Department of Foreign Affairs, had begun a series of volumes on early Australian foreign policy by publishing R. G. Casey's letters reporting to Bruce on political events in London.[2] Having discovered that the Bruce papers included a similar file of letters from McDougall on trade and economic matters, he thought they would make a valuable complementary volume. He wrote to McDougall's daughter Elisabeth, then working in the British Foreign Office, that 'the Bruce papers now make it clear that much of Bruce's most interesting and constructive ideas in fact were your father's or were stimulated by him'. Elisabeth welcomed the proposal for publication and offered to have what papers she possessed copied at Australia House; a set was made for our use and another for the National Library of Australia.[3]

As we worked on the letters, my colleagues and I were awed by the energy and devotion McDougall brought to his work, and saddened to think that he remained so little known in Australia. In 1984 Bill Hudson suggested I write a full biography of McDougall for publication. Work on that progressed slowly in my spare time, but biennial conferences of editors of foreign policy documents, inaugurated by our British counterparts in 1989, provided brief but essential opportunities for research in overseas archives.

Continuing a project over so many years involves difficulties and inefficiencies, but there are compensations. As I worked through a mass of collected material, notes in familiar handwriting reminded me of the help of past colleagues and friends. I particularly want to acknowledge the hard work and enthusiasm of those who worked with me on *Letters from 'A Secret Service Agent'*: Kate Birrell, Rhonda Piggott and Ken Fower. The many able scholars who worked

---

[1] W. J. Hudson and Wendy Way, eds, *Letters from 'A Secret Service Agent': F. L. McDougall to S. M. Bruce 1924–1929*, [hereinafter LFSSA] Australian Government Publishing Service, Canberra, 1986.
[2] W. J. Hudson and Jane North, eds, *My Dear PM: R. G. Casey's Letters to S. M. Bruce 1924–1929*, Australian Government Publishing Service, 1980.
[3] NLA, MS6890. Letters from Hudson to E. McDougall, 1 August, and reply, 30 August 1979, are now in my possession.

in the Department of Foreign Affairs/Department of Foreign Affairs and Trade (DFA/DFAT) Historical Documents Section over the years noted useful material as they trawled through the archives, discussed the project and generally encouraged me. For this support, I thank Pamela Andre, Frank Bongiorno, Damien Browne, Barbara Cooper, Kathleen Dermody, Philip Dorling, Jeffrey Grey, John Hoffmann, David Hood, Ashton Robinson and Martin Sharp, former DFAT archivists Leith Douglas, Elizabeth Nathan and the late Chris Taylor, and diplomatic consultant Jeremy Hearder. I am grateful for the interest and assistance of distinguished scholars and public servants associated with the DFAT Editorial Advisory Board, particularly Geoffrey Bolton, Peter Edwards and Roger Holdich. I owe a considerable debt to David Lee, the present Director of DFAT Historical Publications and Information Section, for his interest over many years, and in particular for his superb biography of S. M. Bruce,[4] which has been of great assistance in my most recent work.

After my retirement from the Department of Foreign Affairs and Trade, a chance meeting with Pat Jalland led her to suggest a doctorate at The Australian National University based on the biography. The suggestion was timely: it had become clear to me that the task was too big to be undertaken without the help and advice of others. I am most grateful to Pat for persuading me to a course that has proved of immeasurable value. The friendship of scholars engaged in various studies has been both enjoyable and stimulating and I thank all members of the History Program for the enlargement of experience and ideas to which they contributed. I am particularly grateful for helpful discussions with Barry Smith and Margaret Steven, and for the assistance of Barbara Dawson, Janet Doust, Karen Smith, John Thompson, Susan Mary Withycombe and Malcolm Wood. Members of my advisory panel, Anthony Low and James Gillespie, provided helpful suggestions and encouragement. Barry Higman was most generous with his time—both as advisor and as supervisor in 2004; he questioned and prodded, impelling me to greater efforts in the scope of the study and in clarity of writing and thinking, and gave frequent assistance with books and suggestions for further reading. I remain very grateful for his important contribution. My supervisor, Tim Rowse, was unfailingly patient, diplomatic, enthusiastic and ready with ideas. He helped me maintain excitement about the undertaking; his own work and thinking were inspiring, and he spared no effort to ensure that my efforts measured up to the standards he set.

Primary historical research relies on the dedication of archivists and librarians in many places. My work has been undertaken in locations ranging from the grandeur of the *Palais des Nations* library by Lake Geneva in a glorious autumn, to the warehouse then housing the archives of the Commonwealth Scientific and Industrial Research Organisation (CSIRO), its roller doors open to the warmth of

---

4   David Lee, *Stanley Melbourne Bruce: Australian Internationalist*, Continuum, London, 2010.

the sun on a Canberra winter morning. It was a privilege to work in the archives of the Food and Agriculture Organization (FAO), close to the heart of Rome, in the building in which McDougall spent the final years of his life. From there one can look to the Colosseum in the northern distance; just across the road is the site of the Circus Maximus, and a short distance to the south, sheltered by the ancient city wall, is the tranquil Protestant Cemetery where McDougall lies, his tombstone commemorating one who 'nobly served his generation'.

On my journey there have been many moments of discovery and the occasional piece of luck. In 1984, Kate Birrell and I visited the offices of the Australian Dried Fruits Corporation in Melbourne, seeking records of McDougall's work for its predecessor, the Dried Fruits Export Control Board. The staff had been clearing out cupboards: ranged along the floor were files going back to their earliest days in the 1920s. 'What should we do with them?', they asked, and happily accepted our suggestion that they contact the Victorian Branch of what was then Australian Archives, where the records are now preserved. I have thus to thank Kevin Walker and Margaret Terrill of the corporation. I also want to record my thanks to former CSIRO archivist Michael Moran, and to his successor, Rob Birtles, who guided me through the difficulties of translating old Council for Scientific and Industrial Research (CSIR) file numbers to their new identities in the National Archives of Australia. I thank Graeme Powell, former Manuscripts Librarian at the National Library of Australia, who was responsible for acquiring the McDougall papers for the library. I am grateful for the assistance of other librarians in Canberra, Melbourne and Adelaide, in Edinburgh and in Washington. I owe even more to the expert assistance of staff in national archives in Canberra, London, Ottawa and Washington, in the archives of the League of Nations, and of FAO. I have been helped by residents of Appleton-le-Moors, North Yorkshire, and of the Renmark district, particularly Douglas and Merridy Howie and David Ruston, who have shared memories of the McDougall family, and I thank Heather Everingham, Chair of the Renmark Branch of the National Trust, for assistance with photographs.

There are few scholars working in the field of 'McDougall studies', but I am grateful for the assistance and encouragement of those who know and write of his work: Michael Roe, Bernard Attard and the late John O'Brien. Sean Turnell has been extraordinarily generous in sharing the results of his own research and writing; sections of this work depend to a considerable extent on that generosity and on Sean's fine writing about McDougall's economics. His infectious enthusiasm for McDougall's work has been a constant inspiration.

I can no longer thank in person the two people who really made this study possible. Bill Hudson introduced me to McDougall and encouraged me to undertake his biography. Bill set a high example in his own biography of R. G.

Casey,[5] and in his other work. During the years we worked together at DFAT, he gave me much practical assistance and wise advice; he remained a mentor and friend until his untimely death in 2002.

Elisabeth McDougall died in 2005. She was, I imagine, very much like her father in her energy, enthusiasm and determination. She too became a friend. In retirement, she devoted herself to the history of her remarkable family, and made all relevant findings available to me. When I visited London, she took me around Greenwich and Blackheath to many sites connected with her family, and provided lunch, which we ate with impressive McDougall family cutlery. We had a considerable correspondence, and many long phone calls after she moved back to Adelaide in her final years. I am saddened that Elisabeth did not live to see my study completed, and grateful for the generous assistance of her nephew, Ian McDougall, since her death.

I want to acknowledge the assistance of staff of The Australian National University and of ANU E Press for their patience and assistance in the process of publication, in particular Duncan Beard, Lorena Kanellopoulos, Karen May, Nic Petersen and Liz Walters, and Nausica Pinar for a splendid cover design.

Finally, I thank my longsuffering family. My daughters and their families have borne without complaint my neglect and distraction from their interests and have assisted with helpful advice, particularly on computing problems. My husband, John, has given me much domestic help and constant encouragement; he has scanned pictures, printed documents and listened patiently and helpfully to chapter after chapter. This book is dedicated to him with gratitude and love.

---

5   W. J. Hudson, *Casey*, Oxford University Press, Melbourne, 1986.

List of Figures

Figure 1	F. L. McDougall in Vienna, 1930s.	vii
Figure 2	Renmark 1907: Spreading fruit to dry on the property of E. E. Hutton.	23
Figure 3	The Murray River 1909: Paddle-steamers moored near stores of the Renmark Fruit Packing Union.	24
Figure 4	McDougall pruning vines, c. 1919: 'full of hope for the next season's crop.'	26
Figure 5	F. L. McDougall, AIF, 1916.	26
Figure 6	Australian Production of Dried Vine Fruits, 1880–1960.	27
Figure 7	Dried Fruit Prices on the London Market, 1900–27.	29
Figure 8	The London Agency of the Dried Fruits Export Control Board meets British dealers and bakers in Liverpool.	56
Figure 9	Declining Percentages of British Exports of Manufactures to Europe and the Americas.	63
Figure 10	McDougall's Hydraulic Metaphor for Imperial Preference.	65
Figure 11	Tariff Levels on Wheat in France, Germany and Italy.	133
Figure 12	F. L. McDougall at the Institute of Agriculture, Rome.	161
Figure 13	Assembly of the League of Nations, 1934.	168
Figure 14	F. L. McDougall with J. L. A. Avenol, Secretary-General of the League of Nations.	192
Figure 15	High Commissioner, S. M. Bruce, and Mrs Bruce with Australian troops in London.	224
Figure 16	John Winant with F. L. McDougall in Geneva.	232
Figure 17	F. L. McDougall presenting a copy of the *Story of FAO* by Gove Hambidge to Eleanor Roosevelt in 1955.	253
Figure 18	FAO Director-General, John Boyd Orr, with Special Advisers F. L. McDougall and S. L. Louwes.	289
Figure 19	Lord Bruce and F. L. McDougall at FAO.	295
Figure 20	F. L. McDougall and John Boyd Orr.	297
Figure 21	F. L. McDougall, Counselor, FAO.	302
Figure 22	F. L. McDougall.	306

List of Tables

Table 1	Australian Exports of Dried Fruits by Weight and Price.	28
Table 2	Wheat Exports of the 'Big Four' as a Percentage of the World Wheat Trade: Five-Year Averages, 1922–37.	131
Table 3	Wheat Exports of the 'Big Four' in Millions of Bushels: Five-Year Averages, 1922–37.	131
Table 4	Export Quotas Allocated under the Wheat Agreement 1933: Actual Exports Shown in Parentheses (in million bushels).	151

Imperial Currency and Weights

Currency

1 shilling = 12 pence/pennies (d)

1/6d = one shilling and six pence

1 pound (£1) = 20 shillings

1 guinea = 21 shillings

Conversion to Decimal Currency in Australia, 1966

1 shilling = 10 cents

£1 = $2

Weights

1 ounce (oz) = 16 drams

1 pound (lb) = 16 oz

1 hundredweight (cwt) = 112 lb

1 ton = 20 cwt = 2240 lb

Metric Conversion

1 lb = 0.454 kg

Abbreviations

Text

AAA	Agricultural Adjustment Administration (US)
ADFA	Australian Dried Fruits Association
AIF	Australian Imperial Force
BAE	Bureau of Agricultural Economics (US)
BEW	Board of Economic Warfare (US)
BMA	British Medical Association
CFB	Combined Food Board
CSIR	Council for Scientific and Industrial Research (Aus.)
DFECB	Dried Fruits Export Control Board (Aus.)
DMC	Development and Migration Commission (Aus.)
ECC	Economic Consultative Committee of the League of Nations
ECOSOC	Economic and Social Council of the United Nations
EMB	Empire Marketing Board
FAO	Food and Agriculture Organization of the United Nations
FBI	Federation of British Industries
IAB	Imperial Agricultural Bureaux
IBRD	International Bank for Reconstruction and Development
IEC	Imperial Economic Committee
IEFC	International Emergency Food Council
IIA	International Institute of Agriculture, Rome
ILO	International Labor Organization
IMF	International Monetary Fund
ITO	International Trade Organization
LCC	London County Council
LEC	Economic Committee of the League of Nations
LEO	Economic Organization of the League of Nations
LNHO	Health Organization of the League of Nations
MAF	Ministry of Agriculture and Fisheries (UK)
NFU	National Farmers' Union (UK)
NSW	New South Wales
OSS	Office of Strategic Services (US)
OWI	Office of War Information (US)

PLB	Paterson, Laing & Bruce
RAF	Royal Air Force
RFC	Reconstruction Finance Corporation (US)
RGC	Research Grants Committee of the Empire Marketing Board
SA	South Australia
TUC	Trades Union Congress (UK)
UK	United Kingdom
UN	United Nations
UNESCO	United Nations Educational, Scientific and Cultural Organization
UNRRA	United Nations Relief and Rehabilitation Administration
US	United States
USA	United States of America
USSR	Union of Soviet Socialist Republics
WFB	World Food Board
WFC	World Food Council

Sources

ADB	*Australian Dictionary of Biography*
CSIR	Records of CSIR held in CSIRO Archives, Canberra
DAFP	*Documents on Australian Foreign Policy*
DCER	*Documents on Canadian External Relations*
DFAT	Department of Foreign Affairs and Trade, Canberra
HLH	Hollis Library, Harvard University
HWD	Diaries of Henry Wallace, microfilm, Library of Congress, Washington, DC
FAO	Archives of the Food and Agriculture Organization, Rome
LFSSA	*Letters from 'A Secret Service Agent': F. L. McDougall to S. M. Bruce 1924–1949*
LN	Archives of the League of Nations, Geneva
MP	*Murray Pioneer*
NAA	National Archives of Australia, Canberra
NAA(CSIR)	Archives of CSIR/CSIRO, now located in NAA
NAAV	National Archives of Australia, Victorian Branch, Melbourne
NAC	National Archives of Canada, Ottawa
NLS	National Library of Scotland, Edinburgh
ODNB	*Oxford Dictionary of National Biography*

SSEA	Secretary of State for External Affairs (Canada)
TTES	*The Times Trade and Engineering Supplement*
UKNA	National Archives of the United Kingdom, London
USNA	National Archives of the United States, Washington, DC

Introduction

The creation of an international organisation bringing together the needs of the hungry and the interests of farmers seems so logical that it is hard to imagine how revolutionary it was in the first half of the twentieth century.

The idea of international organisation was itself new. Late nineteenth and early twentieth-century advances in communications had made it feasible. The advent of professional scientists and other experts meant it could become a means of pooling knowledge and experience. International congresses and societies to share expertise in medicine had existed from the mid nineteenth century. Efforts to stem epidemic diseases brought further cooperation, but it was only after World War I, when the newly created League of Nations Council called for an organisation that would fight disease but also strive to prevent it, that programs to promote better health on an international scale were envisaged. The League of Nations Health Organization (LNHO) undertook work on public hygiene. Scientific studies in various countries increasingly showed poor levels of nutrition in many social groups; the LNHO published its own authoritative report on the problem in 1935.

Agricultural crises in the late nineteenth century provided a challenge, suggesting, for the first time, that international cooperation in agriculture might be desirable. The most notable response was the formation in 1905 of the International Institute of Agriculture, a body conceived as a world federation of farmers collecting and exchanging technological and market information to improve their own profitability. But in the crisis of the depression of the 1930s much more help was needed.

In 1935 the 'marriage of health and agriculture' was first promulgated at the League of Nations. The idea turned traditional policy on its head. Farm incomes had become an issue of public policy, but accepted remedies were restrictive: import quotas, production limits, protective tariffs and pooling schemes to keep prices high, thereby putting high-nutritive, 'protective foods' out of reach of the poor. The marriage of health and agriculture proposed increased production of those foods, to lower prices and thereby to increase consumption. The poor and hungry would benefit from the first, and farmers from the second.

This study traces the evolution of that idea in the thinking of a man almost forgotten by history. Frank Lidgett McDougall was an English-born fruit farmer in the Murray irrigation area of Australia, a self-taught expert on patterns of international trade. He became economic adviser to Stanley Melbourne Bruce, who was Australian Prime Minister in the 1920s and Australian High Commissioner in London in the 1930s and 1940s. McDougall was influenced by the campaign of British scientist John Boyd Orr for government action on nutrition. While Bruce and Orr led the 'nutrition initiative' in public, it was McDougall behind the scenes whose thinking formed its basis. Bruce once said 'McDougall brings me a new idea every morning'.[1] Some of those ideas became the basis for joint action.

The work and ideas of McDougall and of Bruce developed through three phases, reflected in the organisation of this study. The first phase covers the 1920s, when both believed that the British Empire could hold the key to greater rural prosperity in Australia, and that economic problems in Britain could also be solved by imperial economic cooperation. They saw the interests of the empire and its constituent countries as indivisible; London was the centre of their work to that end. Imperial cooperation came into being, to a limited extent, in the Ottawa Agreements of 1932.

By then it had become apparent that some problems were beyond the scope of an imperial solution. In the second phase, the League of Nations at Geneva seemed to offer a forum for international cooperation. There the 'nutrition initiative' succeeded beyond expectations. Nutrition was accepted as a proper subject for international cooperation, and resulting League publications increased public awareness of the problem. Bruce and McDougall went on to extend their thinking beyond nutrition, arguing that measures to implement social justice and higher standards of living could ease the political tensions of the late 1930s. They sought to renew the League as an agency for social reform.

Finally war turned their focus to Washington and 'a new deal for the world'. In the United States McDougall took a significant part in planning for a postwar food and agriculture organisation—eventually to become the Food and Agriculture Organization of the United Nations (FAO)—while Bruce and Orr lobbied in support. All three were to play key roles in the early years of FAO, where McDougall's memory is honoured still in a biennial McDougall Lecture.

A booklet published by FAO in 1956 was called *The McDougall Memoranda: Some Documents Relating to the Origins of FAO and the Contribution made by Frank L. McDougall*.[2] The booklet consists of four documents. Three were written in 1935 for the League of Nations. The first is a memorandum by McDougall called 'The Agricultural and the Health Problems', which urged placing adequate nutrition of populations in the forefront of economic policy. The second, a speech by Bruce on Australia's behalf to the League Assembly, called for 'the marriage of health and agriculture'. The third document is the subsequent Assembly resolution establishing nutrition as a subject for international consideration. The final document in the booklet is a memorandum written in October 1942

---

1 Recorded by Alfred Stirling, *On the Fringe of Diplomacy*, The Hawthorn Press, Melbourne, 1973, p. 148.
2 Copy of booklet in National Library of Australia [hereinafter NLA], MS6890/5/4.

by a Washington group including McDougall. It was essentially a revision of an earlier paper in which McDougall set out points he and Bruce had discussed. The memorandum outlined proposals for an international organisation that would 'make the marriage between health and agriculture a reality'. This study aims to show how these documents and the organisation itself came into being.

though I discovered that the central issue it raised lay in the intersection with other issues, which I had not anticipated or even recognised.

Part A: Visions of Empire

Prologue

The working partnership between Frank Lidgett McDougall and Stanley Melbourne Bruce in the 1920s aimed to ensure the economic welfare of the British Empire. Their partnership also illustrates aspects of the relationship between a dominion and the central imperial power, for both men had family connections to imperial trade. Both spent most of their careers working for what they believed to be Australian interests, yet for much of their lives they were based elsewhere.

McDougall grew up in Blackheath, near the River Thames. He could have watched ships at the docks nearby unloading cargoes from around the world, including grain for the flour mill managed by his father, an expert on imported grains. After arriving in Australia at the age of twenty-five to take up an irrigation block at Renmark, South Australia, McDougall spent less than a decade in the country he adopted as his own. But he had served in the Australian Imperial Forces, and was to represent Australia officially for most of his life. In later years he would work hard to present an Australian persona—one which other Australians sometimes found unconvincing.[1]

As a young child, Bruce lived close to Port Phillip Bay. He probably watched ships bringing goods from Britain into the port of Melbourne for his family's softgoods importing business.[2] Bruce was schooled largely in Australia, then spent the formative years of late adolescence and early adulthood in England and served in the British forces during World War I. He acquired a 'quality of aloofness from the Australian man in the street', leaving him 'a foreigner in his own land, a man out of touch with the people he was leading'.[3] In fact, like many of their time, neither Bruce nor McDougall saw a need to distinguish between the interests of Australia and of the Empire as a whole; serving one should and would benefit the other.

The terms centre and periphery may be used as shorthand for a metaphor of the Empire as a circle with the imperial capital, London, at its centre. From any point on the periphery one sees the immediate surroundings most clearly, while points on the far side of the circle are hardly visible. From the centre the whole periphery is visible, but the central area is clearest. Some change of perspective can thus be discerned as Bruce and McDougall move from the periphery to the imperial centre. This change is not necessarily an indication of less patriotic

---

1 H. C. Coombs, *Trial Balance*, Macmillan, Melbourne, 1981, pp. 41–2; interview with W. Way, 7 April 1995.
2 Lee, *Stanley Melbourne Bruce*, p. 1.
3 Ibid., quoting journalist Warren Denning, p. 9.

goodwill towards Australia. Patriotism for Australians in the early twentieth century was not a simple matter of loyalty to an Australian nation. It involved complex loyalties within Australia itself and beyond it to the Empire.

Empire, Dominion and Nation

By the early twentieth century the British Empire had become a many-tiered institution. Colonies scattered through Africa, the Caribbean, South-East Asia and the oceans were controlled by Whitehall. India, by virtue of size and a variety of polities, held a unique status, somewhere between colony and dominion; final control remained with Whitehall, a matter increasingly subject to controversy. Dominions—the white settler societies of Canada, Australia, New Zealand and South Africa and, for a time, the Irish Free State, Newfoundland and Rhodesia— were by convention largely self-governing in domestic matters, which included the right to impose tariffs on imports. During the 1920s, dominion rights to legislative and executive independence, including external relations, were gradually defined—a process driven by 'radical' dominions: Canada, with trade and security interests increasingly focused in North America; the Irish Free State and South Africa, both concerned with 'equality and separateness'.[4] The 1923 Imperial Conference agreed that separate empire governments could conclude treaties with foreign powers. The 1926 Imperial Conference adopted the Balfour Report, defining dominions as 'autonomous communities within the British Empire, equal in status, in no way subordinate one to another in any aspect of domestic or external affairs, though united by a common allegiance to the Crown, and freely associated as members of the British Commonwealth of Nations'. Expert recommendations on outstanding legal issues were approved by imperial leaders in 1930; dominion legislative independence was solemnised in the Statute of Westminster on 11 December 1931.

Australia did not adopt the Statute until 1942, and then only to expedite implementation of wartime measures.[5] Australian and New Zealand reluctance to accept independence was based on defence and on trade; in neither dominion was there public pressure for change. Official statements stressed pride in empire and its achievements, using the words 'empire' rather than 'British Commonwealth of Nations' and 'self-governing' rather than 'independent'.[6] The Crown, common history and literature drew 'cultural and even emotional attachment to the material representations and rhetoric of Britishness and imperialism' even for

---

4 W. J. Hudson and M. P. Sharp, *Australian Independence: Colony to Reluctant Kingdom*, Melbourne University Press, Melbourne, 1988, p. 86.
5 Ibid., pp. 130–8.
6 Ibid., pp. 64, 126–7.

a prime minister with the Irish heritage of Joseph Lyons.[7] On the other hand, a prime minister like Robert Gordon Menzies—with deep allegiance to idealised, mythologised views of the British Isles and the monarchy—was capable of hard bargaining on matters of trade and Australian national interest. Menzies could revel in a 'romanticised' English countryside, yet privately lampoon pompous officials, make cutting observations of royal princes and treat British politicians and bureaucrats with 'courteous equality'.[8] For all Australian governments of the 1920s and 1930s, cultivation of imperial loyalty, based on allegiance to the Crown, and cultural heritage, was a pragmatic necessity for defence. Tough negotiation on issues of trade was equally necessary.[9]

Australian national sentiment was therefore complex. English-born constitutional lawyer W. Harrison Moore wrote: 'I do not ask myself whether I am English or Australian in any sense which excludes one for the other… to be called on to make a choice…would be deeply painful as a disintegration of my own personality.'[10] Australian-born historian W. K. Hancock described a developing Australian nationalism, but added: 'A country is a jealous mistress and patriotism is commonly an exclusive passion; but it is not impossible for Australians, nourished by a glorious literature and haunted by old memories, to be in love with two soils.'[11]

Allegiance was not merely divided between the new world and the old. A new federal authority seemed of little relevance to the lives of most Australians, for whom State and regional loyalties predominated.[12] Apathetic they might be about their new Federal Government, but Australians had nevertheless adopted the land as their own.[13] Native species became national symbols. The 'Australian legend' portrayed the bush as the real Australia. Distinctive art developed: the Heidelberg School 'shaped the popular Australian view of the land'.[14] McDougall, from youth a lover of German poetry and grand opera, visited the studio of Hans Heysen in the Adelaide Hills and was enraptured by Heysen's treatment of the way 'sunlight streamed through the trunks of the gum trees… not painted sunlight but the very illuminating thing itself'.[15] Australian nature formed part of children's literature and the school curriculum. Yet, like their prime ministers, Australians were also formed by the literature and images of

---

7   Kosmas Tsokhas, *Making a Nation State: Cultural Identity, Economic Nationalism and Sexuality in Australian History*, Melbourne University Press, Melbourne, 2001, p. 136.
8   Ibid., pp. 141–50.
9   See, for example, ibid., pp. 11–12, 19–20.
10 Quoted in Hudson and Sharp, *Australian Independence*, p. 111.
11 W. K. Hancock, *Australia*, reprint 1961 edn, Jacaranda Press, Brisbane, 1930, p. 51.
12 J. M. Powell, *An Historical Geography of Modern Australia: The Restive Fringe*, Cambridge University Press, Cambridge, 1988, p. 20.
13 Thomas R. Dunlap, *Nature and the English Diaspora: Environment and History in the United States, Canada, Australia and New Zealand*, Cambridge University Press, Cambridge, 1999, p. 95.
14 Ibid., pp. 98–102.
15 NLA, MS6890/1/5, letter to Kit, 29 May 1913.

the British Isles, of 'things and places [they] had never seen'. Thomas Kenneally wrote of being 'educated to be exiles'.[16] To see a world they had learned to love at a distance, many left Australian shores. They sailed away with the comfortable security of British passports, and rejoiced to smell the eucalyptus as they neared those shores on return.

Thus it was with both Bruce and McDougall, the one Australian-born and raised, with close links to the Australian business world, but formed from late adolescence in Britain, and having served in the British Army before returning to Australian politics; the other born and raised in England, with family links to British industry, but whose subsequent career was determined by an early adulthood in rural Australia, and service with Australian forces. Their view that the interests of the two were indivisible would be proved wrong in the testing years of the 1930s depression, and the issue of tariffs was a key part of the problem.

Challenging Free Trade

In the 1840s the British Government had adopted a policy of free trade or 'economic liberalism'. Advocates of the policy argued that the economy had its own rule: the 'invisible hand' of the market; governments should not interfere with the benign forces of unrestricted trade. Individual good and common good were one and the same. This policy benefited an economy leading the world in the export of manufactured goods and with a need for cheap imported food. The value of the policy seemed to be confirmed by Britain's early experience. In the decades following its adoption, the worth of British exports and imports trebled. More than half the world's ocean-going tonnage was British; public and private wealth increased by every measure; living and working conditions improved.[17] From 1860 Britain was the pivot of 'a new multilateral trading system' based on a pattern of economic specialisation in three broad trading groups.[18] Two were essentially complementary: industrial, or rapidly industrialising, countries of Western Europe and the United States; and agricultural and plantation economies of Eastern Europe and the tropics. A third group, including Australia, exported primary products, but aspired to industrial development. Britain, with a trade surplus derived from a high level of invisible exports, was able to import from

---

16   Dunlap, *Nature and the English Diaspora*, pp. 108, 115–18.
17   David Thomson, *England in the Nineteenth Century*, Penguin, London, 1950, pp. 138–9, 142.
18   A. G. Kenwood and A. V. Lougheed, *The Growth of the International Economy 1820–1980: An Introductory Text*, George Allen & Unwin, London, 1983, pp. 103–4.

all three groups and export industrial goods and capital to all of them. In 1913, despite decline over some decades, Britain still provided 43 per cent of world investment capital.[19]

Economic instability in Britain had become apparent, however, in the 'great depression' from the 1870s to the mid 1890s. Agricultural prices declined as supplies of cheap overseas wheat increased.[20] Britain's rate of export growth fell; world market share in its traditional industries like heavy machinery and textiles drifted to Germany and the United States; emigration increased; and capital moved overseas.[21] In 1888, when McDougall was four years old, the word 'unemployment' appeared in the *Oxford English Dictionary* for the first time.[22]

The irony was that, in the words of Robert Skidelsky, British export trade could be considered 'largely self-liquidating'. British capital had helped develop overseas industries that would eventually become competitors. It had also enabled New World countries to develop agriculture and to export foodstuffs in return for the sterling they needed to buy British manufactures. Cheap New World wheat and meat 'invasions' 'ruined' British agriculture. British workers paid for their cheap food in potential and actual unemployment as foreign demand for British manufactures declined.[23]

Free trade was therefore questioned. Liberalism and individualism also lost ground as a result of economic and social change with the development of business combines and cartels, organised labour, political parties and urbanised populations. There were opposing solutions to the key problem of creating a market for British goods. Socialists wanted measures to increase domestic purchasing power. Imperialists, focusing on a need for security in a new and uncertain world, turned back to an earlier 'mercantilist' doctrine—that foreign imports destroy domestic employment. Thus, the aim of mercantilists was self-sufficiency, based on an imperial trading system protected by tariffs.[24]

Around the turn of the century, as the young Bruce was at Cambridge and McDougall was contemplating his future in London, Britain debated a 'tariff reform' campaign, led by Colonial Secretary, Joseph Chamberlain. The phrase covered a mix of policies, including a protectionist tariff with imperial preference—that is, lower import tariffs for goods from the empire. The idea appealed to some agricultural and manufacturing sectors. Since foodstuffs formed a large proportion of British imports, a significant objection would always be that abandoning free trade would raise the cost of food. The Conservative

---

19 Ibid., p. 41.
20 W. W. Rostow, *The World Economy: History and Prospect*, University of Texas Press, Austin, 1978, pp. 164–5.
21 Robert Skidelsky, *Oswald Mosley*, second edn, Macmillan Papermac, London, 1981, pp. 49–50.
22 Robert Skidelsky, *John Maynard Keynes: Hopes Betrayed 1883–1920*, Macmillan, London, 1983, p. 38.
23 Skidelsky, *Oswald Mosley*, pp. 51–2.
24 Ibid., pp. 45–53.

Cabinet divided; Chamberlain and others resigned; and the Government lost heavily in a general election on the issue in 1906. Socialists and imperialists remained outside a mainstream not yet willing or able to challenge laissez faire.[25]

Tariff Policies in the Empire

While the mother country and dependent colonies adhered to the principles of free trade, self-governing dominions used tariffs to protect infant industries. Dominion tariff policies were driven by mixed motives of imperial sentiment and self-interest. Dominions rejected late nineteenth-century suggestions for an imperial tariff union, preferring differential tariffs encouraging their own trade at the expense of other parts of the empire. Canada introduced protection, with some preferential treatment for British goods, from 1879. Australian colonies, prevented by 1850s constitutional legislation from imposing differential duties, fought for and won that right from Britain in 1873.[26] Tariff preference beyond the empire was ruled out by a series of most-favoured-nation agreements negotiated in the 1860s and binding on British colonies. In 1897 Britain yielded to colonial pressure by denouncing treaties with Belgium and Germany. In the following decade, while the mother country remained committed to free trade, all self-governing dominions initiated some tariff preferences in favour of the United Kingdom.[27] Ambiguities in relations between dominions and the mother country were apparent. It has been argued that adoption of distinct tariff policies constituted an affirmation of independence from Britain.[28] It has also been argued that a key factor in persuading the Australian Federal Parliament to institute a British Imperial Tariff at a lower level than the general rate was imperial sentiment attaching to Chamberlain's tariff reform campaign. Domestic considerations, however, remained pre-eminent: the interests of Australian industry were not to be compromised.[29]

In Britain, two opposing groups, cutting across party lines, would continue to contest economic policy: laissez-faire free-traders and the group described by Ian M. Drummond as 'Imperial Visionaries'. The principal figure in the latter was Lord Milner, a former high commissioner in South Africa and later a member of Lloyd George's Cabinet. A loyal following of former junior colleagues, known as 'Milner's kindergarten', included Geoffrey Dawson, destined to become Editor

---

25 Ibid.
26 Emmett Sullivan, 'Revealing a Preference: Imperial Preference and the Australian Tariff, 1901–1914', *Journal of Imperial and Commonwealth History*, Vol. 29, no. 1, 2001, p. 37.
27 W. K. Hancock, *Survey of British Commonwealth Affairs. II: Problems of Economic Policy, Part 1*, Oxford University Press, London, 1940, pp. 86–8.
28 Francine McKenzie, *Redefining the Bonds of Commonwealth, 1939–1948: The Politics of Preference*, Palgrave, New York, 2002, p. 14.
29 Sullivan, 'Revealing a Preference', p. 39.

of *The Times*; Philip Kerr, later (as Lord Lothian) Ambassador in Washington; Leo Amery, Colonial and Dominions Secretary in the 1920s; and Lionel Curtis. *The Round Table* was established as a forum for the group's ideas. The visionary agenda aimed to raise total empire output by means of emigration and investment, and to increase British exports to the empire, thereby reducing unemployment and promoting greater prosperity in Britain. A preferential tariff would help make empire development more profitable and the empire more attractive to capital and migrants. It would strengthen sterling against the dollar and tend to replace foreign imports into Britain with empire goods. Tariff preferences would secure and raise Britain's share of trade in manufactures in the overseas empire.[30]

In 1914–18 imperial patriotism and imperial self-sufficiency again came together. Australia's Prime Minister, W. M. Hughes, was hailed in London as 'one of the empire's strong men in her hour of need', as he campaigned to bring the empire's essential supplies under imperial control, urging that Australians 'must get nothing overseas that we can produce ourselves, and we must in particular buy nothing except from Britain and her allies'.[31] A combination of wartime sentiment and pragmatism thus brought small steps towards the imperial vision in Britain. The 'McKenna Duties' of 1915 imposed tariffs on luxury goods. The Imperial War Conference resolved in 1917 that the empire should be as independent as possible in food supplies, raw materials and essential industries. It also resolved in favour of British emigration to empire countries and imperial tariff preference, linking 'in a single logical system the policies of migration, investment, and tariff preference'.[32] Bruce would later call it 'men, money and markets'. The visionary agenda seemed to be in place.

Tariffs were the first aspect of the program to receive attention, but the results fell well short of tariff preference. Conservative leader, Bonar Law, and Liberal Prime Minister, Lloyd George, both promised to honour the 1917 resolution on preference, but would not tax food. The 1919 British Budget included preferences equivalent to one-sixth of low revenue duties existing on tea, cocoa, coffee, sugar, tobacco and dried fruits. The small sums involved were unlikely to make empire goods competitive. This was the situation when McDougall began to argue that the future of the Australian dried-fruits industry depended on a substantial British tariff preference.

---

30 Ian M. Drummond, *British Economic Policy and the Empire 1919–1939*, George Allen & Unwin, London 1972, pp. 36–42.
31 L. F. Fitzhardinge, *The Little Digger 1914–1952: William Morris Hughes, a Political Biography. Volume II*, Angus & Robertson, Sydney, 1979, p. 60.
32 Hancock, *Survey*, pp. 137, 126–7.

1. Renmark

Summary

The story of the 'marriage of food and agriculture' begins with the arrival of twenty-five-year-old Englishman Frank McDougall at Renmark, South Australia, in 1909. This chapter decribes a young man born into a devout, inventive and successful family, but struggling to find his own purpose in life. He was to find that purpose as a fruit-grower in Renmark, an irrigation settlement depending on self-help and technical efficiency. McDougall worried about the prospects for his rapidly growing industry; he studied market possibilities and concluded that the industry's prosperity depended primarily upon gaining preferential tariff treatment in the world's biggest market for dried fruits: Great Britain. His idea was developed in a series of memoranda written in the years immediately after World War I. With the support of press and fellow growers, he lobbied State governments and the Federal Government, gaining support in spite of a daunting political barrier: with a few minor exceptions, the British Government was committed to a policy of free trade.

The Making of 'an incorrigible optimist'

The Murray River meanders for hundreds of kilometres through eastern and southern Australia. In the early twentieth century it formed, with its chief tributary, the Darling, an inland highway, carrying goods and passengers upstream from South Australia through the pastoral inland to settlements in Victoria and the far west of New South Wales.

In the spring of 1909 a paddle-steamer tied up at Renmark, on the eastern fringe of South Australia. It had churned its way past yellow sandstone cliffs topped with desert saltbush and mallee scrub, and skirted mazes of billabong and swamp. At times passengers had been asked to jump ashore and gather firewood to keep its engines going. It had been a long trip for all, but for none more than twenty-five-year-old Frank McDougall. In the past two years he had sailed three times from England to the southern hemisphere, spending several weeks in the tiny crown colony of St Helena and making two visits to the spectacular escarpments around the Umgeni Valley near Pietermaritzburg in Natal, part of the self-governing dominion of South Africa.[1] Finally he had sailed to Adelaide,

---

[1] NLA, MS 6890/1/4, letters to family.

and then travelled by train north to the river port of Morgan and by paddle-steamer to Renmark. As he stepped ashore at Renmark, he may have thought he had reached his destination. He could not have imagined the extraordinary journey he was about to begin.

Such extensive travelling suggests that young McDougall's family was no ordinary one. Members of his father's family managed two successful national industries. Their lives were marked by discipline and a commitment to education and self-improvement, derived from a heritage of devout Methodism.[2] The McDougalls applied their devotion equally to work and to social activism, and took an inventive, inquiring approach to the world and its workings. Frank's grandfather, Alexander McDougall, had been a schoolmaster in Manchester and a keen amateur chemist. Alexander might well be seen as an example of the way in which scientific progress occurred in nineteenth-century Britain, where pioneering developments in both industrial and agricultural production were led by amateurs. Later industrial development was often pioneered in German universities, and American governments established colleges and agricultural extension services to support a massive development of agriculture, but early British innovators were nurtured by the scientific societies.[3] The Royal Society, from 1660, and similar bodies in Birmingham, Manchester, Edinburgh and Liverpool a century later were first founded to cater for wealthy cultured enthusiasts. John Dalton, doyen of the Manchester Literary and Philosophical Society, represented a new kind of scientist: lower middle class, provincial, dissenting—and completely devoted to science. Unlike the earlier amateurs, these 'devotees' had concerns associated with professional science: research at the frontiers of science, publication, 'keeping up' with the output of leaders in their specialities, and scientific communication.[4] Although they were generally self-taught, and in many cases unable or unwilling to make a career of science, their chief purpose and status lay in its pursuit. They worked not only for personal gratification. Like the utilitarians, they believed science to be vital to Britain's progress.[5]

Alexander McDougall had been drawn from his native Carlisle to Manchester by its scientific institutions and organisations and was probably taught by Dalton. Alexander based a successful milling industry on his invention of self-raising flour. McDougall's Flour was merged with Hovis in 1957 and absorbed into the conglomerate Rank Hovis McDougall in 1962. Since 2007 it has been part of

---

2 David Hempton, *Methodism: Empire of the Spirit*, Yale University Press, New Haven, Conn., 2005, pp. 52–4.
3 Peter Alter, *The Reluctant Patron: Science and the State in Britain 1850–1920*, trans. Angela Davies, Deddington, Oxford, 1987, pp. 22–4.
4 Robert H. Kargon, *Science in Victorian Manchester: Enterprise and Expertise*, Manchester University Press, Manchester, 1977, pp. 11–13.
5 Ibid., pp. 35, 78.

Premier Foods.[6] Alexander made an important contribution to the welfare of Manchester, a 'boom town' suffering the inevitable pollution of smoky factories and rivers described as 'stinking foul open sewers'. Edward Chadwick's report of 1842[7] spurred a public health and sanitation movement and a royal commission. With professional chemist and crusader for research into air pollution Angus Smith, Alexander McDougall experimented on deodorisation of sewage. A disinfecting powder, patented by Smith and McDougall in 1854, was subsequently sold as 'McDougall's powder'. Alexander also marketed a liquid of carbolic acid and limewater for use in sewers; later it was used in dissecting rooms and for treating sores, dysentery and footrot. Joseph Lister claimed to have taken the idea of using carbolic acid for surgical purposes from its use disinfecting sewage at Carlisle, where McDougall's process had been adopted.[8] These inventions formed the basis of McDougall's chemical company. By 1926, Coopers, McDougall & Robertson was the leading British supplier of sheep dip and insecticides and a major supplier of disinfectants;[9] it was acquired by the Wellcome Foundation in 1959.

Privately funded regional colleges were gradually established to meet demands from British industry for training. Owens College in Manchester was established in 1851 by a bequest from a nonconformist cotton merchant. Alexander's son, Arthur McDougall, won the Dalton Prize for original research at Owens College, but abandoned science for business. Owens College was to become Manchester University, and Arthur's son, Sir Robert McDougall, one of its significant benefactors.[10] But there were no counterparts in Britain for the large American foundations nurtured in the Social-Darwinist tradition that wealth should be used to benefit society.[11]

The children of Alexander McDougall, however, were encouraged to teach in ragged schools in the slums of Manchester, and as adults they devoted their spare time to church and municipal work. Two of them—Frank's father, John, and young Alexander—eventually sold out of the family business to concentrate on this work. John McDougall left his position as manager of McDougall's London flour mill in 1888 to represent the district of Poplar on the London County Council (LCC), for 25 years. His chief interests were efficiency and social welfare, in particular lunatic asylums and drains; he spent three days a week visiting inmates in the former, and personally toured the latter. During

---

6   *A Matter of History: A Century of McDougall's Self Raising Flour 1864–1964*, in NLA, MS6890/5/4; company sites on Internet; E. McDougall.
7   *Report on the Sanitary Condition of the Labouring Population of Great Britain*.
8   Kargon, *Science in Victorian Manchester*, pp. 108–23; Ellen, Lady McDougall, *The McDougall Brothers and Sisters: The Children of Alexander McDougall*, privately published and printed in Blackheath, UK, 1923.
9   NLA, MS6890/1/9, McDougall to Norman, 3 January 1926.
10 Kargon, *Science in Victorian Manchester*, p. 179; McDougall, *McDougall Brothers and Sisters*.
11 Sidney Fine, *Laissez Faire and the General-Welfare State: A Study of Conflict in American Thought 1865–1901*, University of Michigan, Michigan, 1956, pp. 114–15.

Frank's childhood, John McDougall was pilloried in *Punch* and was a focus of demonstrations against his campaign to clean up entertainment offered in music halls. He was knighted as LCC Chairman in the coronation year of Edward VII. Traditions of disciplined devotion to work and social activism persisted in a third generation including high achievers and dedicated idealists: industrialists and academics, doctors and missionaries.

Frank Lidgett McDougall was born on 16 April 1884, in Blackheath, London, the eldest child and only son of John McDougall's second marriage. Frank's mother, Ellen Lidgett, also had a strong Methodist heritage. Her double first cousin, Dr John Scott Lidgett, 'the grand old man of Methodism', was co-founder of the Bermondsey Settlements, leader of the LCC Progressive Party from 1918 to 1928, and Editor of *Contemporary Review* and the *Methodist Times*. Ellen herself was serious and bookish, a writer of devotional works, some based on the McDougall and Lidgett families.[12]

With this powerful array of examples it is hardly surprising that the young Frank McDougall believed he should devote his life to a worthy purpose. It is equally understandable that he struggled to find his own path. Frank completed his schooling to matriculation level at Blackheath Proprietary School in 1901, but his father had suffered financial difficulties and there was no money for university. There was thought of joining the family chemical business. In 1903 Frank spent time in Godesburg am Rhein learning German, and late in 1905 he was in Darmstadt, preparing to enter the famed University of Technology. He learned to enjoy reading German, but found 'the language does not come very quickly' and 'everyone speaks too fast'.[13] He did not complete any qualification. In 1906 he was back in England, tinkering with chemical experiments.[14] He drifted. Members of his extended family conferred about possible employment. He was sent to investigate possibilities of wattle growing on St Helena, and in Natal where there were Lidgett relatives. Finally the decision was made to invest in a fruit block in South Australia, where there were also family connections. His oldest half-sister, Lucy, was married to Charles Napier Birks, of a prominent Adelaide retailing family, and the two families had other connections in England. The Birks family was active in charitable causes; Lucy later founded the Adelaide Babies Aid Society.[15]

Frank was twenty-five years old when he reached Renmark. Long letters to his younger sister Kit, written between 1906 and 1909, have survived. In these one

---

12   McDougall, *The McDougall Brothers and Sisters*.
13   NLA, MS6890/1/3, letter to Mother, 30 October 1905; letter to Kit, 19 October 1905.
14   Ibid., letters to Kit, 30 September, 10 October and 15 November 1906, 13 May 1907.
15   Information from E. McDougall; Martin Woods, 'Birks, Rosetta Jane (1856–1911)', Australian Dictionary of Biography [hereinafter ADB], online edn, 2006, <http://www.adb.online.anu.edu.au/biogs/AS10040b.htm> [accessed 22 October 2008].

sees a young man desperately searching for a purpose. He lectured her thus: 'You have the job of choosing something to do, well let it be something really worth while'; adding, 'I feel and you too will feel like the young man who went to Jesus to know what must I do to be saved. I think in a few years it will be clearer.'[16]

Extensive reading shaped McDougall's religious and political thinking: 'Books are everything not quite!!' he wrote to Kit in 1907. A later letter provided an impressive list of recommended reading for her, drawn from his own favourites.[17] Gradually he shed his Methodist heritage, briefly considering 'High Puritanism' and Anglo-Catholicism, but subsequently abandoning Christianity altogether. Much remained of his Christian upbringing: he was known ever after for his ability to recite Bible passages and his affection for the *Book of Common Prayer* and the *Authorised Version*. A colleague wrote of him being 'always ready to cite scripture for his agricultural purposes', but the obituary writer for *The Times* recalled that 'there were some who held the view that what he quoted was not always to be found in Holy Writ', adding, 'he nevertheless could produce phrases which sounded as if they could have come from nowhere else'.[18] More significant Methodist legacies were a dedication to disciplined, unremitting work; a preference for empirical, rather than theoretical, understanding of problems; and a belief in the perfectibility of man through his own efforts.[19]

The young McDougall mentioned Immanuel Kant, Henry George and Max Nordau amongst those who led him briefly to consider socialism, and to aver that 'the great standing lie is our present social system and our economic system'.[20] He made no formal commitment to left-wing politics, but as a young man of twenty-three saw much in his own world needing change and, without any sense of direction, he wanted to help. He thought that 'the social question' (a euphemism for prostitution) was in truth much wider, perceiving evils in domestic service, in 'Society' living in idleness on the work of others, in obsession with material possessions, and even smoke pollution.[21] He had difficulty in formulating a social philosophy, partly because of a fundamental aversion to compulsion, and partly in reconciling social activism with his youthful enjoyment of luxuries: 'the hard thing is to get a reasonable line of junction for a love of Beauty and a dislike of luxury.' The individualist 'sees that no real progress for the indiv[idual] can be made without the personal choice, while the Socialist is accused of the absurd belief that we can bring

---

16 NLA, MS6890/1/3, letter to Kit, 18 June 1907.
17 Ibid., letter of 1 June 1907; NLA, MS6890/1/4, letter of 9 February 1909.
18 Coombs, *Trial Balance*, pp. 41–2; Gladwyn Jebb, *The Memoirs of Lord Gladwyn*, Weidenfeld & Nicolson, London, 1972, p. 63; *The Times*, 17 February 1958; Alfred Stirling, *Lord Bruce, the London Years*, The Hawthorn Press, Melbourne, 1974, p. 65.
19 Hempton, *Methodism*, pp. 50, 57–8.
20 NLA, MS6890/1/3, letter to Kit, 1 June 1907.
21 Ibid., letter to Kit, 18 June 1907.

about a millennium by Act of Parliament'.[22] Nevertheless, perhaps with some bravado at home, he was happy to declare himself a socialist. He tried to connect Christianity and socialism by writing. In one proposed piece he was to imagine a Christian socialist parson working in London's East End, falling asleep and dreaming of what England would be like 50 years after socialism had won a parliamentary majority. McDougall's purpose in this project was to crystallise his ideas of 'what we socialists want'.[23]

The result of all this thinking is contained in the last such letter, written some months after reaching South Australia, and summarising his philosophical journey. From it, he wrote, he had emerged with three principles. First, he was, and hoped to remain, 'lacking in reverence', meaning that no belief could be acceptable if it failed the test of human logic. Second, he declared himself an optimist: 'What I mean by that is that I believe that a man has the power to choose to live his own life in the way that seems to him best…"I am the master of my fate; I am the captain of my soul"'. In a much later version, he would write: 'I am an incorrigible optimist or at least feel that it is well worth while to make a great effort to secure an intelligent attitude towards economic problems!'[24] He called his third principle 'the ascent of man'. 'Man is capable by his own efforts of rising'—a belief 'essential to the creed of a socialist as I understand the term'.[25] In essence these principles were applied to all the problems he would encounter over more than 40 years. In very different times and amid calamitous events, McDougall would remain an optimist, pragmatic and independent—which surely adds up to 'lacking in reverence'—and ever labouring for a better world.

Renmark: Self-Help and Dissemination of Knowledge

This, then, was the young man who arrived in Renmark in the spring of 1909. He came to a community and to an industry that had been created in response to ideas of Social Darwinism, progressivism and imperial efficiency.

In the later nineteenth century, ideas described collectively as 'Social Darwinism' were influential in the United States, where William Graham Sumner suggested that without state interference men are rewarded in proportion to their efforts;

---

22   Ibid., letter to Kit, 8 June 1907.
23   NLA, MS6890/1/4, letter to Kit, 23 January 1909. The piece itself does not seem to have survived.
24   National Archives of Australia, Records of the Council of Scientific and Industrial Research [hereinafter NAA/CSIR], A9778, M14/32/5, letter to A. C. D. Rivett, 29 February 1932.
25   NLA, MS6890/1/5, letter to Kit, 20 April 1912.

only the fittest survive.²⁶ In Britain such ideas combined readily with evidence that British industry was falling behind in a competitive world, encouraging scientists and organisations like the British Association for the Advancement of Science to lobby for government support. When British industry took few awards at the Paris Exhibition of 1867, the association successfully called for investigation into science teaching and research. The Devonshire Royal Commission on Scientific Instruction agreed in 1875 that development of research depended largely on the state being willing to 'assume that large portion of the National Duty which individuals do not attempt to perform, or cannot satisfactorily accomplish'.²⁷

Social Darwinist thinking also underlay campaigns for 'national efficiency' triggered in Britain by German and American competition, and reinforced by revelations of the poor health of British recruits for the Boer War, weaknesses of army organisation and German technical superiority. Journalists warned of Britain's imminent defeat in the 'battle for commerce in almost every land on earth'. The British Science Guild was established in 1907 by J. Norman Lockyer, astronomer, Secretary to the Devonshire Royal Commission and founder of the journal *Nature*, which he used as a vehicle for his arguments. The guild held an essay competition on 'the best way of carrying on the struggle for existence and securing the survival of the fittest in national affairs'. In 1903 Lockyer called his presidential address to the British Association 'The influence of Brain-power on History', a deliberate allusion to Alfred Mahan's *The Influence of Sea Power upon History* (1890). Lockyer said 'we have not learned that it is the duty of a State to organise its forces as carefully for peace as for war; that universities and other teaching centres are as important as battleships or big battalions; are, in fact, essential parts of a modern State's machinery'.²⁸

Branches of the British Science Guild were established in Australia. Prime Minister Alfred Deakin was impressed by the British aim to achieve efficiency and by an informal movement called Progressivism in the United States. Progressivism embraced a wide range of measures—democratic reform, regulation of corporations and monopolies, labour rights, social justice—all aimed at mitigating the harsh effects of unregulated large-scale capitalism. Many Progressives campaigned for professionalism, rationality and efficiency in government. In 1907 Deakin expressed what might be called 'imperial progressivism':

> The task of Empire is the…scientific conquest of its physical…problems. [We must] endeavour…to acquire that knowledge in scientific manner,

---

26 Fine, *Laissez Faire and the General-Welfare State*, pp. 79–91.
27 Alter, *Reluctant Patron*, pp. 100–2, 118.
28 Ibid., pp. 81–2, 92–4, 113, 127–30.

and by scientific methods, which shall enable us to appreciate...the vast, the incalculable natural resources which are present in our possession under the Flag—the means of utilising these instruments of material power for the benefit of our race.[29]

The term 'pragmatic progressivism' has been used to describe the coincidence of a new federal bureaucracy and a new interest in applying science to problems of agriculture, health and transport.[30] Yet in the 1880s, before federation, Deakin as a Victorian State Minister had chaired a royal commission on irrigation and led a group investigating irrigation schemes in California. There he met brothers George and W. B. Chaffey, who came to Mildura in 1886 to demonstrate their methods. Deakin subsequently introduced the first legislation in Australia to promote irrigation systems and providing for state-aided local trusts, thereby enabling establishment of irrigation settlements at Mildura and Renmark in the late 1880s.[31]

The family decision that the young McDougall—drifting apparently without purpose—should go to Renmark demonstrates an imperial view: it has echoes of the long tradition of sending 'remittance men' who were an embarrassment to their families to the outposts of the empire, and of another tradition of investing in primary industries overseas. It also reflected a new phase of empire. Irrigation in the newly federated Australian colonies was barely a generation old when McDougall reached Renmark in 1909. Irrigation was in the forefront of consolidation of settlement, a pattern extending throughout the temperate empire, particularly in Canada and Australia, driven by the idea of creating a stable 'yeoman class' of small farmers, of diversifying production and of taming a harsh interior. Immigrants were sought to populate these new areas. New production could be exported to the imperial 'centre', Great Britain, in return for manufactured goods exported to lands on the imperial 'periphery'. Within this pattern lay seeds of conflict between centre and periphery.

Most first-generation settlers at Renmark were genteel middle-class English families, with little knowledge of horticulture. They were attracted by illustrations of river scenes and by tales of healthy outdoor life and easy profits, found in the 'Red Book' produced by the Chaffey brothers' London agent. They purchased orchard, town and residential blocks—all on 10-year terms.[32] Through blisteringly hot summers and miserably cold winters, they camped

---

29 Quoted in Roy M. MacLeod, 'Science, Progressivism, and "Practical Idealism": Reflections on Efficient Imperialism and Federal Science in Australia, 1885–1915', in *The 'Creed of Science' in Victorian England*, ed. Roy M. MacLeod, Variorum, Aldershot, UK, c. 2000, p. 7.
30 Ibid., p. 13.
31 R. Norris, 'Deakin, Alfred (1856-1919), ADB <http://adb.anu.au/biography/deakin-alfred-5927/text10099.htm> [accessed 17 January 2013].
32 Prices in the late 1880s are given as £40 per acre for orchard and town blocks, and £200 for residential. G. Arch Grosvenor, *Red Mud to Green Oasis*, Raphael Arts Pty Ltd, Renmark, SA, 1979, p. 18.

in tents and set about tree clearing, stump removal and channel digging. The soil was fertile, irrigation work proceeded with only minor problems, but both settlements faced early difficulties: poor-quality or incorrect orchard stock, salt seepage from unlined channels, floods in 1890, untimely rains in 1895, and frosts. After years of waiting, there was fruit to harvest, but there remained risks of transporting it fresh to city markets over long distances by the uncertain river, which, unhindered by lock or dam, could shrink to a series of barely connected waterholes in a dry season and in flood spread miles beyond its banks. Improvised drying techniques brought poor results. Refusals to pay rates and bank failures bankrupted the Chaffeys; many settlers walked off their blocks.[33]

Figure 2 Renmark 1907: Spreading fruit to dry on the property of E. E. Hutton.

Source: Photograph by Reiners Studio, image courtesy of Renmark Branch, National Trust.

Those who remained drew on their own resources. A small State Government loan kept the Renmark settlement viable, but the foundations for its future success lay in self-help, cooperation and dissemination of technical knowledge. With no money to keep the irrigation pumps going, settlers cut wood for fuel to run them. The locally elected Irrigation Trust, established by State legislation, managed irrigation and water rights. A Community Trust Hotel sought to combat the sly grog trade that had flourished under the teetotal rule of the Chaffeys;

---

33   Ernestine Hill, *Water into Gold*, Robertson & Mullens, Melbourne, 1946, pp. 89–108; Grosvenor, *Red Mud to Green Oasis*, pp. 14–21.

the town 'drank itself into a state of solvency'.³⁴ Fruit-growers organised to tackle the problems of transport and prices. The Secretary of the Renmark Fruit Packers' Union was pioneer settler F. W. Cutlack, McDougall's future father-in-law. Grower organisations in Renmark and Mildura amalgamated in 1907 to form the Australian Dried Fruits Association, controlling purchasing, packing and marketing. Distilling and citrus packing cooperatives followed in Renmark. Solutions were found for the early problems of cultivation: gypsum would break up heavy soils; wells could drain away salt; sultana grapes brought better prices than the Gordo Blanco (muscatel); the Greek practice of cincturing Zante currants would prevent early fruit loss.

Figure 3 The Murray River 1909: Paddle-steamers moored near stores of the Renmark Fruit Packing Union.

Source: Photograph by Reiners Studio, image courtesy of Renmark Branch, National Trust.

Growers gained their technical information at meetings of the 'Agricultural Bureaux', an initiative of the SA Department of Agriculture in 1889. Each local bureau was run by the growers themselves, gathering in private homes or packing sheds to read papers on everything from 'manuring' to keeping accounts, usually setting out the personal conclusions or research of an experienced member. Bureau rules required constant attendance and annual

---

34 This is the version given in Hill (*Water into Gold*, p. 171). Another is that 'Renmark proceeded, as the year unfolded, to drink itself…into a state of beauty and prosperity'. F. M. Cutlack, *Renmark: The Early Years*, Nancy B. Basey, Eltham, Vic., 1988, pp. 26–7.

turnover of one-third of members, thus helping newer arrivals who were most in need.[35] By 1909 the population of Renmark numbered some 2000, and most aspects of a settler's livelihood depended on these patterns of cooperation and shared technical knowledge. They were to influence profoundly McDougall's future thinking.

With family funds, he purchased blocks totalling some 40 acres (16 ha) in an area close to the river known as 'The Crescent', where he grew vines and apricots. The holdings doubled as he was joined later by two half-brothers and a sister.[36] He was fit and, despite apparent reluctance to accept the family decision, and references in letters to a determination to remain only briefly, he enjoyed the physical and intellectual challenges. In 1910 he reported with satisfaction that he could do three times as much manual work in a day as he had been able to achieve on arrival. Later he wrote that he enjoyed pruning most of all his tasks: 'the work is interesting, requiring judgement and it's full of hope for the next season's crop for which one is pruning.'[37] His neighbours were congenial, many with interests similar to his own, and he participated in the Agricultural Bureau with enthusiasm. In October 1911, with a mere two years' experience as a grower, he joined two prominent members of the Renmark Settlers' Association travelling by river to Mildura to examine the new method of rack drying. He wrote a paper setting out designs they had seen, with advantages and disadvantages and costs. The paper was printed in the *Renmark Pioneer*, whose Editor, H. S. Taylor, was an influential advocate of causes benefiting growers and their organisations.[38]

The block had prospered sufficiently for McDougall to take a holiday in England in the summer of 1914. He married Joyce Cutlack in 1915 shortly after enlisting in the Australian Imperial Force (AIF); their son, John, was born in England in 1918; a daughter, Elisabeth, followed in 1920. McDougall served on the Western Front as a lieutenant in the Cycle Corps, then as battalion quartermaster, and was promoted to captain. He was transferred to the AIF Education Service just days before the Armistice. He attended a school for instructors preparing troops for future careers on the land; then, from 18 December until his departure for home early in April 1919, he participated in a massive effort to keep troops busy and interested while they awaited repatriation. Staff hastily prepared suitable textbooks, with McDougall contributing some pamphlets on fruit-growing.[39]

---

35 *Murray Pioneer* [hereinafter *MP*], 17 December 1920.
36 Registrar-General's Unregistered Document Service, SA Department of Lands: Certificates of Title; Merridy Howie, 'Personalities Remembered: No. 24, Frank McDougall', Radio talk on 5CL, 15 November 1970. I am grateful to Mrs Howie for giving me a copy of her talk.
37 NLA, MS6890/1/5, letters to Kit, 18 November 1913 and 21 August 1910; NLA, MS6890/1/7, letter to Kit, 11 June 1921.
38 *MP*, 3, 17 and 24 November 1911; Malcolm Saunders, 'Taylor, Harry Samuel (1873–1932)', ADB, online edn, 2006, <http://www.adb.online.anu.edu.au/biogs> [accessed 21 November 2006].
39 Resume of military service supplied by Central Army Records Office, Melbourne, 31 August 1983.

Figure 4 McDougall pruning vines, c. 1919: 'full of hope for the next season's crop.'

Source: E. McDougall.

Figure 5 F. L. McDougall, AIF, 1916.

Source: E. McDougall.

Production

Sultanas (Sultana, Sultanina and Thomson's Seedless varieties) are the most important vine fruits, although raisins (*Gordo Blancos*) and currants

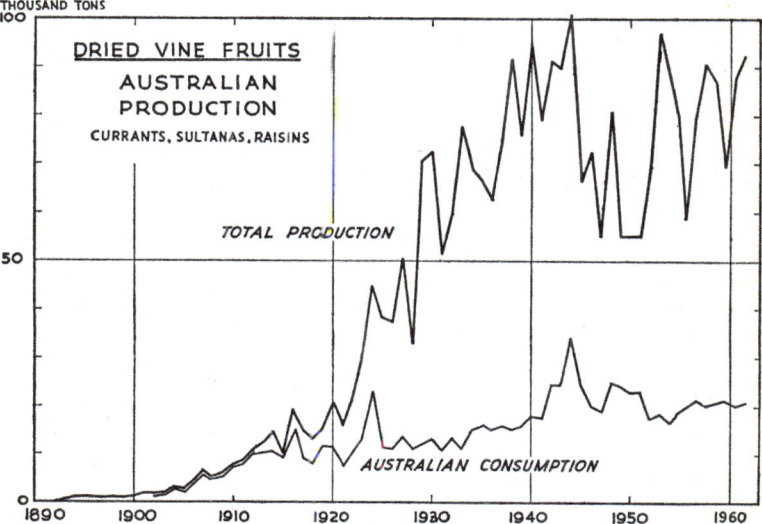

Fig. 42. Production of dried vine fruits

The great expansion of the dried vine fruit industry after World War I was due to soldier settlement, mainly in the irrigation areas of the Murray Valley. Large annual variations in output are typical of the industry.

[2] See R. H. Martin in *The Journal of the Australian Institute of Agricultural Science*, vol.

Figure 6 Australian Production of Dried Vine Fruits, 1880–1960.

Source: S. Wadham, R. Kent Wilson and J. Wood, *Land Utilization in Australia*, fourth edn, Melbourne University Press, Melbourne, 1964, p. 194.

It was an immature Frank McDougall who had arrived in Renmark in 1909. Captain McDougall returned in 1919 as a family man of purpose, with an understanding of the importance of education and the problems of his industry. As a returned soldier, he received financial benefits, enabling him to repay the loans from his family and to plan expansion of his enterprise. Those plans were soon to be overtaken, however, by the first of his important ideas: a program to put Australian dried fruit on a sound export footing.

Australia's Dried Fruit Industry

Cultivation in the Murray irrigation areas had increased gradually in the first decade of the twentieth century, in 1901 yielding 2050 tons of dried fruits (sultanas, currants and lexias—the last popularly called raisins) and 9000 tons in 1910. By 1912 local production was almost sufficient to supply the domestic market. In 1913 Australia exported just over 1000 tons of dried fruits—roughly one-tenth of total production. About half of this, in value, went to the United Kingdom, and one-quarter to New Zealand. A smaller market was being developed in Canada.[40]

Production increased rapidly during the war years, encouraged by high prices at home and overseas. Exports rose gradually at first, and then leapt rapidly as blocks planted in response to high wartime demand began bearing. Between the 1914–15 and 1917–18 seasons Australian exports of dried fruits increased fourfold; returns increased more than sevenfold. Exports had doubled again by 1920–21, and returns trebled. Postwar soldier settlement rapidly expanded the acreage under vines; greater acreage stimulated technological innovation, further increasing output. Production exceeded 40 000 tons in 1924, and reached 55 000 tons in 1927. The local market at best could consume little more than 20 000 tons. With assistance from a wine bounty, that figure included some 6000 tons used by wineries and distilleries.[41]

Table 1 Australian Exports of Dried Fruits by Weight and Price.

Season	1000 lbs	£1000s
1914–15	2314	36
1915–16	8255	244
1916–17	13 460	373
1917–18	9427	266
1918–19	8525	253
1919–20	16 788	606
1920–21	17 811	729

Notes: For clarity, numbers have been rounded to the nearest thousand. Figures for 1919–21 have been consolidated from separate categories.

Source: *Year Books of the Commonwealth of Australia.*

London was the largest market for dried vine fruits in the world, absorbing annually some 125 000 tons before 1914. Its most important source was the Mediterranean region, where labour and transport costs were low. A rapidly growing industry in California, subsidised by a large domestic market, could supply cost-competitive, high-quality products. Prices had risen as Mediterranean supplies were cut off by war, reaching a peak roughly three times the prewar level in 1919–20.

---

40 Development and Migration Commission, *Report on the Dried Fruits Industry of Australia*, Melbourne, 1927, pp. 9–11; Commonwealth Bureau of Census and Statistics, *Trade and Customs and Excise Revenue of the Commonwealth of Australia*, Melbourne, 1914–20.
41 Development and Migration Commission, *Report on the Dried Fruits Industry of Australia*, pp. 6, 9–10.

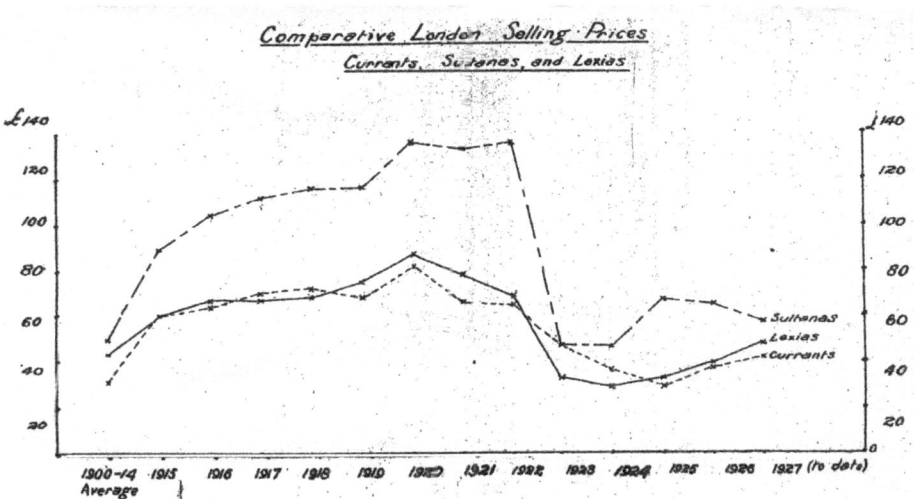

Figure 7 Dried Fruit Prices on the London Market, 1900–27.

Source: Development and Migration Commission [Australia], *Report on the Dried Fruits Industry*, Melbourne, 1927.

On his return to Renmark in 1919, McDougall found evidence of expansion and optimism. New irrigation areas were being planned and block prices were rising. His own property had almost doubled in value. Electric light and power had reached the area. A nearby orchard boasted a new heated drying house with trays carried in and out on a sunken railway. An automatic dip installed on another property processed 800 tins of sultanas an hour; at best a man could manage 2000 a day.[42]

McDougall's optimism was tempered with understanding of the difficulties facing his industry. On his holiday home in the European summer of 1914, he had taken time to investigate the British market, seeking views in London and in northern cities from buyers, brokers and cooperative organisations. He had asked for comparison with Californian dried fruit and about government-controlled marketing of South African fresh fruit. He had learned that Australian fruit was supplied spasmodically and thought to be badly graded, over-dried and unattractively presented compared with the Californian product. He had concluded that Australian growers were ill informed about prices in Britain, with little understanding of the needs and views of British buyers.[43]

During the war, H. S Taylor is reported to have sent free copies of the *Murray Pioneer* to all Riverland volunteers serving overseas.[44] These may have kept

---

42 NLA, MS6890/1/9, letter to mother 2 July 1919; *MP*, 28 February 1919.
43 F. L. McDougall, 'Notes on the Dried Fruits Industry in Australia', *Murray Pioneer*, 23 May 1919.
44 Saunders, 'Taylor, Harry Samuel'.

McDougall abreast of developments in Renmark: higher prices, extended plantings and the prospects of soldier settlers adding to grower numbers.[45] In the rest camps in France he organised several 'Agricultural Bureaux' at which 'good papers' were followed by 'keen discussion'.[46] He had estimated that some 3000 men awaiting repatriation were in the agriculture section of the AIF Education Service; he knew at first hand that many returned men were eager to take up land under soldier-settler schemes.[47] But he also knew that a postwar export market could not be assured.

1918–19: 'Notes on the Dried Fruits Industry in Australia'

McDougall was in his late teens in 1903 when Joseph Chamberlain's tariff reform campaign began. With his strong social conscience and eagerness to find a place in the world, he must have taken some interest in it then. He had occasional home leave in London during the war, and probably had access to British newspapers, as well as mail, at other times. He would have known of renewed imperial sentiment and debate on the tariff as the date for the Budget of 1919 approached. During the last months of the war and its immediate aftermath, he wrote a paper about the problem of finding markets for Australia's increasing dried-fruits crop. He later explained to the Renmark Returned Soldiers' Association that he had some 'difficulty in persuading the adjutant of his unit to allow the manuscript to be typed in the orderly room'.[48] 'Notes on the Dried Fruits Industry in Australia' set out his ideas on marketing, considering every step of the trade from irrigation block to London retailer. He drew on his own experience. The paper was deceptively simple, its language matter-of-fact and unpretentious, but, like the best of the many papers McDougall was destined to write, it was authoritative, logical and persuasive. It was sent from England to be read at the Renmark Agricultural Bureau. Bureau President, F. H. Basey, thought it important enough to keep for the 1919 Conference of River Agricultural Bureaux. The paper was read there by McDougall's half-brother Norman on 16 May.[49]

---

45 Marilyn Lake, *The Limits of Hope: Soldier Settlement in Victoria 1915–38*, Oxford University Press, Melbourne, 1987, p. 32.
46 Remarks by Norman McDougall at River Agricultural Bureaux Conference, 16 May 1919, in *MP*, 23 May 1919.
47 *MP*, 17 June 1919. Total numbers, across agriculture, mechanics and commercial courses, were estimated at perhaps 40 000; Geoffrey Serle, *John Monash: A Biography*, Melbourne University Press, Melbourne, 1982, p. 409.
48 *MP*, 11 November 1921.
49 *MP*, 23 May 1919. The issue includes a full copy of the paper. A separately printed copy is in NLA, MS6890/4/2.

The conference opened on a pessimistic note. A. J. Perkins, SA Director of Agriculture, recalled marketing difficulties of the past and predicted more with postwar expansion. He could see little hope for export markets, given high labour and production costs, even with mechanisation and cooperative methods; he could suggest no alternative crop for irrigation areas except sugar beet. Perkins' speech was followed by McDougall's paper. It also recalled past problems and dismissed some solutions. Export had not been taken seriously earlier and some fruits—currants, peaches and lexias—could not compete in world markets. The irrigation settlements had been saved in the past by establishment of distilleries and there was a view that canning and jam production offered the only hope of absorbing expanded postwar production. But then came a call for a 'radical… change of viewpoint':

> …overseas markets will become of increasing importance…
>
> I want the Dried Fruit Producers to establish an industry in the semi-arid districts of Australia which will be able to hold its own in the world's markets and therefore to welcome expansion of the industry and to look forward to making the Murray Valley a second California…
>
> I want us to maintain that the Dried Fruits Industry is particularly suited to Australia…
>
> I want the industry to be in such a state that we can cheerfully claim that it is the pioneer industry of good working conditions…that it combines the healthy conditions of ordinary agricultural life with the reasonable hours and opportunities for recreation of the urban industries.

McDougall surveyed the industry's problems, including the criticisms encountered during his own investigations in 1914. He proposed solutions under four headings: information, organisation, cooperation and British preference. He wanted efficiency and rational organisation. Growers themselves, with government support, should investigate and disseminate information on marketing conditions. He suggested tighter control by packing houses and use of cooperative organisations, well paid and run, to improve production, packing, finance and marketing, to eliminate middlemen and to maintain consistent quality.

His crucial argument, however, was the need for British tariff preference. As matters stood, Australian fruit could not compete with fruit from California, which was subsidised by an immense home market, or with Mediterranean fruit produced by cheap labour and with low transport costs.[50] Indications were that

---

50 The daily wage for unskilled labour was the equivalent of 1/6d per day in Greece, 11/6d in Australia. Hill, *Water into Gold*, p. 235.

Britain was preparing to abandon its historic free-trade policy; it had already been chipped away by small revenue tariffs on some commodities, including most dried fruits. A British preferential tariff would allow Australian fruit to meet competitors on a level footing. He urged growers to establish exactly how high that tariff need be for a profitable export trade.

The paper drew considerable discussion and praise. It was read just after the British Budget of 1919 had been delivered, but before full details of the new British tariff—whether dried fruits were included in tariff duties and, if so, to what extent—had reached Australia. H. S. Taylor noted that dried fruits had not been mentioned in cabled information about British intentions, and asked whether the Australian Dried Fruits Association (ADFA) had brought the question before the British authorities. As the Agricultural Bureaux were self-help, not political, organisations, the conference passed a motion recommending ADFA 'take such steps as are necessary to secure full British preference for Australian dried fruits'.[51]

The Australian Dried Fruits Association

ADFA owed its existence to W. B. Chaffey's desire that growers cooperate to control marketing. It had grown from separate organisations in Mildura and Renmark to a membership including 95 per cent of all fruit-growers, some packing companies, agents and merchants. In theory it could control the product from grower to retailer. Australia's prewar consumption of currants and sultanas, protected by tariff duties, was almost fully supplied by ADFA, and absorbed nearly all the product. Tentative prewar export consignments had generally proved unprofitable, and talk of developing markets in the Far East and the near north had been little more than that.[52] When war opened up profitable export markets in Britain, ADFA undertook to supply home needs first, but soon found that much higher prices could be gained on the London market. Forty per cent of the 1919 crop was exported.[53]

ADFA's first response to the problem of finding markets for postwar production was to increase home consumption. The 1918 ADFA Conference appointed a young Mildura businessman, C. J. De Garis, as its Director of Publicity. De Garis spent a lavish budget of £20,000 on a nationwide advertising campaign owing much to US inspiration. Richly rewarding competitions were launched on billboards and in newspapers throughout Australia. Slogans, recipe books and gimmicks followed. The *Murray Pioneer* reported campaign progress in detail and

---

51 *MP*, 23 May 1919.
52 Hill, *Water into Gold*, pp. 168–9, 178–81. See also speech by De Garis, *MP*, 30 May 1919.
53 See Figures 6 and 7 and Table 1.

carried its advertisements; it was doing well at the time McDougall's paper was read. De Garis addressed the River Bureaux Conference, urging the importance of brand names, market research, public education about food values and tactics to change buying habits. Admirable as McDougall's paper was, development of the home market offered better prospects than the 'statesmanlike' work for imperial preference.[54]

March 1921: 'A British California on the Murray'

McDougall arrived back in Renmark at the end of May 1919, two weeks after his paper was read. For the next two years, he was busy with family, block and community work. He served a year as President of the Agricultural Bureau and as a member of the Irrigation Trust—effectively the local council. There he had experience of responsibilities for management of public money and control of works.

H. S. Taylor had bought what was then the *Renmark Pioneer* in 1905 and 'transformed it into a major influence' in the irrigation areas of the Murray Valley. Renamed the *Murray Pioneer*, Taylor's paper advocated the cause of irrigation, locks for the Murray, closer settlement, organised marketing and producer cooperation. Taylor himself led deputations to premiers and government ministers.[55] With Taylor's help, McDougall campaigned for State Government support for conservation and cultivation of trees along the river to meet high demands for timber in the area. They achieved support from the 1921 State Conference of Agricultural Bureaux and advice from the State Advisor on Forests, but little in the long term.[56] The campaign did teach McDougall the importance of press support and of expert opinion; it also taught him the need for persistence.

Publicly he did little for the preference cause; privately he talked with friends and rethought his 1919 paper.[57] In 1921 he recast it for a broader audience of political figures and opinion framers in Australia and in Britain by appealing to the burgeoning interest in empire settlement. Its central proposition was now that in return for substantial preference on Australian fruit, irrigation land should be offered free to British returned soldiers. He described the Murray irrigation area, 'thumped the big drum a little, fell back on statistics and finished with a lyrical purple patch over the scheme's Imperial significance…

---

54   *MP*, 30 May 1919.
55   Saunders, 'Taylor, Harry Samuel'.
56   *MP*, 24 September and 8 October 1920, and 19 August, 23 September and 21 October 1921.
57   Recollection of T. C. Angove, reported in *MP*, 10 November 1922.

How I loathe the terms Empire and Imperial but what else can one use without elaborate explanations'.[58] He called the new paper 'A British California on the Murray'.

By year's end McDougall was enthusiastically supporting development proposals, maintaining that half a million people could be settled on the Murray; he later revised that figure to one million.[59] His estimates of potential production were sound and often surpassed in reality; the population prediction was of a piece with over-optimistic contemporary views represented by 'Australia Unlimited'.[60] In 1928 a compilation of essays by various experts included estimates of population potential ranging from 14 to 200 million.[61] Prime Minister, W. M. Hughes, and some experts had endorsed a suggestion of 100 million. Australia's population in 1921 was some five and a half millions; a population of one million in the Murray lands must have seemed a breathtaking figure. The potential of irrigation and other development schemes in Australia was nevertheless widely and enthusiastically embraced; the fragility of inland river systems was scarcely understood at all.

McDougall's idea had taken its final shape by 1921, as had his plan of campaign. First steps would be based on ADFA and on the press in both Australia and Britain.[62] By midyear, the *Murray Pioneer* could claim that preference was 'coming on'.[63] Another rural newspaper, *The Leader*, acknowledged that the dried-fruits industry was becoming increasingly dependent upon successful export. The problem might well be too big for ADFA. By 'pouring settlers by the thousand' into the industry, government had assumed some obligation to those who had already made it a success.[64] The Editor of the *Mildura Sun* wanted the fruit industry put on 'a war footing' to deal with federal and State legislatures. ADFA should buy up his paper and the *Murray Pioneer* and use both to push 'the greatest Imperial idea that has been heard of yet in Australia: the establishment...of a new, self-governing Rhodesia, in the shape of a Sun-Raysed State'.[65] In May, State ministers of agriculture resolved to encourage substantial immigration within the empire and to lobby for increased preference on primary products, including fruits.[66] A speech in the SA House of Assembly calling for preference in return for settling British ex-servicemen on the Murray

---

58 NLA, MS6890/1/7, letter to Kit, 20 March 1921.
59 Ibid., undated letter to Kit, November–December 1921.
60 See, for example, Hancock, *Survey*, pp. 133–4; Powell, *Historical Geography of Modern Australia*, pp. 131–2.
61 P. D. Phillips and G. L. Woods, eds, *The Peopling of Australia*, MUP/Macmillan, Melbourne, 1928.
62 NLA, MS6890/1/7, letter to Kit, 11 June 1921.
63 *MP*, 5 August 1921.
64 *Leader* article of 19 November, reprinted in *MP*, 2 December 1921.
65 Dated 12 September, published in *MP*, 23 September. Taylor, flattered but not tempted, wrote that the proposal was nevertheless worthy of serious consideration.
66 National Archives of Australia [hereinafter NAA], A458, R508/1, part 1. The resolution was forwarded by SA Premier, H. N. Barwell, to the Commonwealth Government on 28 June 1921, and supported by letters from the Premiers of New South Wales and Tasmania.

was greeted with cheers. H. S. Taylor had supported imperial preference for 17 years, 'but it is only recently that it has been taken hold of by men really desirous of bringing the idea to fruition'. Taylor gave McDougall credit for originating the linking of preference and imperial soldier settlement.[67]

'A substantial British preference'

Besides linking preference to settlement, McDougall's 1921 version tackled a question left unanswered in 1919: how much preference was needed to make fruit export viable? The British Budget brought down on 30 April 1919 included preferences equivalent to one-sixth of low, existing revenue duties on tea, cocoa, coffee, sugar, tobacco and dried fruits. Most of the dried fruits attracted duty of 10/6d per hundredweight (cwt), which, reduced for empire products to 8/9d, meant a preference of 1/9d. Currants attracted a duty of only two shillings per cwt, reduced by a four pence imperial preference to 1/8d. The small reductions were unlikely to make Australian fruit competitive in normal times. Currants exported before the war had sold at a loss for 25–30 shillings per cwt; the wartime high of 90 shillings had fallen to 50 shillings by 1921. Lexias had not paid, prewar, at 30–40 shillings per cwt; sultanas had barely broken even at 45–65 shillings.[68] McDougall believed that nothing less than threepence preference per pound (that is, 28 shillings per cwt) could give growers a comfortable income.[69] *The Leader* later found this suggestion 'manifestly extravagant'.[70] It meant the price of foreign fruit in London would rise a further 28 shillings above prices of at least 50 shillings per cwt for empire fruit; it was asking a lot of British consumers and their government. But McDougall had argued to his audience of returned soldier 'blockers'—most of them new to the industry—that price was more important than quantity of sales. New markets in Asia might be developed, as had been proposed, but these, like London markets, would be at risk of dumping from California as other supplies returned to normal. Substantial preference was the only sure protection for adequate returns for growers. The meeting carried unanimously a resolution urging other returned servicemen's groups on the river to lobby the SA Premier to support 'a substantial British preference'. The phrase gained currency: the Renmark Agricultural Bureau passed a motion proposed by McDougall, requesting the SA Advisory Board of Agriculture to recommend 'all possible methods' be used to obtain a substantial preference.[71]

---

67 *MP*, 5 August and 21 October 1921.
68 NAA, A458, K500/2, part 1. The figures were submitted by ADFA to the Premiers' Conference in 1921.
69 Paper delivered by McDougall to Returned Soldiers' Association in Renmark. *MP*, 11 November 1921; copy in NLA, MS6890/4/2.
70 Article printed in *MP*, 19 November 1921.
71 *MP*, 11 and 18 November 1921.

Daily Mail proprietor, Lord Northcliffe, who, McDougall believed, possessed 'greater powers of useful publicity than any man in England', toured the Antipodes in 1921.[72] Both the *Murray Pioneer* and the *Mildura Sun* urged that Northcliffe be invited to the Murray to spark a publicity campaign attracting migration and investment to what was potentially 'a mighty inland State stretching from Renmark far beyond Mildura and with a population of at least a million men'.[73] ADFA representatives asked South Australia's Premier, H. N. Barwell, to seek assurances to Northcliffe from all premiers that Murray lands would be supplied to British settlers in return for preference. Northcliffe did not visit the Murray, but A. E. Ross, a grower from Waikerie 'with a fine presence', was chosen to represent ADFA's arguments to Northcliffe, who promised support from the *Daily Mail* once the scheme was launched in Britain. Barwell assured Renmark growers his government was determined to establish markets for the produce of 'the great Murray lands', and was exploring measures to open up promising trade prospects in the East. State ministers of agriculture had agreed that everything should be done to achieve substantial tariff preference. His own government had contacted the Commonwealth Government and concerted action by the States was contemplated. 'We've progressed a fair way', reflected McDougall, and thought his settlement scheme itself might be named 'A British California on the Murray'.[74]

---

72 NLA, MS6890/1/7, letter to Kit, 3 October 1921.
73 Grant Hervey, Editor of the *Mildura Sun*, in *MP*, 2 September 1921.
74 *MP*, 16 September and 2 December 1921; NLA, MS6890/1/7, letter to Kit, 3 October 1921.

2. Working with Bruce

Summary

The Australian Dried Fruits Association took the campaign for tariff preference, summed up in a memorandum by McDougall, to the Premiers' Conference late in 1921. The Federal Government was persuaded to raise the issue of increasing preference with the British Government, which refused to consider a measure increasing the cost of food. A delegation to help clear stocks of Australian tinned fruits in London included McDougall, who gained the approval of Stanley Melbourne Bruce, a young and very new Federal Treasurer in 1922, for his continuing to campaign for tariff preference while in London. By the time McDougall reached London, Bruce had replaced W. M. Hughes as Prime Minister. McDougall reported regularly to Bruce on the progress of his lobbying, and was asked to assist him prepare for the 1923 Imperial Conference. He helped write Bruce's major speech to the Imperial Economic Conference, calling for imperial economic cooperation through the provision of 'men, money and markets'. Soon after the conference, British Prime Minister Stanley Baldwin sought endorsement of further tariffs at a general election. McDougall assisted Bruce in his support for Baldwin's case and also urged Bruce to ensure that Australia's rural industries were organised for efficient marketing. With the election lost and an anti-tariff minority Labour Government in power in London, Bruce summoned McDougall to Australia in 1924 to help prepare legislation for organisation schemes for industries including dried fruits. It was agreed between the two that McDougall should return to London, paid partly by the Federal Government for lobbying work and partly by the new Dried Fruits Export Control Board, of which he would be the London Secretary. By this time Bruce and McDougall had developed a solid working partnership, the nature of which is discussed at the end of the chapter.

November 1921: 'A California within the Empire'

ADFA delegates from four States conferred in the temporary federal capital, Melbourne, in October 1921, and resolved to seek federal and State help in 'agitating' for a preferential duty on dried fruits in the United Kingdom. As ADFA deliberations concluded, senior ministers from all States were arriving for a premiers' conference. On its first day, Barwell requested consideration of an additional agenda item concerning 'some matter' from the ADFA Conference.

An ADFA deputation wished to see the Prime Minister. H. D. Howie of Renmark and a Mr Victorsen of Clare saw Hughes at his home on Tuesday, 1 November, the Melbourne Cup Day holiday.[1]

Barwell's 'matter' was yet another version of the paper that had occupied McDougall throughout 1921. A preliminary statement submitted to the conference set out the problem: London prices were already falling to half the wartime level, and might reach a point where export would not pay. The present acreage was expected to increase more than sevenfold, and ADFA's expensive three-year advertising campaign had brought sales in Australia to saturation point. With the paper was McDougall's six-page memorandum prepared for Northcliffe, headed 'A California within the Empire'. It presented the case from the British point of view, including likely returns from fruit-growing for a settler and opportunities for British investment. Both depended upon a preference rate of 3d per pound. The paper suggested, as a *quid pro quo*, an invitation to British returned soldiers to work side-by-side with 'old comrades in arms', to enjoy the healthy lifestyle and community spirit of an industry already established for them. It answered objections to taxing food: even at a high price of one shilling per pound, dried fruits would constitute the cheapest form of fruit food. The Imperial Exchequer would benefit from customs duties until 'a great producing centre has been built up within the empire which can supply all its needs'.[2] McDougall already understood the advantage of propounding the vision from a central point of view.

Hughes promised ADFA delegates full support.[3] But next day he told the assembled premiers he held 'no hope whatever' of a British Government agreeing to increase the price of the people's food. Nor did Barwell. The premiers spent much of that day seeking other means to sell more Australian fruit. Stories of poor-quality fruit and packing, inadequate advertising and poor comparison with Californian fruit were shared round the table. All agreed 'immediate action is imperative'. 'We cannot attack a more serious problem', said Victoria's Premier, H. S. W. Lawson.[4] Their decisions formed the major part of Hughes' statement at the conclusion of the conference. The Federal Government and State Governments would work with producers to ensure a standard product through uniform inspection and grading, and would jointly fund commercial representatives overseas. Migration and marketing were linked, said Hughes: 'successful land settlement involves more than merely placing men upon suitable

---

1 *Argus*, 31 October 1921; *MP*, 4 November 1921; NAA, A9504/1, 3, 31 October 1921.
2 NAA, A458, K500/2, part 1.
3 *MP*, 3 March 1922.
4 NAA, A9504/1, 3, 2 November 1921.

areas…We must find remunerative overseas markets for what they produce.' This was particularly true of the fruit industries, whose very success 'now threatens us with disaster'.[5]

Although politicians had written off preference, ADFA persisted. Its conference had appointed a committee of five to pursue the campaign. McDougall, one of the five, explained his vision to his sister. Once the Federal Government was brought 'into line', it might send a full statement of the case, 'nicely printed', to England, followed by representatives 'to push the matter, to lobby in the House, and to try and get the necessary backing…The committee all say that I must be the one to go to England.'[6]

ADFA had discussed the idea of sending irrigation pioneer William Chaffey and McDougall to England early in 1921, but ADFA funds were low after a poor season and the expense of the publicity campaign, so the idea was put aside.[7] Almost a year later the idea remained in doubt. South Australian delegates favoured the trip but Chaffey did not. McDougall wondered, without much hope, whether the Federal Government could be persuaded to share the cost. He was reluctant to go without his family and foresaw jealousies within ADFA.[8] Official travel overseas was a rare and costly privilege, avoided as far as possible by making use of Australians already overseas on other business. It could very easily create resentment.

A two-day Premiers' Conference in January 1922 dealt with matters left over from the November meeting. Barwell, with support from Victoria's Lawson, fulfilled a promise to his State Assembly to continue pressure on preference. Hughes had confirmed with Colonial Secretary Winston Churchill that a British Government would not tax the people's food. Hughes suggested 'a first class business man' could achieve sufficient market share in Britain without preference. Barwell countered that a preference on dried fruits already existed; it was simply not high enough. He recalled the proposal to make fruit-growing land available for settlement in exchange for preference. Still doubtful, Hughes promised to telegraph Churchill again.[9]

Much of the cable sent on 28 January might well have been drafted by McDougall himself. It extolled the 'almost unlimited' possibilities of the dried-fruits industry, the capacity of the British market to absorb the product, its economic importance to the Empire and its direct bearing on settlement of British ex-soldiers. Australian ministers in conference urged British assistance

---
5 *Argus*, 5 November 1921.
6 NLA, MS6890/1/7, letter, probably to Kit. Page one is missing, but the content suggests it was written before the second Premiers' Conference, probably in November or December 1921.
7 Ibid., letter to Mother, 7 March 1921.
8 Ibid., undated letter to Kit of November–December.
9 NAA, A9504/1, 3, 17–18 January 1922.

to the industry in the form of a preferential tariff of one and a half pence per pound on currants and twopence on other fruits.[10] The amount was less than ADFA wanted, but the British Government baulked at such an increase over existing rates. To achieve Australia's request, the existing duty on foreign currants must rise from 2 to 14 shillings per cwt, assuming empire fruit was admitted duty-free. Duty on other fruits would almost double to at least 18 shillings and 8 pence.[11] Britain would not accept so great an increase in the cost of a food largely consumed by the working classes, nor a possible loss of customs revenue if imports of foreign fruits liable to duty were to decrease in favour of free-entry empire produce. Despite 'every sympathy' with the object of the request, it was 'quite impracticable'.[12]

The ADFA committee would not give up. The proposals put to London had been 'half-baked' and bound to fail, claimed the *Murray Pioneer*. McDougall determined to answer the case against preference with more information, better presented. He asked the Commonwealth Statistician to help calculate just what effect an adequate preference would have on the British Exchequer. Armed with a conclusion that revenue would in fact increase by some £1,600,000 per annum, McDougall prepared a new, longer statement: 'The Case for Preference.'[13]

The Fruit Delegation

In May 1922 McDougall was summoned to Melbourne to address 'a special Ministerial Conference' of the Murray States, arranged by Lawson to discuss preference. The conference requested further negotiations with London. It also agreed that the States should prepare 'an attractive immigration offer as a *quid pro quo*', and that States and Commonwealth together should fund a delegation of 'practical men' to put the case before influential people in England. State ministers for agriculture endorsed the proposals in Perth later that month.[14]

A national conference of fruit-growers was called in August 1922 to effect measures to improve standards in the industry. It was attended by ministers for agriculture, more than 100 growers, and Hughes himself on the final day. The conference agenda included establishment of an Australian Advisory Fruit Council and State Advisory Boards, a national trademark, and inspection of

---

10   NAA, A11804, 1922/232; text reprinted in *Argus*, 28 January 1922.
11   Figure 7 shows that London wholesale prices in 1919–22 for lexias and currants ranged between £60 and £80 per ton (60–80 shillings per cwt). Prices for sultanas had been almost double that at their highest, but fell sharply after 1922. McDougall stated in his paper 'A California on the Murray' that the highest retail price was about one shilling per lb (equivalent to 112 shillings per cwt).
12   NAA, A11804, 1922/382, 24 March 1922.
13   *MP*, Editorial, 9 June 1922; report of McDougall's speech to growers, 23 June; a version of 'The Case for Preference' published 7, 14, 21 and 28 July 1922.
14   *MP*, 12 May 1922; NAA, A457, Y300/7, McDougall to conference of fruit-growers, 14 August 1922.

grading and packing, storage and handling. But it was also required to approve 'an urgently pressing matter': the sending of a delegation to Britain in time to negotiate effectively for the new season's deliveries. Arthur Rodgers, Federal Minister for Trade and Customs, dismissed suggestions that 'either the Agent-General or the High Commissioner are the men for this job…it requires men who can go, with the weight of life-long experience and supported by the governments, to fight for every branch of the industry'.[15]

McDougall shared the platform with Rodgers, giving the meeting an account of negotiations with governments since the January Premiers' Conference. Fellow South Australian, ADFA delegate A. E. Ross, countered any lingering reluctance: the real beneficiaries of the venture would be new ex-service growers who, without preference, would 'make an unholy failure' at ultimate cost to the taxpayer'. The motion in support was carried unanimously.[16] The delegation, comprising two irrigation growers, McDougall and Chaffey, and C. E. D. Meares, General Manager of the Coastal Farmers' Co-operative of New South Wales, left for London on 20 November 1922, their work funded jointly by the Commonwealth and the States. Members would be paid a daily living and entertainment allowance of five guineas once they reached Britain and two guineas per day on board ship.[17] McDougall left his block in the hands of his half-brother Norman.

The *Murray Pioneer* credited government support for the delegation to McDougall's 'able advocacy' and 'tireless persistence'. Speakers at a farewell dinner described him as 'a technical expert in the question of preference', a man of 'bulldog tenacity' and 'dogged persistence'. McDougall replied modestly, but did claim to have 'worn holes in about all the ministerial doormats in Melbourne and Adelaide'. The task of 'preparing the British mind for preference' would be an easy one if a Conservative Government were returned in a general election set for 15 November, 'very difficult but not hopeless' under a Labour/Liberal Government: 'if British Labour men meant half of what they said it should not be impossible to convert them to a policy which had for its object the maintenance of a decent standard of living.'[18]

Before the Fruit Delegation left Melbourne, Rodgers handed them formal instructions. They were to advance sales of Australian dried, fresh and canned fruit. They were to consult with the Australian High Commissioner and State Agents-General on a campaign of persuasion and 'judicious propaganda' to

---

15   *MP*, 1 September 1922; NAA. A457, Y300/7, report of conference, pp. 72, 77.
16   Ibid., pp. 77–84.
17   NAA, M111, 1925, statement attached to McDougall's letter to Bruce, 14 December 1925. See also *LFSSA*, 45, 14 December 1925, pp. 129–30.
18   *MP*, 10 November 1922.

merchants, brokers and large eating houses, and on means of disposing of stocks of canned fruit already built up in London. They were to report fortnightly to Rodgers, and the period of the mission was limited to six months.

But then the instructions took a surprising turn. At the fruit-growers' conference Rodgers had explained that the delegation would 'convince the consuming public of Britain, the merchants and the British Government that it was a wise thing to give preference to Australian fruit'. In Rodgers' presence, McDougall had spoken at length on lobbying for preference. Now, the written instructions forbade it:

> You are, under no circumstances, to approach the Imperial Government, British Ministers, or organised bodies on the question of Imperial Preference. This matter, the Commonwealth Government can delegate to no private individual. You may, of course, say that Australia will be represented by a full supply of choice fruits at the British Empire Exhibition…you should make no communications to the press either privately or otherwise.[19]

There was no recorded explanation for the reversal. The letter would have been drafted by Rodgers' department, which had taken a conservative position in regard to preference throughout the preceding debate.[20] It may have been bureaucratic inability to deal with an unorthodox proceeding. McDougall's comment suggests something of the sort: 'Rodgers tried to wind red tape around us but I had an eminently satisfactory interview with Bruce as to aims and objects.'[21]

Bruce

Stanley Melbourne Bruce was then Australia's Federal Treasurer and was shortly to become Prime Minister. Bruce was born in Melbourne in 1883, the youngest child of John Munro Bruce, partner in the prominent softgoods importing firm Paterson, Laing and Bruce (PLB), and his wife, Ann. He was educated largely in Melbourne, though the family spent some time in England during his childhood. After his father's death in 1901, he moved with his mother and invalid sister to England, studying law at Trinity Hall, Cambridge. Bruce's academic achievements were outweighed by his success in rowing: he was a member of the winning crew in 1904 and subsequently coached Cambridge crews. He was admitted to the English Bar and developed a practice in company

---

19 NAA, A458, 1500/2, part 1, Rodgers to Meares, Chaffey and McDougall, 9 November 1922.
20 NAA, A458, R508/1, various examples.
21 NLA, MS6890/1/8, letter to Norman, 10 November 1922.

law. PLB was substantially funded by British shareholders; Bruce became its London Chairman in 1906, while his brother Ernest managed the Australian operation.

In 1913 Bruce married Ethel Anderson, elegant daughter of a prominent Melbourne family. The marriage was a close and devoted one. The couple had no children, but took a keen interest in their many nieces and nephews. Bruce was commissioned in the British Army and fought in the Gallipoli campaign. He was awarded the Military Cross for leading a rescue of 42 isolated comrades and the *Croix de Guerre avec Palme* for support his battalion had given to the French. In October 1915 he suffered a severe knee wound and spent 18 months in hospitals before returning to Australia in 1917, still on crutches, to take over PLB after Ernest enlisted. After Ernest's death in 1919, he became responsible for both the London and the Melbourne offices. He oversaw PLB's increasing expansion and profitability, and also ensured a profit-sharing scheme was extended to junior staff.

During the war the Australian Labor Party had split on the issue of conscription. Prime Minister W.M. Hughes left the party with almost one-third of its MPs, and joined with the pro-conscriptionist Liberals to form the Nationalist Party. In 1917 the Federal seat of Flinders on the outskirts of Melbourne became vacant and Bruce was persuaded to stand for the Nationalists at a by-election. Despite a myth that he stumbled unwillingly into politics, it has been argued that he had a clear agenda and, using the 'guise' of a businessman who disliked politics, he was able to criticise the Government and yet accept most of its legislation.[22] He claimed to stand for 'efficiency and economy in the conduct of affairs'; he urged 'business methods' in government development of industry and government promotion of immigration and irrigation. He was elected to the House of Representatives on 11 May 1918, one of very few returned soldiers at that time. In 1919 he was sent by the Government to investigate repatriation systems in Britain, the United States and Canada. In the general election campaign at the end of that year he claimed to stand for maintenance of the British Empire, a living wage, a health insurance scheme, soldier settlement, development of markets in Asia and tariffs to create new secondary industries.

In the 1919 general election, the Nationalists won 37 seats and Labor 27; 11 Country Party members generally supported the Nationalists, without formal agreement. In 1921 Bruce was obliged to spend several months in London on PLB business. While there, he was appointed senior Australian representative at the League of Nations, where he spoke graphically of the horrors of war, saying: 'If the League of Nations goes, the hope of mankind goes also.' As Bruce returned

---

22   Heather Radi, 'Bruce, Stanley Melbourne [Viscount Bruce] (1883–1967)', *ADB*, <www.adb.online.anu.edu.au/biogs/A070460b.htm>

to Australia in October, Hughes offered him the post of Minister for Trade and Customs. Bruce refused, but did accept a subsequent offer of Treasurer. Members of the National Union, backers of the Nationalist Party, wanted the appointment of a businessman. Bruce resigned from PLB when his appointment was announced on 21 December. His only budget, in 1922, proposed reduced income and company taxes, reduced tariffs on imports of galvanised iron, wire and tractors, and a bounty for local production of these essential products for farmers. He established a sinking fund to redeem the domestic war debt within 50 years, based on the practice of other developed countries. He claimed that his measures were inspired by resolutions of the International Financial Conference in 1920. He also paid special attention to war service homes and soldier settlement.[23]

Many members of the National Union were critical of Hughes' policies as extravagant and socialist. The leader of the Country Party, Earle Page, disliked him. At the general election in December 1922—the first to use preferential voting widely—Nationalist numbers fell to 28 seats. Labor rose to 30, leaving the Country Party with 14 seats and holding the balance of power. That party decided not to support any government led by Hughes. Bruce, aged thirty-nine, became Prime Minister on 9 February 1923. He has been called the 'architect' of the enduring coalition between Nationalists (later United Australia Party and subsequently the Liberal Party of Australia) and the Country Party (later the National Party and now The Nationals). His policies appealed to interest groups including business and returned servicemen; his speeches, though workmanlike, impressed with solid content; one observer described him as 'ponderous and measured…at his best when addressing Chambers of Commerce or trade meetings'.[24] But no doubt his courteous and reserved demeanour, enhanced by a commanding stature, contrasted favourably with that of the mercurial and sometimes irascible Hughes.

Although he would come to be respected and influential, and to have wide access at high levels in British and international politics, Bruce was not naturally gregarious. For much of the time he and Ethel lived quietly, enjoying travel, golf, theatre and the company of family and close friends. It has been suggested that his obduracy and apparent aloofness may have masked 'a sense of insecurity and melancholy', and even at times depression. His family life had been marked by loss: his father committed suicide in 1901, as did his brother Ernest in 1919. His sister died in 1908 and his mother four years later.[25]

---

23 I. M. Cumpston, *Lord Bruce of Melbourne*, Longman Cheshire, Melbourne, 1989, pp. 21–2.
24 Lee, *Stanley Melbourne Bruce*, p. 20.
25 Ibid., p. 14.

Bruce had been marked out from his schooldays as a leader; in his military career and at PLB he demonstrated a natural ability. He valued 'order, tradition, structure and discipline'. His primary loyalties were to nation and empire, the interests of which he believed to be inseparable. As a leader, he could look beyond political boundaries and personal oddities to make use of talented individuals. He had formed a government with Page and worked constructively with him despite Page's 'habitual incoherence and tendency to giggle'.[26] While he could inspire loyalty and confidence in his supporters, and respect even more widely, his lack of a 'common touch' and refusal to bow to opinion created enemies. Opponents would characterise him as a conservative Anglophile and representative of big business. He 'saw himself as a modern man' looking to international cooperation to prevent future wars and build a more prosperous world. As a businessman depending on international trade himself, he would always aim to facilitate and encourage its expansion. As a political leader of a country dependent on primary exports, he understood the needs of the rural sector and the importance of marketing its products.[27] In the later years of his prime ministership, his determination to take a hard line with maritime unrest strengthened the forces against him.

Lobbying for Preference

As Federal Treasurer, Bruce had visited Renmark with members of the Murray Waters Commission, a few months before his meeting with McDougall in November 1922, and just after a period of record rains. Despite being bogged and spattered with mud en route, Bruce was said to be impressed with the construction works in progress, and promised early attention to a telephone connection between Mildura and Renmark.[28] There is no evidence that he met McDougall then, but the experience and a shared vision of the area's potential perhaps gave them common ground. Bruce was persuaded to endorse McDougall's plan of action. Writing just after Bruce's appointment as Prime Minister, McDougall tactfully reminded him of their agreement:

> …you were good enough to clearly state your views as to the best methods to adopt in England. Very briefly stated your views were that the main duty of the Delegation was to prepare the ground for further Preference proposals by the Commonwealth Government, and that the

---

26 Ibid., pp. 9, 33.
27 Ibid., pp. 29–31.
28 *MP*, 12 May 1922.

best method to adopt was by personal interviews with the right people, and other quiet unobtrusive methods. My own views entirely concurred with your own.[29]

McDougall thus had two objectives: sales promotion with the Fruit Delegation as instructed by Rodgers, and covert lobbying for preference sanctioned by Bruce. He welcomed the return of a Conservative Government days before he sailed. By the time he reached England, however, Prime Minister Bonar Law had repeated his earlier pledge not to tax food. Undaunted by this difficulty, or by Rodgers' instructions, McDougall used every spare moment from a busy round of sales promotion and investigation, 'going very hard…seeing people morning noon and night' about preference. No-one 'uttered the word impossible, except the High Commissioner [former Prime Minister Sir Joseph Cook] who did not understand the scheme', although in the City 'one and all emphasise the very great difficulties that will have to be overcome'.[30]

This early period in London was effectively an apprenticeship, where McDougall applied himself to learn the skills of the lobbyist: putting a persuasive case, whether in person or in writing; mastery of his subject matter; and an understanding, even empathy, with those he sought to persuade. Talking, especially informally, came easily, but he claimed to find writing difficult. It was to become integral to his work, in a constant flow of memoranda and in the press. He began gradually, drafting replies, refutations and corrections to letters and public statements criticising Australian tariff policy, to be published over the signature of Australia House officials as he maintained a policy of anonymity. He cultivated journalists of like mind, particularly at *The Times*, and at first was content to 'inspire' articles. Very soon he was 'doing much more journalism than I ever imagined would fall to my lot…I have no style and mightily little ear or eye for bad faults, and a 1500 word article is some joke to me'.[31] With experience, and later the confidence of a well-received book behind him, he was to become a contributor of major articles, often in series of three or more. He spent a great deal of time in editorial offices; his articles in *The Times* were usually part of a coordinated campaign, timed to advantage and frequently echoed by editorials.

Within a few weeks, McDougall could report to Bruce sympathetic views from importers of dried fruits; a promising contact within the Colonial Office; an interview with the President of the Board of Trade; approaches to the Federation of British Industries, numbering 200 Members of Parliament amongst its

---

29   NLA, MS6890/1/8, copy of McDougall's letter to Bruce, 8 February 1923, sent to Norman for circulation in ADFA.
30   Ibid., letters to Norman, 4 January and 8 February 1923.
31   Ibid., letter to Norman, 29 August 1923.

membership, and to Lionel Curtis of the *Round Table*; editorial support from the *Morning Post* and *Daily Mail*; and favourable views at *The Times*. British commercial treaties with Spain and Greece, however, remained an obstacle.[32]

Bruce's elevation to the prime ministership delighted him: 'I hope…that it will be a new bright page on the history of Australia.' He quickly sent a brief letter of congratulation with a longer letter, which included the reminder of their agreement quoted above, a report on progress and assurance that as the names of the delegates were not associated with preference, Rodgers' 'very hurriedly given' instructions were being observed. But he also explained that the campaign had been 'forced open'. Press support of Australia's preference proposals had followed the 'unexpected' publication by *The Times Trade Supplement* of McDougall's report to the most recent ADFA conference. He confessed to Norman: 'I did not exactly write either article but I entertained the writers of both to separate dinners!! I hope I shall not sacrifice my figure on the altar of preference.'[33]

The Imperial Conference, 1923

McDougall began his mission hoping to stay longer than the six months allowed for the Fruit Delegation. Before leaving Adelaide, he had made tentative arrangements for his family to join him in the English spring. An Imperial Conference, expected in June, would provide the opportunity for pressure at the highest level to seal the preference campaign. He believed there would be work to do following up any broad decision, and hoped the task would be his. When the conference was delayed until autumn, beyond the six-month limit of the delegation, the NSW and Victorian Premiers, both then in London, were persuaded to send almost identical cables to the Prime Minister, stating that McDougall was doing 'excellent work', he should remain for the conference and the cost could be covered by financial arrangements made for the Fruit Delegation.[34] The South Australian Government, lobbied by ADFA, joined in the request; South Australian winemaker T. C. Angove saw Bruce; and Meares, by then back in Australia, also dealt with the Federal Government. All governments and departments involved assented and the High Commission in London was formally notified of approval on 8 June. Chaffey departed and McDougall, still officially the Fruit Delegation, remained to prepare for the Imperial Conference.

An Australian Prime Minister could expect to return with some reflected glow from the glories of empire. But his electorate, and a modest federal budget,

---

32 Ibid., letters to Norman, 4, 12 and 24 January 1923.
33 Ibid., letters to Norman, 2 and 8 February; copy of Letter to Bruce, 8 February 1923.
34 NAA, A458, 1500/2, part 1, cables from Fuller and Lawson, 28 April 1923.

could not accept undue squandering of funds for accompanying officials. The party of advisers accompanying Bruce would be small. He might well supplement his team with any likely Australian who happened to be in Europe, and would make good use of his long and expensive journey by staying on for several weeks, publicising and attending to Australia's interests. All of this had occurred to McDougall. He wrote to Meares suggesting he join Bruce's ship in Naples, to help brief him for the conference, but was told that his presence would not be required. Then suddenly, early in September, he was delighted to be ordered to meet Bruce's ship at Port Said: 'This will give me five days with Bruce, Wilson, Oakley and Fred and a splendid opportunity to press home the points I particularly want to make.'[35] Senator R. V. Wilson, Assistant Minister, was the only other politician in the party travelling from Australia. Officials were the Comptroller-General of Customs, R. McK. Oakley, Solicitor-General, Sir Robert Garran, an adviser on defence and another on foreign policy. There were also secretaries to Bruce and Wilson and a publicity officer, F. M. [Fred] Cutlack, lawyer, journalist and author of a much-publicised volume of the official war history, who was also McDougall's brother-in-law. Four prominent businessmen acted as advisers: Herbert Brookes and Claude Reading, both members of the Commonwealth Board of Trade; London-based John Sanderson, who represented Australia on a committee advising the British Board of Trade; and Walter Young, General Manager of Elders and an active promoter of orderly marketing, particularly of wool and wheat.

McDougall was never so overawed by the great and powerful as to lose sight of their use to his cause. As he prepared papers and arguments aboard the *Orsova* steaming towards Port Said, he sensed the approach of a momentous leap for his cause and for his career: 'I suppose the next six weeks are going to be immensely interesting, perhaps decisive as to Preference, Inter Empire Trade, Baldwin and Bruce's governments and perhaps for FLMcD!!' FLMcD had his objectives in order of priority and, while he placed Australian interests— preference on dried fruits, on canned fruit and on wine—first, Bruce should express his policy 'in harmony with' British agriculture and British consumers, and with broader imperial interests: the proper organisation of empire trade, giving empire producers competitive advantage over foreigners and eliminating price fluctuations.[36] McDougall had prepared the ground with an article in the *Yorkshire Post*, written by 'an Australian correspondent who has made a special study of the problems to be discussed at the Imperial Economic Conference'. It argued for reciprocal tariff preferences and access to British markets for Australian producers labouring in a sparsely populated continent of enormous potential,

---

35   NLA, MS6890/1/8, letter to Norman, 5 September 1923.
36   Ibid., letter to Norman, 26 September 1923.

one unlikely in the foreseeable future to become a significant manufacturing economy. Australia needed men and markets.[37] The day before Port Said, he confessed to apprehension about the fate of his cause and his future:

> Everything is so dependent upon what is forthcoming next month. Tomorrow and the few days following will show me whether Bruce will be the man for the occasion. The occasion is here, I perhaps have done a little mite to prepare and to bring the occasion but Bruce can if he will bring it to the climax.[38]

The Imperial Conference of 1923 was in some ways the first of a new era. No longer overshadowed by the Great War and subsequent treaty-making, it was in fact two conferences, held concurrently over some six weeks, seeking to define how the British Empire would meet the political and economic challenges of a difficult postwar world. The Imperial Conference opened on 1 October, chaired by the British Prime Minister, and dealt with foreign policy, defence, communications and legal issues. The President of the Board of Trade, Sir Philip Lloyd-Greame, chaired the Imperial Economic Conference, beginning the following day. Its agenda included financial assistance for imperial development, technical research and communication, commercial facilities, currency and exchange, forestry, livestock and the Imperial Institute. Bruce had worried that absence threatened his tenuous hold on office; he tailored parliamentary sittings to include a brief but busy session before his departure after which Parliament would not sit again until his return. The arrangement meant that Bruce could not reach London until a week after the conference opening; its program was modified to allow him to speak at the Economic Conference as soon as he arrived. He was met on arrival by Lord Milner, inspiration for the 'imperial visionaries', who became a close associate and encouraged him to promote imperial economic cooperation.[39]

Bruce proved indeed to be the man for the occasion. McDougall had 'worked up' the opening address with him as they sailed through the Mediterranean. Its theme was an imperial vision very much along the lines of McDougall's *Yorkshire Post* article, summed up in the words 'Men, Money and Markets'. Its thesis was that markets were paramount; without them migration and development would at best be slow and limited. About one-quarter of it was devoted to figures drawn from McDougall's increasing interest in proving the value of Australia's tariff preference to British manufactures and of dominion markets to British employment. Bruce countered arguments against tariff preference on the grounds of harm to British trade relations, surveyed tariff practices of

---

37 Ibid., cutting from *Yorkshire Post*, 7 September 1923, with letter to Norman, 11 September 1923.
38 Ibid., letter to Norman, 26 September 1923.
39 Lee, *Stanley Melbourne Bruce*, pp. 37–8.

other colonial powers and recommended his audience study, as McDougall undoubtedly had, the report of the US Tariff Commission. He quoted Cobden: 'I doubt the wisdom, I sincerely doubt the prudence, of a great body of industrial people to allow themselves to live in dependence on foreign Powers for the supply of food and raw material.' He described the threat from products of cheap labour and recalled the 1917 resolution on imperial preference: 'the last expression of the view held by a Conference of this character.' Australia would again subscribe to a similar resolution, but would prefer 'something practical to give effect to what we actually believe in'. He described the potential of the Murray irrigation scheme and its increasing production dependent on markets: 'if we have no markets we cannot have great migration, we cannot have great development in the near future.' Interestingly Bruce altered McDougall's order of objectives, placing assistance to the British farmer first, before dominion producers, with the British consumer last. He surveyed possible methods: tariffs, subsidies, import licensing or import control, and recommended an imperial royal commission to make recommendations on these alternatives. Finally he apologised for the length—26 typed pages—of his speech.[40]

McDougall sat watching the faces of British ministers as Bruce delivered the speech they had written together 'extraordinarily well'. It was 'a day of some personal triumph'. That afternoon, Lloyd-Greame, who less than a year earlier had declared increased preference impossible, announced a reversal of government policy, to allow full preference—10/6d per cwt—on all dried fruits except currants, about which there would be further discussion, and a new duty giving a preference of 5 shillings per cwt on canned fruit.[41] The campaign, it seemed, had been won.

The General Election of 1923

It was a brief triumph. Bruce learned next day that the decision on tariffs had been made by the British Government well before his arrival and that the new Conservative Prime Minister, Stanley Baldwin, long committed to tariff reform, intended to take the country to a general election on the issue. Baldwin believed a wider tariff policy offered a solution to the 'gaping wound' of unemployment—his government's most pressing problem—as well as falling exports and dominion demands for secure markets.[42] With further depression predicted for the approaching winter, Baldwin and his ministers sought means to stimulate industries. Suggestions included the visionary program: increasing

---

40   NAA, CP103/3, Volume 6, 'Imperial Conference 1923, Stenographic Notes', Fourth Meeting, 9 October 1923, pp. 3–22.
41   NLA, MS6890/1/8, letter to Norman, 11 October 1923.
42   Keith Middlemas and John Barnes, *Baldwin: A Biography*, Weidenfeld & Nicolson, London, 1969, p. 219.

the 1919 preferences and allocating more funds for empire development.[43] Having agreed in principle to extending preferences on empire sugar and dried fruits in March, Baldwin suggested delaying the announcement until the Imperial Conference: 'if it were given in advance the Dominions would accept it and ask for something more.'[44]

Proposals to extend the McKenna and 1919 duties and to impose 'safeguarding' measures against dumping had been approved by Cabinet in August. Wider reform would break the 1922 pledge not to extend tariffs without another election. As imperial leaders gathered, the question in government had become not whether, but when, that election might be held. Besides aiding the economy, tariff reform might also heal a split in the Conservative Party and forestall a rumoured declaration of support for tariffs by Lloyd George. Senior Conservatives consulted Bruce as to 'how far we can go at the Conference and also what we are to go to the country upon and when'. Cabinet members did not support immediate action; even keen tariff reformers like Leo Amery urged cautious and slow preparation. Baldwin hesitated, giving a 'somewhat Delphic' speech to his party conference on 25 October and choosing, on the final day of the Imperial Conference, not to deliver a prepared speech explaining a decision to go to the polls. Three days later, on 12 November, he sought a dissolution.[45]

Those who had advised caution were proved right. The party did not reunite; its case for tariff reform was unprepared; there was substantial press opposition and too little time to educate the electorate. Although the Liberal Party could expect to attract free-traders who normally voted Conservative, and to hold much of their working-class support for fear of 'dear bread', there was little else to distinguish Liberal policy from Labour, and Labour reaped the benefit. At the general election on 6 December, the Conservative Party lost its majority, but remained the largest party in the House of Commons, with 258 seats. Labour won its largest number to that date, 191; the Liberals were left with 159.[46]

Baldwin chose not to resign immediately, forcing the Liberals to decide which party to support. They chose Labour, possibly because the Liberal Party was more likely to survive a brief alliance with the left, whereas alliance with the Conservatives might encourage a move for fusion.[47] Thus, a censure motion on 21 January 1924 brought a minority Labour Government into power. Labour tended to oppose tariffs to keep food cheap, and was now dependent

---
43   John Barnes and David Nicholson, eds, *The Leo Amery Diaries 1896–1929. Volume I*, Hutchinson, London, 1980, entry for 10 July 1923, p. 333.
44   David Dilks, *Neville Chamberlain: Pioneering and Reform 1869–1929*, Cambridge University Press, Cambridge, 1984, p. 339.
45   Barnes and Nicholson, *Amery Diaries*, Vol. I, p. 348; Dilks, *Neville Chamberlain*, pp. 341, 344; Middlemas and Barnes, *Baldwin*, p. 23.
46   T. O. Lloyd, *Empire to Welfare State: English History 1906–1985*, Oxford University Press, Oxford, 1986, pp. 127–8.
47   Ibid., p. 129.

on free-trade Liberal support. Chancellor of the Exchequer Philip Snowden, a determined free-trader, abolished the McKenna duties and reduced revenue tariffs on sugar, tea, coffee and cocoa. The preference increases announced at the Imperial Conference, which involved no increase in duty, were to be put to a free vote in Parliament in June 1924.

McDougall had expected the election campaign to be a full-scale, measured debate to which he could apply 'all the propaganda we can work to disarm hostility to our preference'. When the election date was uncertain, he suggested he should remain in England long enough to ensure the duties announced at the Imperial Conference survived the budget. Bruce agreed that he should stay at least until the election, and later extended his approval until the budget, due in May 1924.[48] In the weeks after the Imperial Conference, Bruce toured the country, accompanied much of the time by McDougall, who drafted speeches and articles on the tariff question, expounded his own views on the needs of the dried-fruits industry, and developed the partnership begun on the *Orvietto*. The announcement of a December election, and its outcome, shifted their focus. The preferences announced at the Imperial Conference might now depend on Labour support, and McDougall congratulated himself on his foresight in establishing early links with Labour. At his first shipboard meeting with Bruce and his economic advisers, he had argued that agreement with Conservatives could not be permanent without Labour support, and wrote ruefully of the 'rapid and cruel vindication of my prophecy', adding, 'but again, it's interesting'. 'I am now all out on propaganda for Labour. It was a damned wise move on my part to get in touch with Labour people before the Conference, as a result I have a leg in with them now.' Rapidly developing as a skilful lobbyist, he was learning to tailor his arguments to a solution to the listener's own concerns.

> I have had [J. R.] Clynes [then considered a potential Chancellor of the Exchequer, but who became Lord Privy Seal] to lunch during the last week and also [F. W.] Pethick Lawrence and G. D. H. Cole. I hope to see Sidney Webb, [J. H.] Thomas and other lights, I am stressing to them the importance of Labour putting itself right with the nation on Empire matters, showing them the wonderful purchasing power of Australia and New Zealand for British Goods and the value of our preference.[49]

He found little sympathy for the imperial vision amongst Labour intellectuals, and nothing developed from a tentative approach based on common problems of agriculture. He had more success on the issue of labour conditions, stressing the importance of supporting products of good working conditions against those of sweated labour. He found 'among Trade Union MPs a clearer realisation

---

48   NLA, MA6890/1/9, letters to Norman, 30 October and 1 November 1923.
49   Ibid., letters to Norman, 10 and 24 January 1924.

of economic fundamentals than among either Liberals or the Labour intelligen[t]sia'.[50] But he did make some headway with the Liberals by focusing on their perceived need for an 'empire policy'.[51] Early in 1924, McDougall determined 'to bring every argument and every battery to bear upon Parliament. When Bruce leaves I must start to see Chambers of Commerce, Trade Union leaders and represent the danger of losing the Australian preference on their own lines of commodities.'[52] Bruce lunched with leading Labour figures on 19 December and was 'convinced that they are most anxious to put themselves right with the Empire', although their election speeches had also convinced him they would find 'preferences hard to swallow'.[53]

In the event the free vote taken on the preferences announced in 1923 was narrowly lost. 'Just fancy', wrote McDougall, 'a Free Trade majority of 80, the McKenna duties defeated by 62, and we get within six of our goal. It's damnable but in some respects a triumph and at least argues irresistibly that in the near future we shall succeed.'[54]

Bruce's 'secret service agent'

McDougall wrote that Bruce's biggest failing was of 'imagination'—a term that for McDougall may well have included empathy with his own position. He knew from the beginning that he could not expect warmth or personal understanding from Bruce, whose 'main disadvantage' was a 'lack of general knowledge and experience, powerfully offset by commonsense and a very considerable power for close thinking'. Yet, wrote McDougall, 'one could work for him with enthusiasm, because he thinks and is mentally progressive'. On Bruce's last evening in London, he and McDougall talked until 2 am. In that cosy intimacy, Bruce answered a question about the nature of the position McDougall was to hold while he remained in London: 'Well McDougall, in your more uplifted moments you can call yourself the confidential representative of the Australian Prime Minister, when less inflated a secret service agent.' Amusing and flattering as that reply might have been, there was no such designated position in the sprawling bureaucracy of Australia House and no comfortable line of direction for McDougall once Bruce departed. Bruce's authority would be carried by Senator Wilson, who was remaining to oversee the British Empire Exhibition. Wilson was not well disposed to McDougall. Nor did Bruce's joke hold any solution to the problem that had been nagging McDougall and his wife for many

---

50   Ibid., letters to Norman, 2 February, 2 and 29 April, 10 and 17 May 1924.
51   *LFSSA*, 2, pp. 4–5.
52   NLA, MA6890/1/9, letter to Norman, 24 January 1924.
53   NLA, MA6890/1/8, letter to Norman, 20 December 1923.
54   NLA, MA6890/1/9, letter to Norman, 19 June 1924.

months. Dwindling funds allocated for the Fruit Delegation could not last much beyond his fortieth birthday in April. He knew now that 'active fruit growing' would not 'be my calling unless as a stop gap...I do want to be in the big things and I want to help Australia'.[55]

The Dried Fruits Export Control Board

Before the Imperial Conference McDougall had organised a committee in London to discuss an organisation scheme for the dried-fruits industry. A 27-page memorandum recording their views was sent to Bruce, McDougall privately claiming responsibility for 65 per cent of the ideas in it. Contributors were Sir James Cooper, A. H. Ashbolt, Agent-General for Tasmania, and M. L. Shepherd, Official Secretary at the High Commission.[56] McDougall had personally urged Bruce that organisation was essential: 'it will be impossible to get very much improvement in the manner of marketing our fruit in London until we control the fruit itself...until [ADFA] has been re-organised and is able to finance its own shipments.' Bruce had been 'very much struck' with the suggestion and asked for more detail to be mailed to reach him on the journey home.[57] At Sydney's Royal Easter Show, he announced that some tariff revenue would be used to assist struggling primary industries with marketing, freight subsidies, export bounties and transport—all conditional on efficiency and organisation.[58] The next day a cable instructed McDougall, through Wilson, to return to Melbourne. Despite a second cable two days later, Bruce's intentions were not clear, but McDougall assumed, correctly, that he was to help establish an organisation scheme. It was a request he could not refuse, reluctant as he was to leave London before the parliamentary vote on preference in June.[59]

Having left his family in England, McDougall helped prepare short-term assistance proposals for the dried-fruits industry and legislation for grower-funded control boards for dried fruits, canned fruits and dairy products. Progress was slow, giving the lonely McDougall leisure to worry about his own future. Bruce agreed he might be of greatest use in London and promised a proposal, but was slow to make a definite offer.[60] There were other options. Before leaving England, McDougall had been offered a position with the Conservative Party, presumably as part of a short-lived policy secretariat to be organised by Leo Amery after the election loss, its purpose to prepare propaganda for preference

---

55   Ibid., letters to Norman, 10 and 24 January, 7 February 1924; for Wilson, see letters to Norman, 13 July and 4 September 1924.
56   NLA, MS6890/1/8, letter to Norman, 30 July 1923.
57   NLA, MS6890/1/9, letters to Norman, 24 January and 24 April 1924.
58   *Sydney Morning Herald*, 17 April 1924.
59   *LFSSA*, 7 and 8, pp. 18–21.
60   NLA, MS6890/1/9, letters to Norman, 19 June and 10 July 1924.

and 'safeguarding'.⁶¹ A firm offer for a management position in Australia with ADFA remained open.⁶² After a delay of two months, Bruce could only offer employment in Melbourne for the rest of the year, for remuneration considerably less than that of the Fruit Delegation. He explained that while 'he recognised my value Cabinet did not and that I must impress them. I said that as I had to hide my light under a bushel of obscurity how could I?' McDougall believed Wilson, now Minister for Markets, was against him; the ADFA job would not do for that reason, and was, in any case, 'off my track'. 'What was I? Answer: an expert on Empire Trade and an artist at propaganda.' He would best serve his cause by being in London 'to work for Preference and for Empire Development'. Bruce accepted his suggestion therefore that he return to London to work part-time for the Dried Fruits Export Control Board (DFECB) being established by the new legislation. His remuneration would be shared between the grower-funded board and the Commonwealth Government, on behalf of which he would continue to lobby. It would be £650 per year less than he had received on the Fruit Delegation and inadequate, McDougall would later argue, in view of his lack of tenure, lobbying expenses and in comparison with salaries in other dominion marketing agencies in London—a sore point that would persist throughout the 1920s. The lack of permanency meant that his family lived in a succession of rented houses, often at considerable distance from London, and this may well have contributed to the disintegration of his marriage towards the end of the decade. 'What we all want is a home and garden', he would lament to his brother, less than a year after the arrangement began.⁶³

The London Agency of the Dried Fruits Export Control Board

The London Agency was responsible for ensuring that fruit was sold profitably, having regard to the state of the markets and the quality of the product. Its Chairman was London-based businessman Sir James Cooper. The agency employed technical staff to appraise fruit consignments and supervise provision of physical needs such as storage and fumigation. It liaised with importers, supervised traders' records of sales, promoted the product, and sent a steady stream of reporting back to Melbourne. As its part-time secretary from April 1925, McDougall spent two busy days each week in the agency's office at 2 Talbot Court, Eastcheap, close to the docks, until 1928, when members of the board in Melbourne became concerned that he was overstretched. Prompted to action by

---

61   NAA, M111, 125, statement enclosed with letter from McDougall to Bruce, 14 December 1925.
62   NLA, MS6890/1/9, letter to Norman, 19 June; NAA, M111, 1924, Bruce to McDougall, 12 May 1924.
63   NLA, MS6890/1/9, letters to Norman, 13 July 1924 and 3 January 1926; *LFSSA*, *16*, *20* and *45*, pp. 45–6, 55–6, 131–2.

an illness which, unusually, kept him away from work for some weeks, the Board Chairman, W. C. F. Thomas, in 1929 proposed a new position for McDougall of Deputy Chairman, freeing the board to employ 'a real secretary' and still provide McDougall with financial support and a status 'more appropriate to his present functions' and to 'the higher work which he might yet be called upon to perform for the Government'.[64] McDougall was to be Chairman of the agency from 1936 to 1947. He maintained some financial involvement with the Renmark blocks, which continued to be worked by Norman McDougall, but was never to visit Australia again after 1924.

Figure 8 The London Agency of the Dried Fruits Export Control Board meets British dealers and bakers in Liverpool, date unknown. Sir James Cooper is third from left in front row; J. S. Scouler, Agency Secretary, on far right. McDougall, then Vice-Chairman of the Agency, is third from left in back row; agency member A. E. Gough fourth from left; A. E. Hyland, Director of Australian Trade Publicity, is on the far right.

Source: E. McDougall.

---

64 National Archives of Australia, Victorian Branch [hereinafter NAAV], B4242, Thomas to Cooper, 8 February 1928.

The Partnership

McDougall's first important idea—to secure an export market for the dried-fruits industry by more efficient organisation and by persuading the British Government to institute substantial tariff preference—thus had a mixed result. He almost certainly played a major role in persuading Bruce to establish and control the industry on a basis of efficiency. By his work in Melbourne and as London secretary, he contributed much to putting the idea into effect. The idea was in accord with those of efficiency and 'practical progressivism'; Bruce's ready acceptance is not surprising. The move seems to have been successful: both industry and organisation have survived.[65]

The tariff preference campaign was, in the event, only marginally successful. The Conservative Government returned to office at the end of 1924 removed duties on empire dried fruits, giving a preference of 10/6d per cwt on fruits other than currants, which were left at 2 shillings. It was below what the Federal Government had requested in 1922, and well below what McDougall had calculated as necessary. Sales, however, continued to be good, at least until 1927 when severe frost affected yield. The work of the agency on the periphery and publicity from the Empire Marketing Board at the centre helped focus public attention on the products of empire, until general tariff protection was achieved by the Ottawa Agreements of 1932.

For McDougall, the preference campaign had brought conviction that his place, physically, and to an increasing degree intellectually, was not on the periphery, but at the empire's centre. He saw himself still as a servant of Australia, but knew he could not flourish in the Australian political arena. To be in 'the big things', which would help Australia and his own industry, he had to be in London and he returned assuming, probably correctly, a personal responsibility to the Prime Minister who had accepted an arrangement placing him outside the lines of conventional bureaucracy. He could perhaps be seen as a prototype of the modern-day political staffer. Certainly Bruce's good-natured but vague designation as 'personal representative of the Prime Minister…secret service agent' set the pattern for their future cooperation.

During the 1920s McDougall wrote a least one letter to Bruce each week, and often more. The letters reported on his activities, but also contained gossip and frank comments and suggestions on policy. He presumed a special relationship and did not hesitate to approach Bruce with ideas, or indeed with complaints. Bruce responded in kind. He wrote infrequently, but took care when he did to

---

65   The Australian Dried Fruits Board became a subsidiary of the Australian Horticultural Corporation in 1987.

comment on every letter; one reply runs to 37 pages.[66] He assured McDougall that 'every one of your letters is read by me', and he hoped McDougall would not be discouraged by the infrequent replies to letters that were 'of the utmost interest and value to me'. He made use of some material McDougall enclosed in speeches. After a quick first reading, Bruce saved McDougall's letters until he had time to read them carefully, usually at weekends; he often reread them months later. During a two-month tour of outlying States, he 'read many of them while travelling in trains, and one even when travelling in an aeroplane'.[67] His replies reported candidly on the political situation in Australia and his own activities and plans, occasionally sending copies of his correspondence with others. He happily joined in the gossip and sometimes vented, to the point of indiscretion, his dissatisfaction with events and people, at one time complaining about the 'hopeless incompetence' of the Prime Minister's Publicity Department, at another describing fruitless efforts 'to get some useful information from the Customs Department'.[68] At times he asked for more information on particular issues; he engaged in discussion and referred to views and aims that he and McDougall shared. This extended to broad economic and social philosophy:

> I read with considerable interest what you say with regard to the continuous harping upon the question of economy as if that was going to solve the economic problems the nations are up against. I agree with you that it will not: that the world is now so far advanced that we have to recognise we must face great expenditures upon social amelioration, and the only way to solve our problems is to adopt the same course that every modern business has been forced to, and that is of expanding our turn-over rather than imagining we can solve our difficulties by reducing our expenses.[69]

Bruce frequently sent the suggested cable or took other action recommended by McDougall: 'I am taking up the question you raise about the position of the standard of living in Australia and Great Britain…I will let you have the information…when I have received it.' If he did not agree, he often explained why, or at least let McDougall down gently. He commended one memorandum McDougall had sent him, urging British cooperation in Australian development, as 'very useful', but rejected a suggestion that it be sent to British ministers: 'on reading it carefully I think the time is hardly ripe for us to let the British Government have copies of it.' On another suggestion, which McDougall had pressed more than once, Bruce was more severe:

---

66   NAA, M111, 30 April 1929. Bruce's letters are filed chronologically with McDougall's in this series.
67   Ibid., 21 May and 15 September 1927, 16 January 1928.
68   Ibid., 10 February 1929, 26 September 1927.
69   Ibid., 30 April 1929.

> You will have grasped from my not having taken any further action after receiving your letters, that I do not want to deal with this matter at the present time. I gave a lot of thought to it when your suggestion came through by cable, and I think for the present it would be better for you to drop the idea.[70]

He was generous in his appreciation:

> I have made a point of advising [the Minister for Trade and Customs] of the different actions you have taken…in the first place to bring home to him the value of the services you are rendering…

> I read the Debate in the House of Commons with regard to Empire trade, and your trail could be seen very distinctly through the whole of the discussions.

> …your work is extraordinarily interesting and extremely valuable, but you have to remember that everything depends upon your maintaining your health.

> …after I have read [McDougall's letters] I am certain that my mind subconsciously goes on thinking over the points that you have raised, and the conclusions that are arrived at are unquestionably very considerably influenced by what you write to me.[71]

Bruce did, however, urge McDougall to keep enclosures to a minimum, because 'the time which I have at my disposal for such a purpose is extraordinarily limited'.[72]

In sum, the correspondence shows a solid partnership between men with common aims and mutual respect, despite their difference in status. Yet that difference was to be maintained over more than 20 years of working together. They could share a joke, but McDougall would never presume to address Bruce as anything other than 'Mr Bruce'; Bruce called him 'McDougall', not the familiar 'Mac' used by his friends. The partnership, close and effective as it was, did not develop into easy friendship. The gulf between prime minister/high commissioner and 'secret service agent'/economic adviser was never to be bridged by intimacy.

---

70   Ibid., 12 December and 7 March 1927, 16 January 1928.
71   Ibid., 30 April and 27 August 1928.
72   Ibid., 21 May 1927.

3. A Vision of Empire

Summary

In London McDougall pursued his vision of imperial cooperation through the second half of the 1920s. His book, *Sheltered Markets*, published in 1925, argued the case for imperial tariff preference, based on an analysis of patterns of trade. Imperial preference would not be introduced until 1932; the returning Conservative Government in 1925 acknowledged that public opinion was opposed to the idea. An Imperial Economic Committee was established instead, to advise on improvements to production and marketing of empire goods, and an Empire Marketing Board was to encourage 'voluntary preference' by means of marketing and scientific research into problems. McDougall was an active member of both. He also acted as London liaison for Australia's Council for Industrial and Scientific Research (CSIR) and the Development and Migration Commission (DMC). Representation on all these bodies brought him into contact with leading scientists, notably nutritionist John Boyd Orr, and A. C. D. (David) Rivett, Chief Executive of CSIR.

The new influences led McDougall to devise forms of imperial cooperation in science, agriculture and industry. Seeking efficiency, he suggested pastoral improvement in Australia should be directed by Orr and based on British research; Rivett argued that Australia must develop its own research capabilities. Similarly McDougall was influenced by British criticism of rising Australian tariff protection to argue for an imperially rationalised approach to industrial development: Australia and other dominions concentrating on simpler manufactures. Bruce was anxious to moderate Australian tariffs, but his efforts to bring about economic reform faced increasing hostility; he lost government at the end of 1929.

'Sheltered Markets'

The idea of writing a book about empire trade occurred to McDougall soon after Bruce left London early in 1924. Amery was enthusiastic. The enforced idleness of some of McDougall's time in Melbourne and the return voyage to London provided the opportunity. *Sheltered Markets: A Study of Empire Trade* was published on 30 June 1925, a volume of some 150 pages selling for 5 shillings a copy. McDougall was promised a foreword by Lord Milner, but the mentor of imperial visionaries died in May and Sir Robert Horne, businessman, philosopher and former Conservative minister, took his place.¹ Horne recommended the work as providing 'the most likely means—if not the only means—of redressing the precarious problem in which we stand today', meaning unemployment. McDougall had 'gone at once to the root of the problem in discussing the markets in which British manufactures are most likely to find their market'.²

Sheltered Markets argues the case for imperial preference. It is more varied and broader in vision than the simple message of the preference campaign waged from Renmark three years earlier. McDougall writes as before of the benefits of closer settlement, in the Murray Valley in particular. He argues, as always, the value of Australia's tariff preference to the United Kingdom; he preaches the importance of organised marketing of empire commodities for both producers and purchasers. Preference is needed to compensate for disadvantages that empire produce suffers on the British market: the costs of transport over long distances and of decent living wages.

These familiar points are argued in the context of a sophisticated analysis of patterns of trade and their relative value to the economy. They are argued from the perspective of Britain and its empire, not from the perspective of Australia alone, still less of one Australian product. McDougall's thesis is that British trade with dominions and colonies is more robust, and of more value to the British economy, than trade with foreigners. To support it he explains the importance of what have come to be called 'value adding' and 'sustainability': exports of finished manufactures are more profitable than raw materials. Coal, a major British export, is a non-renewable resource. Although its export to Europe keeps miners in work, it is of limited value when compared with the manufactured goods that make up almost the entire export to the empire. Tables demonstrate Britain's balances of trade: unfavourable with the United States and Europe; better but declining with Canada; and favourable with other dominions. He considers factors thought to influence the buying patterns of Britain's trading partners: prosperity and wealth distribution, British loans, British buying patterns and the availability of shipping. Two factors predominate: there is 'no doubt as to the very great advantages that national sentiment and tariff preference give to British trade.' He devotes several pages to the advantages of the Australian tariff to that trade.³

---

1 NLA, MS6890/1/9, letter to Norman, 24 January 1924; *LFSSA*, *2*, *15* and *23*, 20 February, 2 April and 11 June 1924, pp. 4, 41, 63.
2 F. L. McDougall, *Sheltered Markets: A Study of Empire Trade*, John Murray, London, 1925, p. v.
3 Ibid., pp. 68–74, 89.

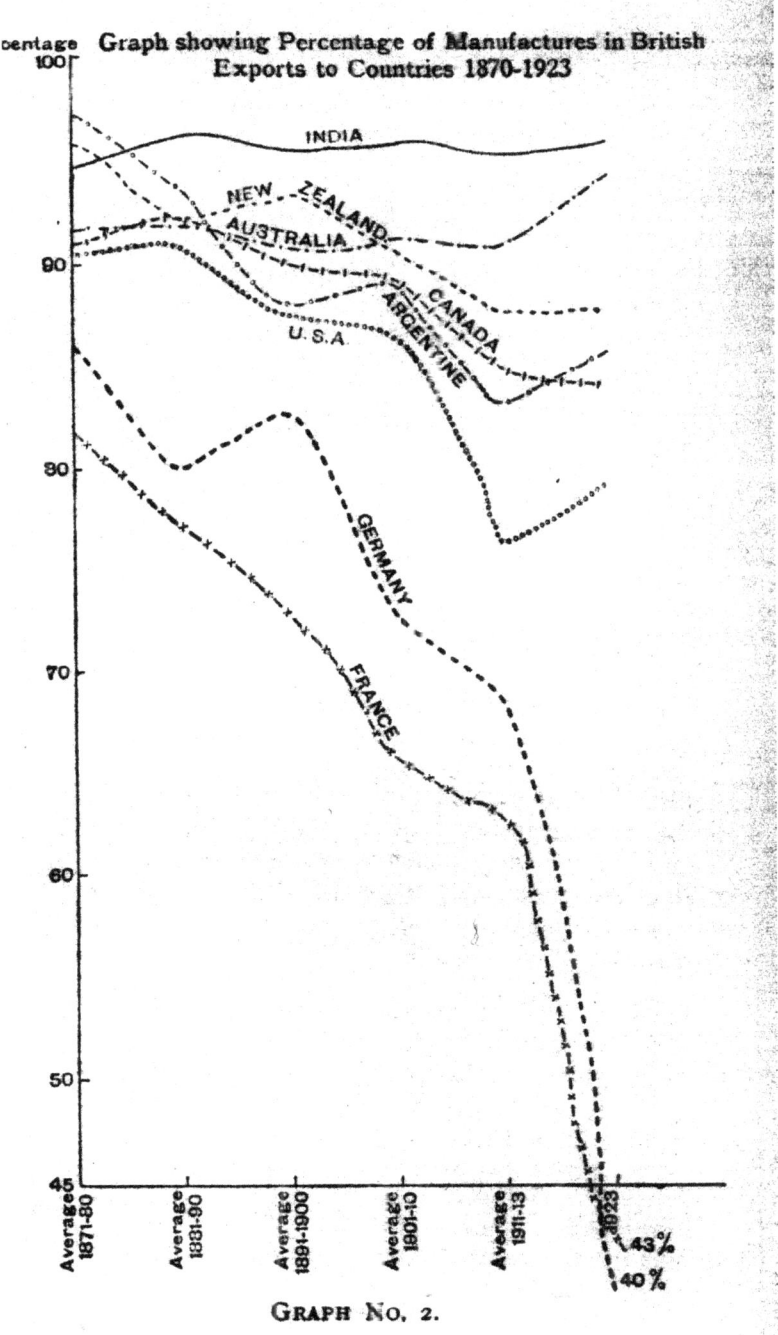

Figure 9 Declining Percentages of British Exports of Manufactures to Europe and the Americas.

Source: F. L. McDougall, *Sheltered Markets: A Study of Empire Trade*, John Murray, London, 1925, p. 46.

For the first time McDougall tackles free trade head on. It had not been an unmitigated benefit even in the nineteenth century. It effectively encouraged sweated labour in other countries by disregarding the growth of foreign protective tariffs and ignoring improved wages brought about in Britain, Australia and New Zealand by trade unionism. It destroyed the colonial sugar industry and developed a dangerous reliance upon foreign foodstuffs and raw materials. Lancashire had been brought to near starvation by blockades preventing cotton leaving the United States during the Civil War—a lesson forgotten until 1914–18. In 1925, McDougall writes, consumers face formidable competition on all sides: combines and cartels, US farm politics, the Soviet Government combine, agricultural cooperatives in Denmark and organised marketing in the dominions. The day of laissez faire and cheap food has passed; 'unrestricted competition no longer exists'.[4]

McDougall does not advocate immediate abandonment of free trade, nor does he push one simple solution. He examines the difficulties of all major food imports and concludes that the solutions are as varied as the Empire itself. Empire buying power must be enhanced by soundly based migration, particularly to closer settlements; empire industries must become viable with organisation, technical aid, London advisory bodies to cooperate in price stabilisation and publicity to create a demand for empire goods. Part of the solution might well be some tariff preference.

Sheltered Markets deals with difficult and complex material. The fact that it reads easily and persuasively is tribute to McDougall's maturing skills as a writer, still more as an educator. As the book moves to a new phase of argument, he summarises the previous one, reinforcing his points. He provides tables and graphs that are easy to understand. He uses simple metaphor with effect: a preferential tariff system is like a weir. Foreign trade must flow over the top 'while British goods flow through the preferential sluice gates' lower down.[5]

The book was well received. An editorial in *The Times Trade and Engineering Supplement* urged 'every thoughtful citizen of the empire' to read it

> for it deals in a broad, comprehensive and lucid fashion with the essential economic problem that confronts the whole British race…facts are marshalled so admirably that even the reader who may open this book with strong preconceived opinions of an opposite character will find it difficult to dissent from the argument.[6]

---

4   Ibid., p. 133.
5   Ibid., pp. 66–7.
6   *Times Imperial and Foreign Trade and Engineering Supplement* [hereinafter *TTES*], 4 July 1925.

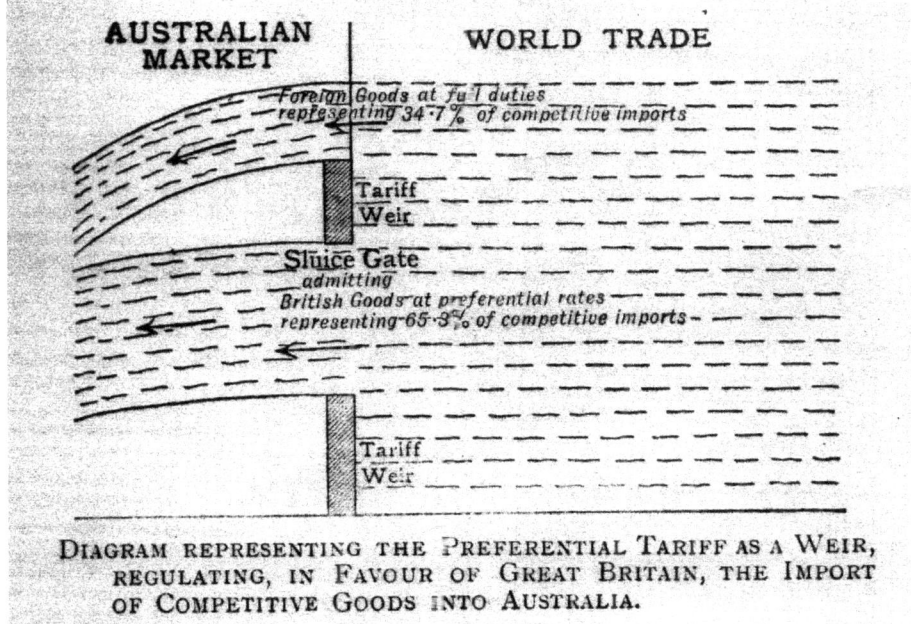

Figure 10 McDougall's Hydraulic Metaphor for Imperial Preference.

Source: F. L. McDougall, *Sheltered Markets: A Study of Empire Trade*, John Murray, London, 1925, p. 77.

Bruce wrote that it was 'timely and useful and will accomplish a considerable amount of good'. It 'continues to get a most remarkable press and equally remarkably small sales considering the press', complained its author.[7]

Conservative Party headquarters took 1000 paperback copies at 2 shillings each, and McDougall's friend Brooks Crompton Wood, MP, sent a copy to each member of the House of Commons. By August, 750 copies had been sold to the general public. McDougall stood to receive only 7d per copy of the regular edition. It would not end his financial worries but it did give him something he needed. He had written before publication that 'if the book causes real discussion, I shall feel that it has served its purpose'. It had achieved that, but it also 'let people know that F. L. McD is a person who exists'.[8] No longer was he an anonymous 'secret service agent' of the Australian Prime Minister; he was the author of *Sheltered Markets*.

---

7 NAA, M111, 1925, Bruce to McDougall, 31 August 1925; NLA, MS6890/1/9, letter to Norman, 18 August 1925, and copy of Bruce's letter; MS6890/5/1, copies of many reviews.
8 NLA, MS6890/1/9, letter to Norman, 25 June 1925; *LFSSA*, 24, 2 July 1925, p. 67.

The Imperial Vision by Other Means

Substantial imperial preference was ruled out after Stanley Baldwin's election loss on the issue late in 1923. The case for change then had been poorly prepared and opposed by a formidable combination of forces: Liberal and Labour politicians, some Conservatives, financial and commercial interests, much of industry, the union movement, some agricultural interests and the popular press. Baldwin commented ruefully: 'the people of this country cannot be shaken out of their fear of high prices.'[9]

The Conservative Party returned to government under Baldwin's leadership included former Liberal and free-trader Winston Churchill, who was appointed Chancellor of the Exchequer. Joseph Chamberlain's son, Neville, the new Minister of Health, was principal author of a restatement of the party's principles and aims, produced during its opposition in 1924, and the basis of its manifesto for re-election. It included a statement of the imperial vision, recognising the importance of strengthening and developing the empire, fiscal autonomy of the dominions and the assistance tariff preference could give them: 'In this way we can provide through our markets the opportunity which they need in order to develop their plans of land settlement, and they, in their turn, will absorb our surplus population and provide new outlets for our manufactures.'

But while the party undertook to defend British trade against unfair foreign competition, 'proposals for a general tariff will not again be submitted except upon clear evidence that on this matter public opinion is disposed to reconsider its judgement'.[10] Thus, although the small preferences announced in 1923 were finally confirmed in Churchill's first budget in 1925, an effective imperial preference was shelved for the remainder of the decade.

Joseph Chamberlain had recognised the link between imperial tariff reform and Social Darwinism. He had been Deputy President of the Science Guild; its purpose was described as 'making the Empire strong and secure through science and the application of scientific method'. World War I demonstrated the need for British Government support for science. The Committee of the Privy Council for Scientific and Industrial Research, created in 1915, was enlarged in 1916 as the Department of Scientific and Industrial Research. It was to be a 'scientific bureaucracy for the administration of research grants, scholarships for research training, and the provision of scientific advice to the government'.[11] In 1923 the Imperial Economic Conference recognised the significance of science

---

9 Dilks, *Neville Chamberlain: Pioneering and Reform*, pp. 358–9.
10 Ibid., p. 381. The statement was published as a pamphlet, entitled 'Looking Ahead', and was also printed in *The Times*, 20 June 1924.
11 Alter, *Reluctant Patron*, pp. 92–4, 97; C. B. Schedvin, *Shaping Science and Industry: A History of Australia's Council of Scientific and Industrial Research 1926–49*, Allen & Unwin, Sydney, 1987, p. 25.

to economic progress, approving establishment of an advisory body on imperial cooperation and a resolution commending cooperation in science: 'All possible steps should be taken to encourage the exchange of scientific and technical information between the various parts of the Empire and the co-operation of the official and other organisations engaged in research for the solution of problems of common interest.'[12]

In 1925, the Baldwin Government turned to these as safer alternatives to an imperial tariff, creating an Imperial Economic Committee, and granting £1 million to be spent by an Empire Marketing Board. As part of the 'visionary' scheme to strengthen imperial trade and cooperation, the board was to spend much of its £1 million annual grant on research stations throughout the Empire, studying problems of science as applied to production and transport of foodstuffs and other exports. In Australia the Commonwealth Council of Scientific and Industrial Research and the Development and Migration Commission were seen as complementary bodies. McDougall made important contributions to all of them.

The Imperial Economic Committee

The new Conservative Government undertook immediate establishment of the Imperial Economic Committee (IEC). Its terms of reference were

> to consider the possibility of improving the method of preparing for market and of marketing within the United Kingdom the food products of the oversea parts of the Empire with a view to increasing the consumption of such products in the United Kingdom in preference to imports from foreign countries, and to promote the interests both of producers and consumers.[13]

In deference to Canadian opposition to creating anything resembling an empire secretariat, it was to be an ad-hoc body and these terms of reference were to be reviewed at the next Imperial Conference.[14]

The IEC was an advisory body responsible to the several governments of the Empire, though it reported to the Board of Trade—giving, in McDougall's view, a misleading impression that it was subsidiary to the board.[15] Four

---

12   NAA, CP103/3, vol. 6, IEC 1923, Stenographic Notes, Eighteenth Meeting, 1923, p. 24.
13   *Reports of the Imperial Economic Committee on Marketing and Preparing for Market of Foodstuffs Produced in the Overseas Parts of the Empire. First Report—General*, Cmd. 2493, pp. 2, 4.
14   Stephen Constantine, 'Anglo–Canadian Relations: the Empire Marketing Board and Canadian National Autonomy between the Wars', *Journal of Imperial and Commonwealth History*, Vol. 21, no. 2, 1993, pp. 360–3.
15   *LFSSA*, 38, 5 November 1925, pp. 11–12.

members, including the Chairman, represented the United Kingdom; there were two each for the six dominions (at this time including the Irish Free State and Newfoundland) and India. Leo Amery, then Colonial Secretary, chose as Chairman Sir Halford Mackinder, the geographer who had inspired Amery's own view of the imperial vision. Representatives nominated early in 1925, with the exception of those from Canada, were senior and notable: chairmen of associated chambers of commerce and of the Co-operative Wholesale Society, high commissioners, trade commissioners and the London Manager of the New Zealand Producers' Board. Canadian representatives were 'humble technicians with specialist expertise only in the areas of immediate inquiry': a livestock expert from the Department of Agriculture and a fruit expert on the staff of the Canadian High Commission in London. They were instructed not to act on matters of high principle without consulting the High Commission, whence reference would be made to Ottawa.[16] The 'senior' Australian representative was Sydney businessman Sir Mark Sheldon, a former chairman of the Associated Chambers of Commerce and holder of several government appointments. According to a newspaper report, F. L. McDougall, 'well-known orchardist on the Murray River', who had 'rendered splendid service in conducting a campaign in Great Britain on behalf of the Australian fruit industry' and at the 1923 Imperial Conference, was to be the 'other representative during initial portion work of [the] committee'.[17]

McDougall had hoped for such an appointment. Immediately after the 1923 Imperial Conference, he had written: 'if the Imperial Economic Committee comes to a head and if Bruce offered me a salaried post as Australia's official representative thereon I think it would be well worth doing.'[18] But he was disturbed by the apparent short-term nature of his appointment and by the distinction in seniority. He suggested to Bruce that Australia would be served best by one permanent member stationed in London and a second representative chosen for expertise relevant to a particular inquiry who could maintain close, 'educational' contact with Australia.[19] This pattern did occur, for a time. Sheldon was succeeded after a year by W. H. Clifford, General Manager of the North Coast Co-operative Company Limited, New South Wales, and a member of the Dairy Produce Control Board. Sir James Cooper attended during 1928. Thereafter, McDougall often served as sole Australian representative. He was to chair the committee during the 1930s.

The IEC first met on 17 March 1925. Its method of supplying advice was to publish voluminous reports on commodities approved by empire governments.

---

16 Constantine, 'Anglo–Canadian Relations', p. 363.
17 *Sydney Morning Herald*, 25 February 1925; cable quoted in *LFSSA*, *11*, 26 February 1925, p. 28.
18 NLA, MS6890/1/9, letter to Norman, 20 December 1923.
19 *LFSSA*, *42*, 19 November 1925, *107*, *109* and *113*, 11 and 25 May and 16 June 1927, *143*, 11 January 1928, pp. 120, 358, 68, 78–9, 495.

At first limited to foodstuffs, its terms of reference were amended by the 1926 Imperial Conference to include raw materials, and its membership augmented by representatives of British agriculture and the colonial empire.

The first commodities to be examined were meat and fruit. McDougall was appointed to the fruit subcommittee and to its three-man drafting committee. He quickly demonstrated determination to see the IEC functioning effectively and producing reports reflecting his views: 'I think that it is essential to cultivate the personal acquaintance of the members of the Committee so that we may be able to get them privately to see the full importance of this committee and thus…obtain general support for comprehensive proposals for assisting Empire Trade.' He began the process by lunching with the Chairman on the first day. Mackinder also chaired the Imperial Shipping Committee and was a member of the Royal Commission on Food Prices. McDougall and Sheldon worried that his commitment to the IEC would be limited.[20]

No member worked harder than McDougall to ensure the IEC lived up to the hopes of empire visionaries. When the 250-page fruit report[21] was published, in June 1926, he organised wide publicity for its appearance, contacting 'all the Editors that I know personally', arranging special treatment in *The Times* and *Daily Telegraph*, and interviews for four Conservative MPs with correspondents of provincial newspapers.[22] He spent long hours in the tedious tasks of drafting and revising this and most subsequent reports. The committee worked by hearing witnesses and inviting submissions. This itself was a burden: 63 memoranda were submitted for an inquiry on tobacco, each averaging about four typed foolscap pages, so that 'one's weekends are fairly well employed in reading them'.[23] He viewed the job as an imperial, rather than simply an Australian, responsibility, and he served at times on drafting committees for commodities of little interest to Australia.

McDougall spent much time compensating for what he considered Mackinder's shortcomings as Chairman, correcting his thinking, smoothing difficult situations and undertaking extra work. He told Bruce that 'in private negotiations I rather get forced into the position of representing the oversea representatives to (a) the Chairman (b) the Members of the Government'.[24]

After the appointment of Sir David Chadwick as Secretary in 1927, the problem of Mackinder diminished. Chadwick included McDougall in consultations on new procedures, which limited the full committee to hearing a few important

---

20 Ibid., *14*, 19 March 1925, pp. 34–5.
21 *Reports of the Imperial Economic Committee on Marketing and Preparing for Market of Foodstuffs Produced in the Overseas Parts of the Empire: Third Report—Fruit*, Cmd. 2658.
22 *LFSSA*, 77, 10 June 1926, p. 248.
23 Ibid., *150*, 23 February 1928, p. 520
24 Ibid., *37*, 29 October 1925, p. 107.

witnesses on each new inquiry, discussing its scope and then handing over to a subcommittee. Definition of the scope of an inquiry proved difficult, so McDougall joined a three-man committee to review that problem.[25] Chadwick and Mackinder later asked McDougall to join them on a standing committee 'to get the proper Imperial aspect suitably expressed in each report':

> I was reluctantly forced to agree that, for this particular type of work, there was no other overseas representative who could usefully function. It is a rather sad comment on the type of man whom the other Dominions have appointed…that experience has shown that, when the general Dominion point of view needs to be taken into careful consideration, they have to turn to me.[26]

The Empire Marketing Board

The first task of the IEC was to advise on a general matter. Having renounced tariffs, Baldwin's government promised an annual grant of £1 million, calculated as the value of preferences that might have been given to empire goods in 1924, to encourage empire trade by other means.[27] The first report of the IEC recommended the money be administered by an 'executive body' subject to the IEC and used to encourage 'voluntary preference': persuading the British public to purchase empire goods by means including advertising and research into problems of production and distribution.[28] While British agreement to the recommendations was delayed for some months—one major reason being the constitutional problem of a body including representatives of other empire governments spending British taxpayers' money—McDougall, virtually alone amongst IEC members, lobbied ministers, the Cabinet Secretariat and Members of Parliament on its behalf. Bruce cabled London twice at his suggestion. After lengthy exchanges of suggestions and reference to a cabinet committee, the Empire Marketing Board (EMB) was constituted as an advisory committee to the Dominions Secretary, Leo Amery's title after the creation that year of the Dominions Office. The board met for the first time on 20 May 1926.

Before the board's composition was finally decided, McDougall, already one of the busiest members of the IEC, was eager to be a dominion representative of the IEC on the board: 'No one here has a wider knowledge of the point of view of the various sections of producers' opinion in all the Dominions for I have

---

25   Ibid., *107*, 11 May 1927, pp. 357–8; *150*, 23 February 1928, pp. 520–1.
26   Ibid., *156*, 29 March 1928, pp. 544–5.
27   NAA/CSIR, A9778, M14/27/9, undated memorandum by McDougall, 'The Empire Marketing Board and Empire Economic Affairs', received 5 December 1927, p. 2.
28   *Reports of the Imperial Economic Committee on Marketing and Preparing for Market of Foodstuffs Produced in the Overseas Parts of the Empire: First Report—General*, Cmd. 2493.

made that question a special study for the last three years.'[29] In the event he was one of five IEC representatives who joined four British ministers—Leo Amery as Chairman; Colonial Secretary William Ormsby-Gore as his deputy; Minister of Agriculture Lord Bledisloe; Under-Secretary of State for Scotland Walter Elliot—and representatives of the Board of Trade.

The EMB formed two temporary subcommittees to recommend methods of tackling research and publicity.

> It was very far from being my wish that I should have to have a say on both these subjects but in the discussions it had become rather obvious that I had perhaps done rather more constructive thinking about how the Economic Committee's recommendations were to be put into operation than anybody else present at the meeting and the Board felt that in devising methods I could give useful assistance both on publicity and research. Personally I should have preferred to be mainly connected with the research side but I am rather afraid that the Secretary of State and other ministers will particularly want me to serve on the Publicity Sub-Committee as they seem to regard me as being particularly expert on educational publicity.[30]

Ormsby-Gore had agreed to chair the Publicity Committee on condition that McDougall was a member.[31] Both committees became permanent—'Research' later being renamed 'Research Grants' to reflect more accurately its function as a dispenser of funds to approved research projects and institutions. The EMB report for 1931–32 lists six main committees; McDougall sat on four or them, including Agricultural Economics, which he chaired, and the Film Committee.

The EMB developed into a substantial institution. It was a 'constitutional oddity': technically an advisory committee, but effectively possessing executive authority, with a civil service staff of some 120 at its height.[32] Amery prevailed over Treasury to ensure that the £1 million annual grant was non-returnable and free from close Treasury control, although the battle to preserve it continued throughout EMB's seven-year existence. The board was also unusual in its non-partisan membership: J. H. Thomas represented Labour from a very early stage, and Archibald Sinclair the Liberals. In 1929 the Labour Government continued its work with enthusiasm.

McDougall's closest ally on the board was Walter Elliot, a doctor who had also gained a DSc for a study on pig nutrition, undertaken at the Rowett Research

---

29 *LFSSA*, 55, 17 February 1926, p. 162.
30 Ibid., 76, 3 June 1926, p. 246.
31 Ibid., 70, 11 May 1926, p. 225.
32 Stephen Constantine, *Buy & Build: The Advertising Posters of the Empire Marketing Board*, HMSO, London, 1986, p. 3.

Institute, Aberdeen, headed by his friend John Boyd Orr. In 1935 Elliot would be elected a fellow of the Royal Society. He held somewhat independent, 'centrist' views, and McDougall was impressed by his book *Toryism and the Twentieth Century* (1927), arguing, in the tradition of Progressivism, for 'a conservatism that would make use of applied and social sciences and of government intervention'.[33] Another ally was Ormsby-Gore, a former Parliamentary Private Secretary to Lord Milner at the Colonial Office, with unusually wide knowledge and experience of the colonial empire.[34] Elliot and Ormsby-Gore were both 'visionary' advocates of 'science for development', believing that research, particularly in agricultural sciences, and an organised colonial agricultural service held the key to developing the resources and potential markets of the colonial empire.[35] McDougall wrote that he, Elliot and Ormsby-Gore 'envisage the problems of Research and Publicity from the same angles and together I feel sure that we shall be able to shape a policy for the Board which will be effective'.[36]

EMB staff were of high calibre. McDougall was impressed by the Secretary Amery had chosen. Stephen Tallents had wide experience as a civil servant, in wartime food rationing and postwar relief, as Imperial Secretary in Northern Ireland and as Secretary of the cabinet committee dealing with the 1926 General Strike. It has been suggested that, as a result, Tallents was, for his time, 'unusually sensitive to the need to assess and massage public opinion', aware that as government functions broadened it would be necessary to use publicity as 'a managerial tool' and 'to obtain public consent by persuasion'.[37] McDougall claimed to have arranged the appointment, as Assistant Secretary, of E. M. H. Lloyd, 'the stabilization expert', who had worked in the wartime Ministry of Food, in the Economic and Financial Section of the League of Nations from 1919 to 1921, and in the Ministry of Agriculture.[38]

The EMB was 'a multi-media event'. It provided public lectures, materials for use in schools, books and pamphlets—some serious studies written or inspired by McDougall, others more light-hearted such as 'A Book of Empire Dinners'. It spent £364,280 on press advertisements over its seven-year life and indulged in occasional publicity stunts, such as hiring aircraft bearing the slogan 'Buy

---

33 Gordon F. Millar, 'Elliot, Walter Elliot, (1888–1958)', *ODNB*, 2004, <http://www.oxforddnb.com.virtual.anu.edu.au/view/article/33003> [accessed 22 November 2005].
34 K. E. Robinson, 'Gore, William George Arthur Ormsby, Fourth Baron Harlech (1885–1964)', *ODNB*, 2004, <http://www.oxforddnb.com.virtual.anu.edu.au/view/article/35330> [accessed 7 May 2007].
35 Joseph M. Hodge, 'Science, Development and Empire: The Colonial Advisory Council on Agriculture and Animal Health, 1929–43', *Journal of Imperial and Commonwealth History*, Vol. XXX, no. 1, 2002, pp. 3–5.
36 *LFSSA*, 78, 24 June 1926, p. 253.
37 Constantine, *Buy & Build*, p. 4.
38 *LFSSA*, 78, 24 June 1926, p. 253. Lloyd published *Stabilisation* in 1923 and *Experiments in State Control* in 1924. Frank Trentmann, 'Lloyd, Edward Mayow Hastings (1889–1968)', *ODNB*, 2004, <http://www.oxforddnb.com.virtual.anu.edu.au/view/articles/34/34566-nav.html> [accessed 20 November 2004].

British' and the baking of a seven-foot-high 'King's Empire Christmas Pudding'. It arranged Empire Shopping Weeks, exhibitions and broadcast talks, and had a film library and a small film unit.[39]

McDougall had not been keen to use the £1 million grant for publicity. He did not believe that voluntary preference could achieve the results possible from tariff preference.[40] But if money had to be spent on publicity, he knew what that publicity should be. He constantly used the word 'educational' because 'what we had to do was to develop…an Empire consciousness in the people of Great Britain and that appeal must be directed to the reason and not to the emotions'. He had to begin by educating the Publicity Committee itself, as it became clear 'that what was urgently needed as a preliminary to more effective advertising was a clear realisation, on the part of members of the Board and of the Publicity Committee, of the essential facts about the importance of Empire Trade to Great Britain'. He therefore prepared a statement of those facts for them.[41]

McDougall was an active member of the EMB poster subcommittee. Its most distinctive posters were series, mounted on large, multi-section frames, a format suiting McDougall's 'educational' principles. The panels could position lists of export and import figures between striking illustrations commissioned from notable artists.[42] McDougall devised a special 'educational' series for use in British factories. 'The idea is to demonstrate chiefly to persons engaged in some of the most important industries of the United Kingdom the great advantage that will accrue to those industries if Empire development is supported by Empire purchasing.' The posters specified how much that firm or factory had sold to a particular part of the empire, or that a contract was now in hand, and asked: 'How can you help to secure further contracts from the Empire? Answer: By buying, and by getting your wife to buy, the produce that the Empire sends to us.'[43]

Research Grants

EMB publicity activities drew some criticism as an inappropriate function for government, but the research side of its work was widely accepted. The

---

39 Constantine, *Buy & Build*, p. 5; Constantine, 'Bringing the Empire Alive', in *Imperialism and Popular Culture*, ed. John M. Mackenzie, Manchester University Press, Manchester, 1986, pp. 205–10.
40 *LFSSA*, *9* and *27*, 22 January and 6 August 1925, pp. 23, 75–7.
41 Ibid., *61*, 31 March 1926, p. 193; *95*, 3 March 1927, pp. 318–19.
42 Copyright restrictions prevented their use in this publication, but illustrations of many posters are included in Constantine, *Buy & Build*. They may also be viewed online at: <http://www.manchestergalleries.org/the-collections> Works by Charles Pears, including *Gibraltar*, *Aden* and *Bombay*, provide a good example of a multi-frame series.
43 *LFSSA*, *171*, 28 June 1928, pp. 599–600; Constantine, 'Bringing the Empire Alive', p. 216. The poster campaign is discussed in the EMB's *Report* for 1928–29, EMB 19, HMSO, July 1929, pp. 24–5.

EMB Research Grants Committee (RGC), chaired by Walter Elliot, made recommendations to the full board for the funding of scientific investigations. It can be argued that this was the most important of all McDougall's activities in this period: it was central to his work in many ways both then and in the future. As with publicity, McDougall was appointed to a subcommittee to consider 'machinery and methods' of putting IEC recommendations into effect and, as in the IEC, he undertook the task of ensuring the RGC fulfilled its proper purpose.[44] He aimed to ensure RGC grants were used effectively in research work of value to the Empire and worked hard to keep the EMB and its committees true to his vision of their role.

EMB scientific staff investigated applications for grants, but RGC members were often actively involved. The committee agreed at its first meeting that advice on the existing state of knowledge should be sought before consideration of any application, and that all bodies likely to be interested should be notified of any inquiry. At that meeting proposals for tropical research centres were referred to the Colonial Office and the Development Commission; investigation of transport and refrigeration of fruit to the Department of Scientific and Industrial Research; and packing of fruit and arsenic sprays to the Ministry of Agriculture and the Imperial Bureau of Entomology. A proposed study of vitamins in dried fruits was referred to the Medical Research Council; one on tainting of dairy produce during transport to the National Institute for Research into Dairying; and another on sugaring of raisins to the Royal Society. A year later, a similar meeting considered a broader range of proposals including the timber industry in Guiana, egg marketing in Scotland, the teaching of economic geography at Cambridge, calf-rearing in Palestine and transport of chilled beef.[45] All of these topics, with the bodies proposing research and those to whom proposals were referred, came within McDougall's ambit. A conscientious member of the RGC needed to understand, at least in basic terms, something of the fundamentals of many branches of the biological and physical sciences, economics, transport and storage, not to mention the geography of the empire itself. It was a mind-stretching undertaking, but it called for general rather than specialist skills. It is difficult to think of any sort of training better than McDougall's: a combination of some basic science, practical experience of production, marketing and organisation, travel in two dominions and a tiny colony, an intensive study of the trade and produce of the empire, and dedication to cooperative development of the empire's resources.

---

44 *LFSSA*, 76, 3 June 1926, pp. 245–6.
45 National Archives of the United Kingdom [hereinafter UKNA], CO760/21, Minutes of EMB RGC, First Meeting, 1 July 1926; Twelfth Meeting, 25 May 1927.

The EMB published more than 50 papers—mostly the work of experts.[46] Many publications arose from research grants. The long evening meetings McDougall spent drafting and redrafting IEC reports were unnecessary for the EMB, although he was the author of one of its publications, *The Growing Dependence of British Industry upon Empire Markets* (1929).[47] McDougall's role on the EMB was an executive and creative one, of policy and liaison between and on behalf of dominions and the British Government, and with research establishments.

Scientific Liaison

In Australia, the founding of a 'national research laboratory' had been announced amid the imperial enthusiasms of wartime. The Institute of Science and Industry was not actually established until 1921, and then limited by its budget to small-scale research. Substantial Federal Government support awaited 'identification of the major economic role' of science. Bruce identified that role as imperial scientific cooperation. In creating the Council for Scientific and Industrial Research (CSIR) in 1926, he made clear its responsibility to concentrate on 'a limited number of major investigations of national importance'. In view of balance-of-payments uncertainties and Bruce's commitment to promoting trade within the empire, this meant the problems of the major export industries. It was, writes C. B. Schedvin, 'a by-product of the Indian summer of neo-mercantilism and the idea that the British nations should combine economically to yield a high level of self sufficiency'.[48] Its goals included cooperation with imperial arrangements for government-sponsored scientific research.

During his visit to London for the 1926 Imperial Conference, Bruce asked McDougall to act as London representative for both CSIR and its twin organisation, the Development and Migration Commission (DMC). Established primarily to examine settlement schemes proposed by State Governments, DMC was also to conduct economic surveys and facilitate establishment of new industries. Introducing the Bill for DMC's establishment, Bruce 'declared his purpose to lift Australia's population to a level that world opinion would either respect or fear, while integrating that process with planned economic expansion and thus maintaining living standards'.[49]

McDougall's appointment could have seemed an odd one. Whatever reputation he had in 1926 rested upon his mastery of empire trade statistics. But Bruce

---

46   A list of 52 publications, to May 1932, is in EMB 53, *Empire Marketing Board May 1931 to May 1932*, HMSO, June 1932.
47   EMB 23, HMSO, December 1929.
48   Schedvin, *Shaping Science and Industry*, pp. 14–18, 25.
49   Michael Roe, *Australia, Britain and Migration 1915–1940: A Study of Desperate Hopes*, Cambridge University Press, Cambridge, 1995, p. 67.

understood the place of science in the scheme of imperial development. He knew that substantial funds were to be had from the EMB. With the RGC's sanction, McDougall had written offering EMB assistance in the form of 50/50 grants for CSIR and coordination to prevent overlapping within the empire.[50] Privately he had written to Bruce at length about the application of scientific research to Australia's rural problems. Bruce must have understood that McDougall—already close to many key political figures and civil servants—was well placed to extend his contacts in science. The appointment probably had its actual genesis in informal discussion during Bruce's visit. It accorded well with McDougall's new interests and his membership of the RGC.

Liaison with DMC

H. W. Gepp, metallurgical engineer, founding manager of the Electrolytic Zinc Company of Australasia, industrial relations pioneer, believer in welfare capitalism and propagandist for scientific agriculture, was appointed to head DMC and, given the imperial dimension envisaged for his work, accompanied Bruce to London for the 1926 Imperial Conference.[51] Gepp's brief was impossibly broad, and Gepp himself unrealistically ambitious for the commission. He proposed to begin with an economic survey of 'the whole of the present resources of Australia' to establish, inter alia: the effects of the seasonal nature of work in Australia, and of migration, upon unemployment; the commercial relationships between industries to be encouraged; and the causes of cyclic economic depressions. He aimed to cooperate with every agency upon which this work impinged and proposed early investigation into development of Tasmanian resources, the goldmining and tobacco industries, rural housing, development of fisheries, and production and marketing problems of the dried-fruits industries. DMC would be 'the national clearing house for all ideas and schemes bearing upon economic development'.[52]

McDougall took the intense, almost hyperactive Gepp under his wing, arranged introductions and inspections with the many institutions in which Gepp expressed interest, and generally displayed his enthusiasm and his range of contacts. Gepp's schedule in Britain was as punishing as his vision was ambitious. He presided over a committee of the Colonial Office to investigate mechanical transport for undeveloped terrain (a type of road train); he visited research stations seeking information on shale-oil production, low-temperature

---

50 *LFSSA*, 82, 22 July 1926, pp. 268–9.
51 Roe, *Australia, Britain and Migration*, p. 69.
52 NAA, A11583, 16, 'Imperial Conference 1926, Unbound Individual Papers, (E[E]1 to E[E]52)'; H. W. Gepp, 'The Development and Migration Commission, Commonwealth of Australia, its constitution and functions', 17 November 1926, Imperial Conference 1926, E(E) 51.

carbonisation of brown coal, producer gas, a physical and chemical survey of Australian coal resources, liquid and pulverised fuels, geophysical prospecting, transport of foodstuffs, dehydration of vegetables, canned foodstuffs, fisheries, goldmining, grass improvement, forestry and the placing of research students.[53]

All these lines of inquiry had to be followed up by McDougall. The new appointment gave him more formal links with the Australian bureaucracy and a regular, albeit officially temporary position in Australia House. He gained, at Bruce's instruction, a clerical assistant, A. W. Stuart Smith, and a technical assistant, Dr A. S. Fitzpatrick, the latter because McDougall doubted his own ability to cope with the technical demands of work relating to the physical sciences.[54] In the first of regular monthly reports, McDougall emphasised the care he was taking to ensure that other officers in Australia House were consulted and informed. He had begun immediately to establish formal liaison with every relevant research board and institute in the United Kingdom, including some 30 bodies under the control of agricultural authorities, and with government departments. McDougall planned to visit the head of each organisation personally.[55]

Liaison with CSIR

McDougall did not meet the chief figures in CSIR for some time and was uncertain about what was expected of him. He wrote to George Julius, Chairman of the Council and of its Executive Committee: 'As I see this job…it is to keep your Council in touch with the work done in this country in the application of science to primary and secondary industries, in so far as these applications may seem to be of interest to Australia.' He went on to explain the interest of the RGC in coordinated research throughout the Empire.[56]

Julius, gifted engineer, inventor and businessman, and CSIR's Chief Executive, David Rivett, both replied courteously, but warily. Rivett had been persuaded to leave the Chair of Chemistry at the University of Melbourne to fill this new position. He was an able administrator and distinguished physical chemist who believed that empirical work must be supplemented and guided by thorough theoretical analysis.[57] Like McDougall, he came from nonconformist stock with a strong sense of social responsibility. Rivett was personally gracious, approachable and patient, but he faced a daunting task. Every area of primary

---

53 NAA/CSIR, A8510, 220/39, 'Reports from F. L. McDougall 1927', 'Statement re Development and Research Activities of Chairman D&M Commission whilst in London'.
54 Ibid., McDougall to Julius, 26 January 1927.
55 Ibid., Report, 10 February 1927.
56 Ibid., McDougall to Julius, 26 January 1927.
57 Schedvin, *Shaping Science and Industry*, pp. 32–3.

industry had pressing problems in urgent need of solution. CSIR was also required to assist secondary industry. Unlike Gepp, Rivett and his fellow councillors saw a clear need to establish priorities, to avoid spreading resources too thinly. In establishing those priorities, handling political pressures, relations with other institutions and practical problems of staffing and accommodation, Rivett had his hands full. He was reluctant to accept advice or assistance from London. Apart from his view that Australia should pay its own way, there was concern that such assistance might come with conditions.

For some months McDougall felt that he and his staff were 'simply improvising, without any clear idea as regards your wishes'.[58] Rivett was reluctant to accept EMB funds—the chief service that McDougall could give. But when Julius visited London in mid 1927, he and McDougall established an immediate rapport. Julius reported favourably to Rivett and explained Rivett's funding scruples to McDougall. McDougall wrote to reassure Rivett, reminding him of Bruce's pivotal role in establishing the EMB, adding what was effectively a short memorandum explaining that any action taken by Britain to aid empire development repaid Britain at least as much as the empire gained. Rivett replied that while this was 'very logical',

> I cannot refrain from suggesting that the case for acceptance of British money for Australian work would be much clearer if Australia were putting into general scientific research an amount of money commensurate not only on a population basis with what Britain is finding but also commensurate with the magnitude of our local problems. We are hardly doing the former and most emphatically not the latter. [Nevertheless] the policy of co-operative work with the EMB is settled and I only hope we shall not appear at any time to be leaning too heavily upon its funds.[59]

Thus began an important correspondence and friendship, and McDougall's role, not just as a liaison officer, but as an adviser, was established. Rivett wrote that 'you have been the guide, philosopher and friend [to Julius, who] knows quite well how impossible his task would have been without you and your staff. I wish we were as happy about everything as we are about our liaison work in London.'[60] He continued to write frankly, sometimes using the correspondence as a means of thinking through problems, and as though McDougall were on a par with the three-man Executive Council. In a sense he was: in these first years of CSIR's existence, his liaison work became an important lifeline in funds, recruitment and contacts, in links to the network of empire. McDougall also used Rivett as a sounding-board and a point of distribution in Australia for his own

---

58   NAA/CSIR, A9778, M14/27/9, McDougall to Rivett, 4 August 1927.
59   Ibid., McDougall to Rivett, 30 June and 7 July; Rivett to McDougall, 15 August 1927.
60   NAA/CSIR, A10666, [1], Rivett to McDougall, 20 February 1928.

ideas and memoranda. The wide-ranging nature of the correspondence is shown by the fact that after 'some difficulty' in sorting the personal from the official material in CSIR, McDougall divided his letters into three categories: those for Rivett's private information, often sent to his home address; confidential letters which Rivett might share with others if he wished, sometimes including memoranda; and official letters to be filed. Rivett seems to have adopted a similar practice.[61] Rivett's son believed that their 'close friendship [was] based on mutual admiration' and a meeting of minds beyond their working concerns, particularly as McDougall's

> insights into British politics were sharp and spiced with humour. This delighted David who had always corresponded on these lines with his own family but seldom received much on the lighter side from his scientists. He and McDougall could see an amusing side in even the grimmest cuts and setbacks and helped each other to keep frustration and genuine grievance in proportion.[62]

CSIR's Executive believed that effective working relationships should be established with research and other establishments, both in Australia and overseas.[63] Even before the understanding about his role had been reached, McDougall was becoming the eyes and ears of the organisation in London: 'I am writing today to McDougall to ask him if he can start any enquiries going amongst the British biologists as to the way in which the ["dingo pest"] question can be attacked.'[64] He was able to inquire informally from Sir Henry Tizard, head of the Department of Scientific and Industrial Research, about British views on the payment of academic scientists seconded to government projects and on the tricky problem, involving copyright of results, of CSIR membership of British research organisations.[65] McDougall's office took care of the finances and progress of young scientists sent to Britain under the studentship scheme. He took this fatherly role seriously, giving practical assistance, arranging placements and doling out advice. He arranged wider experiences for very promising students.[66] On occasion students failed to measure up to the demands imposed upon them. McDougall would counsel, arrange conclusion of the placement and then placate the institution, protecting CSIR's reputation for future students.

McDougall helped recruit scientists. Rivett wanted his research divisions headed by world-class figures to enhance the standing and attraction of the

---

61 Ibid., McDougall to Rivett, 8 December 1927.
62 Rohan Rivett, *David Rivett: Fighter for Australian Science*, Melbourne, 1972, pp. 107–8.
63 Schedvin, *Shaping Science and Industry*, p. 58–9.
64 NAA/CSIR, A9778, M14/27/7, Rivett to Julius, 7 July 1927.
65 NAA/CSIR, A9778, M14/27/9, McDougall to Rivett, 25 August 1927; Rivett to McDougall, 27 September 1927; McDougall to Rivett, 5 October 1927.
66 *LFSSA, 171*, 28 June 1928, p. 598; NAA/CSIR, A9778, M14/28/7, McDougall to Rivett, 3 May 1928; A10666, [1], McDougall to Rivett, 18 September 1928.

organisation. With few experienced research scientists in Australia in some disciplines important to the work of CSIR, there was a view that the advice of leading overseas scientists should be sought. South African veterinarian Sir Arnold Theiler visited for six months in 1928 and recommended an ambitious program, including a large central laboratory. McDougall helped arrange the visit and was subsequently much involved, at Bruce's personal instruction, in unsuccessful efforts to persuade Theiler to head the Division of Animal Health—a field considered crucial to CSIR's success.[67] He assisted in recruitment of other senior men and a young Francis Ratcliffe, then a junior EMB employee, to work on flying foxes.[68]

The work did not always accord with McDougall's views on imperial cooperation. A request that the EMB contribute half the cost of an ambitious plan for research into biological control of various plant and insect pests—devised by the brilliant but difficult head of CSIR's Division of Economic Entomology, Robin J. Tillyard—was supported by Federal Cabinet and by a personal cable from Bruce to Baldwin. In McDougall's view, the proposal threatened imperial cooperation and coordination by demanding too heavy an expenditure on entomology, for which a considerable sum had been committed elsewhere, and it involved a disproportionate allocation to Australia. He persuaded Elliot, nevertheless, and the grant of £36,000 was one of the largest ever made.[69]

Both as EMB member and in liaising for CSIR and the DMC, McDougall visited research stations throughout Britain and came to know some leading scientists well. Chief among them was Elliot's friend John Boyd Orr, 'a very remarkable man of about 45 years of age', wrote McDougall after their first meeting, impressed both by Orr's scientific achievements and his war record: serving as a medical officer, Orr had been awarded both a Distinguished Service Order and a Military Cross.[70] After graduating in medicine, Orr had received a Carnegie Fellowship to study physiological chemistry, and in 1913 had been appointed Director of a new institute of nutrition at Aberdeen. After the war he expanded what became the Rowett Institute, studying nutrition of farm animals and of human populations.[71]

McDougall reported constantly to Bruce on his regular visits to the Rowett Institute; to the Fruit Research Station at East Malling near Maidstone in Kent,

---

67   Schedvin, *Shaping Science and Industry*, pp. 82–4.
68   NAA/CSIR, A9778, M14/27/9, McDougall to Rivett, 7 September 1927.
69   Schedvin, *Shaping Science and Industry*, pp. 99–101; *LFSSA*, *171*, 28 June, *174* and *179*, 12 and 26 July 1928, pp. 599, 608–9, 662; NAA/CSIR, A9778, M14/27/9, Rivett to McDougall, 20 October 1927; A10666, [1], McDougall to Rivett, 24 November 1927, 21 June and 5 July 1928; Archives of the Commonwealth Scientific and Industrial Research Organisation [hereinafter CSIRO], Series 3, AN1/6/7, McDougall to Rivett, 19 and 26 July 1928.
70   *LFSSA*, *124*, 6 September 1927, p. 422.
71   K. L. Baxter, 'Orr, John Boyd, Baron Boyd Orr, (1880–1971)', rev., *ODNB*, 2004–05, <http://www.oxforddnb.com.virtual.anu.edu.au/view/article/31519> [accessed 22 November 2005].

where the Director, R. G. Hatton, headed pioneering work on rootstocks; to the Rothamstead Experimental Station on Soil Science at Harpenden in Hertforshire, directed by Sir John Russell, a pioneer in organising modern agricultural research;[72] and to the Welsh Plant Breeding Station at Aberystwyth, led by Professor R. G. Stapledon, who believed that 'productive grasslands lay at the heart of productive agriculture'. Stapledon's 'ecological' approach influenced farmers, scientists and politicians throughout the British Commonwealth.[73] McDougall was much impressed.[74] He was also impressed by Stapledon's young assistant, Elspeth Grant, who had grown up in Kenya, studied agriculture at Reading and Cornell universities, and was planning to return to the United States after some work experience. McDougall persuaded her to apply for a position as junior press officer with the EMB, popularising its scientific work.[75] She undertook this task with great success, later married Gervas Huxley, Secretary of the EMB Publicity Committee, and, as Elspeth Huxley, achieved fame as a novelist. McDougall became a member of her circle of friends, which included his relative agricultural scientist A. N. (Jim) Duckham. There was frequent correspondence between McDougall and Elspeth for some years after she left the EMB and travelled extensively. She, like Rivett, was one of those to whom he sent ideas and memoranda for comment throughout the 1930s.

Spurred by the views of both Walter Elliot and McDougall, the RGC adopted a role beyond mere dispensation of funds, assuming responsibility for coordinating research throughout the empire in the fields related to its role, avoiding duplication and publishing results. To harness science effectively to the cause of empire development, it was not sufficient simply to supply funds. The Imperial Agricultural Bureaux (IAB) were attached to leading institutions to disseminate results of research through abstracting services. McDougall played a significant role in ensuring the IAB were not tied to any British ministry. He represented Australia and was Vice-Chairman on the IAB Executive Council.[76]

Efficiency in Primary Industries

McDougall had argued since 1919 that rural efficiency would rest on producer organisation, fewer middlemen, improved transport and provision of economic and market information. Now he sought to achieve that efficiency through

---

72   N. W. Pirie, 'Russell, Sir (Edward) John (1872–1965)', rev., *ODNB*, 2004, <http://www.oxforddnb.com.virtual.anu.edu.au/view/article/35877> [accessed 8 May 2007].
73   Elizabeth Baigent, 'Stapledon, Sir (Reginald) George (1882–1960)', *ODNB*, <http://www.oxforddnb.com.virtual.anu.edu.au/view/article/36255> [accessed 8 May 2007].
74   *LFSSA*, *122* and *123*, 18 and 25 August 1927, pp. 415–16, 418.
75   Gervas Huxley, *Both Hands: An Autobiography*, Chatto & Windus, London, 1970, p. 149; C. S. Nichols, *Elspeth Huxley: A Biography*, paperback edn, Harper Collins, London, 2003, pp. 91–5.
76   *LFSSA*, *195* and *198*, 21 November and 6 December 1928, pp. 679–80, 686–8.

imperial cooperation. In 1927 he established, single-handedly it seems, the EMB Committee on Agricultural Economics. He brought together leading academics C. S. Orwin and J. A. Venn in what was then a very new field, with representatives of the EMB, the Ministry of Agriculture and Fisheries and the National Farmers' Union.[77] McDougall had the group list all subjects 'that might usefully be regarded as falling within the scope of agricultural economics' and then select those that could be 'dealt with on an Imperial basis'.[78] The findings included 'economic geography, marketing, co-operation, transport, fiscal and even sociological factors as they touch the agricultural sphere'. He hoped work in some of these fields would be undertaken by the EMB in cooperation with the universities.[79]

He also urged revision of a view that an Australian development rate of 2 per cent per annum was sufficient. 'You have in Australia 6 million really virile people who, in the year of grace 1928, are armed, or ought to be armed, with all the inventions of science to assist them in the more rapid development of their country.' He questioned a suggestion by a committee inquiring into the Australian tariff that, unlike manufacturing, agriculture was subject to the law of diminishing returns. 'It seems to me that for many years to come the application of brains, capital and energy to Australian agricultural and pastoral pursuits will give increasing returns, i.e. returns larger than expenditure, and I personally doubt whether under undiscriminating protection the same will hold good for secondary industries.'[80]

The task of pasture improvement provided, in McDougall's view, the ideal opportunity for scientific coordination on an imperial scale. He had learned of the potential of grassland improvement using superphosphate and subterranean clover from SA grazier W. S. Kelly, whom he had first met as a fellow AIF Education Officer in 1919, and Victorian Agent-General in London, George Fairbairn, who had run successful small-scale experiments on Victoria's Mornington Peninsula. McDougall believed the method offered a means of converting pastoral lands to closer settlement, increasing carrying capacity up to tenfold, and replacing wool with stock fattening and dairying in suitable parts of southern Australia.[81] Orr visited Australia in 1928 and was convinced that 'given a great organized drive, it would be possible' to correct mineral deficiencies and thus, 'within five years, to add at least £10 million to the value of Australia's exports of pastoral products without increasing by one acre the areas at present devoted to

---

77 H. A. F. Lindsay, R. R. Enfield and J. B. Guild.
78 *LFSSA*, *120*, 4 August 1927, p. 407.
79 NAA/CSIR, A9778, M14/27/9, McDougall, undated memorandum, 'The Empire Marketing Board and Empire Economic Affairs', received 5 December 1927.
80 *LFSSA*, *173*, 5 July 1928, p. 604; *168*, 19 June 1928, pp. 587–8. McDougall seems to have been placing his own interpretation on a subject of debate between academic economists at the time. See William Coleman, Selwyn Cornish and Alf Hagger, *Giblin's Platoon*, ANU E Press, Canberra, 2006, pp. 63–5.
81 *LFSSA*, *65*, 16 April 1926, pp. 208–11.

pastoral production'. Orr thought that 'firstclass teams' of veterinary scientists could be formed in Australia to work on the 'practical aspects' of the problem since 'already ascertained scientific knowledge—ascertained in many parts of the world including Australia—only requires to be collated…for an immense advance in pastoral conditions to occur'.

McDougall suggested a joint arrangement between CSIR and the EMB under which Orr might be seconded for a period to get such a program under way: 'the matter should be considered to be a great national enterprise…in order to bring the whole Empire conception into the picture and indeed in order to get the necessary intellectual assistance from men such as Orr.' If CSIR and the EMB assisted State agricultural departments, progress that could normally take 20 years might be made in five. Orr gave authoritative confirmation of McDougall's own view that 'armed with weapons forged for us by modern science, we ought to be able to make much more rapid progress and that, in the first instance, progress should be on our already settled areas'.[82]

Views on Research

The difference in viewpoints of centre and periphery posed problems for cooperation in science. Orr's idea of a great campaign for pasture improvement, enthusiastically supported by McDougall, contradicted the approach of T. Brailsford Robertson, CSIR's Chief of Animal Nutrition. Robertson planned fundamental research to determine the exact nature of amino-acid deficiencies in leaf proteins of fodder plants upon which Australian sheep depended in times of drought, followed by field trials of stock licks and mineral supplements added to water. British scientists were critical: it had not yet proved possible to extract all protein from fibrous plant material; empirical field trials would yield quicker results. Orr even questioned 'the value of any basic nutrition research in Australia. His views were those of the economic agriculturalist…he thought that an adequate increase in production could be achieved by the application of existing knowledge.' The differences raised 'fundamental issues about the choice of research projects, about the balance between basic and applied work, and about the relationship between the ideals of natural science and often conflicting socio-economic reasons'.[83]

CSIR's historian, C. B. Schedvin, suggests that neither Robertson nor Orr had a satisfactory answer. Robertson's investigation of a 'challenging biological problem' was hard to justify on economic grounds. But Orr's 'sweeping dismissal of the need for fundamental research in agrostology bore the imprint

---

82   Ibid., *176*, 18 July 1928, pp. 613–16.
83   Schedvin, *Shaping Science and Industry*, p. 80.

of scientific imperialism'; it assumed that the Rowett Institute could provide all the necessary theoretical knowledge. It was 'a sharp rebuff' to Rivett's views.[84] As Rivett put it to McDougall, economies in the conduct of research might result, but at the risk of neglecting the potential of Australian researchers: 'if we make it a practice to send our severer problems elsewhere…we shall run the risk of definitely lowering the standard of ability in our own workers.' He drew a parallel with the idea called industrial rationalisation—that dominions should not aim to develop higher levels of technology:

> …however economically unsound Australia's present tariff policy may be, the main idea underlying it is not lightly to be set aside. We must have in our midst industries demanding high skill and a good intellectual standard…we will be well advised to let our own people face our own problems, however difficult they may be. Otherwise we shall tend to lose the strength that comes only from exercise.[85]

In this case, neither solution prevailed. Despite McDougall's strenuous efforts, Orr was unable to spend extended periods in Australia, and Theiler, the preferred alternative, declined. Brailsford Robertson, whose work had continued with Rivett's strong support, died suddenly in January 1930. Subsequent recruitment policies were 'more modest and pragmatic' and attention was paid to 'local knowledge and experience'.[86]

Thus ended one grand scheme of imperial science. The entomological scheme, for which McDougall had subordinated his own view of imperial priorities to Australian demands, also met an inglorious end. It was overly ambitious, poorly thought out and spread limited resources over a range of complex problems. Despite some useful pioneering research, expectations of more spectacular successes, like the eradication of prickly pear by caterpillars of *Cactoblastus cactorum* following its introduction in 1926, were not fulfilled.[87]

Defending the Tariff

One of McDougall's greatest assets was, as he told Bruce, that 'I am, like Elijah, "very zealous"'.[88] His account of a typical day bears this out: writing until lunchtime, lunch 'almost invariably of a propaganda nature', more office work and then seeing people in the House of Commons until 7 pm. He worked at weekends, and would remain in London for much of the holiday month of

---

84 Ibid., pp. 79–81.
85 NAA/CSIR, A10666, [1], Rivett to McDougall, 12 April 1929.
86 Schedvin, *Shaping Science and Industry*, p. 84.
87 Ibid., pp. 100–1.
88 *LFSSA*, *37*, 29 October 1925, p. 107.

August. Vacations were rare and short.[89] The letters to Bruce in the 1920s list extraordinary numbers of meetings and other activities. Clearly he possessed considerable stamina, as well as determination and the dedicated discipline of his heritage. But his zeal may well have been a factor in the disintegration of his marriage in the late 1920s. His daughter remembered often seeing her father only as a hand reaching from behind the newspaper for his cup at the breakfast table.

Although he made considerable use of the press, memoranda remained the chief vehicle for McDougall's ideas. Both Elspeth Huxley and her husband Gervas joked about his 'rather touching belief that every problem could be solved by a memorandum'.[90] His memoranda could be short briefing notes limited to a page or two on a very specific topic: many of these were produced, for example, before an Imperial Conference. Others could state a case, be intended as discussion papers or even a means of crystallising a new idea, running through several drafts to reflect the contributions of those asked to read and comment. This process might result in a major memorandum of 20 pages or more, covering a new idea from every possible angle, and designed to be circulated to a carefully chosen list of recipients by both Bruce and McDougall as a catalyst for further discussion and action. McDougall had also amassed an encyclopedic knowledge of trading patterns—literally, in fact, as he contributed an entry on empire trade to the *Encyclopedia Britannica*. Most of his knowledge was based on intensive and self-directed study of statistics, for which, he told his brother, he had developed 'a mania'.[91] In December 1925, he was elected a Fellow of the Royal Statistical Society, proposed by Sir Sydney Chapman and Henry Macrosty, both of the Board of Trade, but he does not appear to have taken any active part in the society.[92]

McDougall was busier than ever in the late 1920s, serving imperial interests on the IEC and the EMB, but also working directly for Australia, and that work included promoting Australian policy. He could not admit to any difficulty in reconciling the two sets of interests, but there were some ominous signs. In the years following publication of *Sheltered Markets*, he was increasingly obliged to defend the Australian tariff in the light of his own published arguments, and to answer the protests of aggrieved British manufacturers. The *Manchester Guardian Commercial* pointed out that McDougall's case in *Sheltered Markets* rested on a complementary trade: Britain supplying goods manufactured from raw materials supplied by Australia. The reviewer pointed to 'the inconsistency between the propaganda in favour of Imperial Preference and the actual

---

89 NLA, MS6890/1/8, letters to Norman, 2 April 1924, 29 August 1923.
90 Elspeth Huxley to W. Way, 15 April 1986; Gervas Huxley, *Both Hands*, p. 127.
91 NLA, MS6890/1/9, letters to Norman, 10 September 1925 and 3 January 1926; *LFSSA, 138*, 14 December 1927, p. 483.
92 Information from RSS Archivist, Janet Foster, 14 and 20 August, and 10 September 2007.

tendencies in Imperial trade as revealed by the recent actions of Australia'. Imperial trade was 'every day threatened by the avowed and adopted policy of complete Protection pursued since 1920 by Australia'.[93]

McDougall's response to such criticism was to demonstrate with figures the benefits of Australia's preferential tariff to British industry, overall. He provided for Sir William Larke, Director of the National Federation of Iron and Steel Manufacturers, a memorandum showing increasing British iron and steel exports to Australia, compared with exports to the rest of the world: 'I may say that he was frankly surprised at the results.'[94] In the hostile territory of Manchester, a manufacturing city harbouring vigorous opponents of Australia's protection policy, McDougall upbraided 'the most influential Chamber of Commerce' in Britain for 'such loose thinking and such loose writing as to make the sort of statements, of which they had been guilty', giving examples and correcting errors.[95] He defended Australian policy by differentiating between established industrial powers and young developing nations. In advance of the 1927 International Economic Conference, certain to be dominated by free-trade views, McDougall provided for *The Times Trade and Engineering Supplement*, anonymously, dominion views on this 'somewhat dangerous conference'. He claimed the 'rights of nations to safeguard their own living standards': 'young and virile nations on the brink of great economic developments' relied on protective tariffs and orderly marketing of agricultural produce.[96]

The 1927 International Economic Conference, called by the Economic Organization of the League of Nations, carried resolutions for reducing tariffs. An Economic Consultative Committee (ECC) was appointed to supervise execution of those resolutions. McDougall was one of 60 ECC members, representing League member states, but chosen for individual expertise. The ECC met for about a fortnight in 1928 and again in 1929. Aware of his government's anxiety about the possibility of League interference in its tariff policy, McDougall expected 'it may become necessary at some stage…to warn the Committee of the dangers to the League of Nations of its interference in the economic affairs of…the younger nations of the world, who are developing their industrial status and may resent dictation from the older industrialised countries'.[97]

Thus, he began a self-appointed role as spokesman for developing economies outside Europe. Impressed by the value and range of economic statistics collected by the League, McDougall wanted to divert League activity from efforts to impose commercial policy to 'provision of full, clear comparable information

---

93 NLA, MS6890/5/1, review of *Sheltered Markets* in *Manchester Guardian Commercial*.
94 *LFSSA, 137*, 8 December 1927, p. 477.
95 Ibid., *152*, 6 March 1928, pp. 528–9.
96 Ibid., *90*, 26 January 1927, p. 296. The article was published in *TTES*, 29 January 1927.
97 *LFSSA, 163*, 9 May 1928, p. 569.

and statistics', thereby assisting planning for empire cooperation.[98] He attended the League Assembly as a substitute delegate for Australia in 1929 and, just as he had on the ECC, he lobbied the British delegation to oppose the 'extreme anti-tariff attitude' of the League.[99]

Rationalising Imperial Industry

McDougall's advocacy on behalf of Australian policies would always be tempered, nevertheless, by the wider imperial interest. He believed that full imperial economic development could be realised only through cooperation and coordination. He therefore supported the idea, current in some sections of British industry, of complementary trade based on imperially rationalised industries. In a memorandum on empire secondary industries, written at the request of Sir Horace Hamilton, permanent head of the British Board of Trade, and circulated within the board, he wrote that dominions lacked the economies of scale to attempt complex industrial development. They should selectively develop, and encourage with protection, simpler forms. They might assemble and manufacture vehicle parts such as chassis and wheels, while British industry supplied engines and gears. A dominion electrical industry might produce its own domestic appliances, small motors and lamps.[100] McDougall favoured industry-specific talks to work out rational arrangements; there were moves in the late 1920s to hold such discussions between the British iron and steel industry and Australia's Broken Hill Proprietary Limited (BHP). A re-examination of inter-imperial tariffs should determine 'reasonable limits' of dominion industrial development for a five or ten-year period, after which 'tariff mongering' should be left alone.[101]

As the need to accommodate and respond to British opinion on tariffs formed an important part of his work on behalf of Australia, that opinion influenced McDougall's own thinking. His understanding of Australian views, based as it was on comparatively brief and narrow experience, never renewed after 1924, faded. Although he worked loyally for what he believed to be Australia's interests, he was blunt in private criticisms expressed to Bruce. He argued that Australia's tariff policy threatened the imperial vision: tariffs should be used for preference, not as protection for inefficient industries. There would be little to criticise in a tariff of 35 per cent, but when it exceeded 50 per cent, 'a howl of indignation' arose in areas producing woollen textiles. Amery agreed:

---

98 Ibid., *155* and *169*, 15 March and 21 June 1928, pp. 542–3, 595.
99 Ibid., *255* and *256*, 4 and 11 September 1929, pp. 885–8.
100 NLA, MS6890/4/2, undated memorandum, 'The Distribution of Secondary Industries within the Empire'; *LFSSA*, *145* and *155*, 25 January and 15 March 1928, pp. 502, 540.
101 *LFSSA*, *106*, 4 May 1927, p. 354.

in 1927 hosiery manufacturers had protested against Australian tariff increases by boycotting an Empire Shopping Week in Nottingham, and McDougall reported Amery's hope that Australia would consider the effect of its tariff on the voluntary preference campaign. He had also urged special consideration for British industries suffering severe hardship, as were cotton tweed manufacturers whose output was geared to specific needs of Australian rural workers.[102] McDougall criticised Tariff Board policy 'encouraging every possible secondary industry to start for itself in Australia'. Tariff rates had doubled from an average of some 10 per cent in 1918 to 20 per cent by 1927.[103] As the increases continued, he grew more insistent on efficiency as a precondition for protection, in both secondary and primary industries. No industry should be supported if it required more than a 25 per cent tariff to be competitive, unless it was essential for defence. Australia should concentrate on large-scale production in enterprises for which it had a natural advantage, lowering overheads and assisting export prospects. Full protection, to the point of import embargo, could be given to selected industries, and even to items within industries. Duties upon other products should be lowered, with a declaration that tariff changes would be kept to a minimum, and the previous 'empirical, experimental, inclusive policy be abandoned for a scientific selective policy based upon the interest of Australia as a whole and without consideration to the special interests of groups of employers or of trade unions'.[104]

McDougall warned Bruce 'how widespread and, therefore, serious, is the distrust of Australian economic policy in very large sections of influential opinion in this country. It is by no means confined to financial circles but affects the attitude of many men who are essentially friendly to us.' A company chairman had told him his board was urging realisation of substantial Australian securities because of such doubts. McDougall could demonstrate the benefit of the tariff to Great Britain and 'the immense possibilities of Australian primary production, yet it is impossible to inspire very much confidence in the immediate future among those who have made any considerable study of recent happenings'.[105]

The Australian Tariff

McDougall urged Bruce to initiate a study of a possible 'vicious circle in Australia...forcing up both wages and the cost of production'.[106] Bruce did

---

102 Ibid., *157*, 4 April 1928, pp. 550–1; *98*, 16 March 1927, pp. 324–5; *30* and *31*, 10 and 17 September 1925, pp. 89–91.
103 Coleman, Cornish and Hagger, *Giblin's Platoon*, p. 66.
104 *LFSSA*, *53*, 1 February 1926; *168*, 19 June 1928, pp. 154, 588; *108*, 23 May 1927, pp. 361–5.
105 Ibid., *149*, 16 February 1928, pp. 517–18.
106 Ibid., *129*, 12 October 1927, pp. 440–1.

establish a tariff inquiry, initially by economists L. F. Giblin, E. C. Dyason and Commonwealth Statistician, C. H. Wickens. J. B. Brigden and D. B. Copland were coopted soon afterwards and the first draft presented to Bruce early in 1928. Bruce asked Brigden to rework it; the final version, known as the 'Brigden Report', was published in mid 1929. Bruce admitted 'the question bristles with difficulty owing to the present policy having been pursued to such a point as to render the calling of any halt difficult'.[107] McDougall was disappointed that the inquiry's first draft report failed to consider the effect of 'indiscriminate' high protection in limiting the effectiveness of more promising industries, but was pleased to find an argument for more discriminating protection.[108] The final report has been described as 'a compromise document', a 'comprehensive manifesto for moderate protectionism' and, although it did include a 'trouncing of Imperial Preference', as 'an attempt to be all things to all men'. It was welcomed by both the protectionist Melbourne *Age* and its rival, the free-trade *Argus*.[109]

The views of Bruce and McDougall were similar in many respects; McDougall often conveys a certainty in his letters that Bruce agrees with his view. While both men wanted Australia's high living standards to be maintained, with emphasis on development, especially in primary industry and on a more rational tariff, McDougall saw things from the centre in London, where he encountered hostility towards Australia's economic policy, British Treasury scepticism about the 'imperial vision', and factions in Whitehall. He understood the threat to Australian borrowing that those negative British assessments posed and he was acutely conscious of the fragility of empire cooperation. Bruce understood those worries, too, but his immediate concern as Prime Minister was to hold together a coalition government subjected to lobbying from rural, manufacturing and importing interests.[110] An example of their similar thinking from different standpoints is found in a pair of letters, crossing in the mail in 1927. McDougall expresses frustration with the increasing Australian tariff, which 'must retard Anglo-Australian cooperation in Australian development [and] *gravely hinders the campaign of Imperial Economic education* on which so much time, energy and money is being spent'. Bruce wrote at the same time: 'the whole question of the Tariff, and its *effect upon Australia's progress* is now very much occupying my mind, and I shall be interested to have any thoughts of yours which you choose to send me.'[111]

---

107   W. H. Richmond, 'S. M. Bruce and Australian Economic Policy 1923–29', *Australian Economic History Review*, Vol. 23, 1983, pp. 249–51; NAA, M111, Bruce to McDougall, 15 September 1927.
108   *LFSSA*, *168*, 19 June 1928, pp. 385–9.
109   Coleman, Cornish and Hagger, *Giblin's Platoon*, pp. 68, 71–2.
110   Kosmas Tsokhas, 'Protection, Imperial Preference and Australian Conservative Politics, 1923–39', *Journal of Imperial and Commonwealth History*, Vol. 20, no. 1, 1992, pp. 67–70.
111   My emphases. *LFSSA*, *108*, 23 May 1927, pp. 361–5; NAA, M111, Bruce to McDougall, 21 April 1927.

The Brigden Report did not 'produce a simple formula' for what Bruce had hoped would be 'the clean up on the Tariff question'. It gave Bruce no more satisfaction than the hope that it would lead to 'a really heavy controversy on the subject'.[112] He worried, like McDougall, that the tariff sheltered inefficiencies, and feared it could harm the competitiveness of rural industries. Yet Bruce was unable to resolve the inherent difficulties of identifying industries deserving protection and of fixing appropriate levels for that protection.

Bruce also agreed with McDougall on the need for efficiencies in rural industry. At McDougall's urging, he had implemented organisation schemes. Although bounties were granted, Bruce 'regarded them merely as necessary palliatives to relieve the distress experienced by growers'. He would tell industry deputations seeking such assistance: 'The government will help you if you help yourselves… there is so much that can be done to improve returns and cut costs but it needs you to take action…go away and come up with a scheme.' It was this view that had led Bruce to establish CSIR and DMC: 'We must…recognize that if we want to maintain that [very high] standard of living and those social conditions we can do it only by adopting the most modern and efficient methods in the conduct of the whole of the industries of our country.'[113]

The 1929 General Election

Bruce, like McDougall, was a 'rural optimist'. It has been argued that his economic policies must be understood in the context of his 'unyielding belief in rural development as the basis of economic progress'.[114] He had identified two factors undermining that progress: the tariff and the complex system of industrial arbitration, divided between State and Commonwealth jurisdictions. Having reached an impasse on the tariff, he determined to use other measures to reduce costs of production and the cost of living, in order to shore up the economy for a depression he believed to be likely. Thus, costs of shipping and industrial disputes had to be curtailed.[115] He had alienated the labour movement and its supporters in shipping disputes in 1925; he saw union militancy as a threat to Australian interests, to the British Empire and to the 'peaceful, prosperous, scientific and rational domestic and international order he wanted'. Thereafter he had become 'a polarising and deeply unpopular figure', 'the best-hated man of his day'. He nevertheless won an 'emphatic' electoral victory later in that year, as the economy recovered from recession; he had succeeded then in 'positioning his government as the genuine representative of the national

---

112 Richmond, 'S. M. Bruce and Australian Economic Policy', pp. 250–1.
113 Ibid., pp. 245–6.
114 Ibid., p. 239.
115 Lee, *Stanley Melbourne Bruce*, pp. 81–8.

interest'.[116] But in 1928, still dogged by industrial unrest, Bruce's government barely scraped back, with a majority insufficient to give it a free hand. Bruce himself feared it might be defeated on a censure motion.[117]

In the early months of 1929, timber workers contested an award reducing wages and increasing hours, and Bruce's government prosecuted the Secretary of the Melbourne Trades Hall Council, E. J. Holloway, for inciting workers. Coalminers in New South Wales objected to an agreement by Bruce and NSW Premier, Thomas Bavin, to reduce the price of exported coal, in an attempt to prop up the industry, with owners to receive reduced profits and workers less pay. Owners closed some mines, but Bruce refused moves to prosecute a coal owner, John Brown, for locking out workers. The decision, though based on sound reasoning, demonstrated the flaws in Bruce's political skills. A 'personification of the capitalist bosses', he had shown, according to a growing number of opponents, 'that he had one law for the rich and another for the poor'. He seemed to have compromised his government's 'moral legitimacy'. The Government narrowly survived a censure motion in the Parliament. Bruce thereupon determined to reduce costs by reforming the arbitration system, either by taking it over completely as a federal activity or by abandoning it to the States.[118] In September he introduced a Maritime Industries Bill, establishing a new type of tribunal to deal with workers in the sea transport industry, but also devolving industrial relations largely to the States. Hughes moved an amendment that the Bill not be passed without approval by a general election or referendum; the amendment was carried by one vote. Bruce had suffered this defeat because of his 'increasing propensity to antagonize those whose support he needed', writes David Lee, who also suggests that at this period he may have been suffering from depression. He need not have resigned, but chose to call an election, hoping to bring coalition dissidents into line. The election on 12 October 1929 resulted in a landslide to the Labor opposition. Bruce lost his own seat to E. J. Holloway, the unionist who had been controversially prosecuted by his government.[119]

---

116 Ibid., pp. 50–1, 57.
117 Ibid., pp. 80–1.
118 Ibid., pp. 82–5.
119 Ibid., pp. 89–92; see also p.14.

4. The Vision Ends

Summary

The years of the Great Depression brought anxiety and disillusion for Bruce and McDougall. Both faced an uncertain future. McDougall's position at Australia House was threatened by the search for economies and by Labor suspicion of his views on tariffs. He was saved—albeit at a considerably lower salary—by his work assisting the new Prime Minister, J. H. Scullin, at the 1930 Imperial Conference. He supplemented his reduced income by writing some paid articles, and a brief period writing on imperial trade for Lord Beaverbrook. He was tempted by various proposals for work on imperial cooperation in London, and by one for economic research in Australia. None of these materialised, but the soul-searching involved in considering them confirmed his determination to remain at the centre in London and to work for the interests of the empire as a whole. As political pressure for a British general tariff increased, he contributed to a ferment of ideas for means to transcend the narrow political process with machinery for broad-based consultation drawing upon industry, commerce, finance and labour.

Bruce was returned to Parliament at the general election in 1931 in a United Australia Party Government headed by Joseph Lyons, who appointed him Assistant Minister. He led the Australian delegation at the Ottawa Imperial Conference, which established an imperial preferential tariff system in 1932. In the months preceding the conference, McDougall had negotiated with British officials on tariff treatment of individual commodities, while Bruce oversaw a similar process in Canberra. McDougall had also worked on plans to transform imperial bodies in London into an organisation promoting imperial economic cooperation. Debate continues about the advantages of the Ottawa preferential system established after hard bargaining. For McDougall, one unexpected outcome, the demise of the Empire Marketing Board, was a heavy blow. It marked the end of his vision for cooperative imperialism.

The Scullin Government: 'A pretty hard fight'

Shocked by Bruce's election loss, McDougall determined to carry on as usual and to report regularly to Scullin.[1] Anxiety for himself and for his cause was

---

1   NAA/CSIR, A10666, [2], McDougall to Rivett, 22 October 1929.

clear as he wrote consoling Bruce, urging him to 'make [Scullin] aware of the amazing opportunity which he can take, if he chooses, at the next Imperial or Imperial Economic Conference'. Otherwise, there could be 'no one effectively to state the case from an overseas point of view in regard to Empire development'. He added: 'I have…no idea what the attitude of the new Government will be in regard to my work. I can only hope, from a personal and from a public point of view, that they will be prepared to allow me to continue my work without any very substantial change.'[2] Bruce replied that he had tried to persuade Scullin, but could give no indication that he had succeeded: 'In any event you have now so established yourself in London that opportunities will present themselves for you to continue your work for the great cause you have espoused, even if it is in a new sphere of activity.'[3] It was, perhaps, a warning.

McDougall knew little of the new ministers. His support base in Australia was small and he had no sympathetic press. His friends in DMC and CSIR were worried. Bruce warned them he feared it would be 'a pretty hard fight to persuade the new Government of the real value of McDougall's services'.[4] DMC was doomed. With few tangible results to show for its efforts, support had been rare even in the former government. Scullin made plain his view that it was a waste of money. In November 1929, he decided to stop most assisted migration in view of Australia's economic difficulties, effectively removing DMC's *raison d'être*. In February 1930 other DMC functions were moved to a 'Development Branch' of the Prime Minister's Department.[5]

CSIR fared better. Its new minister, Senator J. J. Daly, proved approachable and keen to extend its work. But when Rivett and Julius tried to persuade him of McDougall's value, especially as a member of the EMB, 'the Minister said very little, but…there was a slightly antagonistic feeling which was probably due to the fact that you held more positions than one and received more than the basic wage. You know what I mean.' Rivett assured McDougall that there had not been 'the faintest suggestion' of any threat, but late in January he was less optimistic: many ministers did not seem to appreciate the value of CSIR's work and the financial situation was deteriorating rapidly. There would nevertheless be no difficulty in finding McDougall's annual £500 payment, which represented 'a remarkable bargain from CSIR's point of view'.[6]

On 28 February, Daly explained the full extent of the financial crisis to CSIR's Executive. Rivett wrote 'a thoroughly miserable letter' to McDougall, thinking

---

2   *LFSSA, 259*, 17 October 1929, p. 894.
3   Ibid., *260*, 17 November 1929, p. 896.
4   NAA/CSIR, A10085, vol. 3, Julius to Rivett, 13 November 1929.
5   Michael Roe, *Australia, Britain and Migration*, pp. 96, 119–20, 140–5; NAA/CSIR, A9778, M14/29/10, Rivett to McDougall, 15 October 1929.
6   NAA/CSIR, A10666, [2], Rivett to McDougall, 18 and 28 November 1929, 29 January 1930.

through the consequences as he wrote. CSIR must help as best it could. Development and building would be put on hold, contributions to overseas organisations suspended and imperial activities reduced. The Prime Minister's Department had requested justification of McDougall's appointment. It would be 'answered in very emphatic fashion', it was not personal, 'just an evidence of the intense desire of Cabinet to reduce expenditure in every direction'. 'Drastic reductions are the order of the day', but McDougall must remain: 'I shall have to rely upon you to a tremendous extent.'[7]

Scullin's questioning was personal, prompted by a letter from Hume Cook, Secretary of the Australian Industries Protection League. Cook referred to a press report that an EMB paper by McDougall had been hailed by the *Daily Express* as strongly supporting empire free trade, for which *Express* proprietor, Lord Beaverbrook, was waging a vigorous campaign. At the time McDougall had described the review as 'an embarrassing amount of effusion'.[8] Cook's letter recalled his earlier complaint about an apparent attack by McDougall on the Australian tariff and the 'unsatisfactory' response of the Bruce Government. Scullin called for information: 'what salary is being paid by the Commonwealth to Mr McDougall, and what services are rendered by him?'[9] A four-page departmental report dealt with Cook's two complaints and McDougall's response to the earlier one. It noted an account in the *Sydney Morning Herald* of McDougall's statement to the League of Nations Economic Consultative Committee that 'there was a feeling in some quarters that certain aspects of the Australian tariff policy hindered the growth of certain secondary industries and, further, the incidence of the tariff was a handicap to agriculture'. McDougall had explained that he had gone on to say that the Australian Government realised these were 'highly complex' questions, and the Government had therefore decided to establish a Bureau of Economic Research.[10] He had mentioned the decision, he wrote, in order to 'indicate the way in which the League of Nations could, through the provision of a service of economic information, assist countries such as Australia in the assessment of the comparative advantages which they enjoy in regard to primary and secondary industries'. The official report also listed amounts of remuneration received by McDougall: £1000 from the Commonwealth Government for 'general representation' and a further £500 each for liaison on behalf of DMC and CSIR. He received another £500 from the

---

7   Ibid., Rivett to McDougall, 3 March 1930; NAA/CSIR, A9778, M14/30/2, Rivett to McDougall, 14 March 1930.
8   Empire Marketing Board, *The Growing Dependence of British Industry upon Empire Markets*, December 1929, HMSO; Report in Melbourne *Age*, 5 December 1929; NAA, M111, McDougall to Bruce, 11 December 1929.
9   NAA, A981, ECOC 1, Cook to Scullin, 6 December 1929, Scullin to Prime Minister's Department, 10 December 1929; NLA, MS6890/3/2, copy of cable requesting McDougall's explanation, 20 December 1929.
10 A Bill to realise Bruce's aim to provide an independent body to undertake economic research, along the lines of CSIR, was submitted to Parliament in March 1929. It was opposed by Labor, chiefly because it seemed to threaten tariff protection. Efforts to find a suitable director for the bureau lapsed when Bruce lost office. See Coleman, Cornish and Hagger, *Giblin's Platoon*, pp. 73–9.

DFECB, making a total £2500 per annum.[11] McDougall's further response on 20 December explained that the EMB publication *The Growing Dependence of British Industry upon Empire Markets* had already been sent to Scullin, and that it had no connection with the empire free trade campaign; the *Daily Express* must have found the figures in it useful for their own purposes. McDougall pointed out that a recent article of his own, defending the right of younger Commonwealth nations to use protection, condemned empire free trade as 'politically hazardous' for Britain.[12] A reassuring letter was sent to Cook, but McDougall's value to a high-protectionist Labor government was inevitably questioned.

Reorganisation at Australia House

Costs at Australia House in London had increased fivefold, from £24,225 to £124,841 per annum, between its establishment in 1913 and 1930. The vague terms of the *High Commissioner Act* (1909), the pressures of war and changes in the dominion relationship increased the functions of the High Commission and brought uncontrolled growth. By 1926 employee numbers were given as 256, most without permanent tenure.[13] A search for savings focused on trade-related activities; the High Commissioner, Sir Granville Ryrie, was asked for suggestions. He described McDougall as 'a very efficient officer' and 'capable economist' who would be 'of great assistance to the Prime Minister at the Imperial Conference on economic questions'. Either McDougall or A. E. Hyland, Director of an Australian Publicity Scheme established in 1926, would fill a proposed post of Director of Trade and Publicity admirably, but Ryrie would prefer to appoint Hyland. McDougall might be retained on a consultative basis for a year to continue his representation on the IEC and EMB.[14]

Gepp and Rivett responded vigorously to the request to justify McDougall's continuation. His work 'called for the services of one who had special scientific and technical experience and a unique standing with official circles in Great Britain', and 'we esteem Mr McDougall's services…very highly indeed, far beyond the relatively small sum of £500 which we pay for them'.[15] They failed to persuade Daly, who advised Scullin that arrangements initiated by Bruce should not continue. McDougall's communication with DMC and CSIR was controlled neither by the High Commissioner nor by the responsible minister:

---

11   NAA, A981, ECOC 1, Report, 11 December; McDougall's response, 14 May 1929.
12   Ibid. The article was published in the *English Review* in October 1929.
13   NAA, A461, H348/1/8, P. E. Coleman, 'Report Upon the Organization of the High Commissioner's Office, London, and the Activities Associated Therewith', Commonwealth of Australia, Canberra, 1930, p. 5; *Commonwealth Parliamentary Debates* [hereinafter *CPD*], Bruce's reply to question, 19 August 1926.
14   NAA, A461, G348/1/8, part 2, High Commissioner's Office to Prime Minister's Department, 14 March 1930.
15   Ibid., Gepp to Daly, 19 March; Rivett to Prime Minister's Department, 3 March 1930.

'from the point of view of efficient administration I am unable to discern any feature which commends it.' Daly calculated the full cost of McDougall's office as £5000 per annum, and recommended DMC and CSIR liaison work be carried out by regular High Commission staff. Parker Moloney, Minister for Markets and Transport, had no 'detailed knowledge' of McDougall's work on the EMB, but believed it could be done by the officer in charge of trade and commerce at Australia House.[16]

P. E. Coleman, Chairman of the Commonwealth Parliamentary Joint Committee of Public Accounts, then attending an International Labour Conference in Geneva, was appointed to report upon possible savings at Australia House, and recommended economies involving abolition or amalgamation of positions. He paid considerable attention to the branch known 'for want of a better name… as "Mr McDougall's"'. McDougall's total remuneration equalled that of J. R. Collins, Financial Adviser to the High Commission and a former head of Treasury. Collins was paid in part for liaison work on behalf of the Commonwealth Bank. He and McDougall were the highest paid officers at Australia House after the High Commissioner, who received £3000 and use of an official residence valued at £2000. McDougall had three assistants: a technical assistant, Dr A. S. Fitzpatrick; an economic assistant, Miss Pitts; and a clerk, A. Stuart Smith. His office also employed two typists. Collins had one clerk. Like some others, McDougall's branch operated with virtual autonomy; its records were not kept in the central registry and 'the control exercised by the High Commissioner over Mr McDougall's activities is purely nominal'. There was general recognition of the value of his work and claims that he had been 'a valuable propagandist in the development of Empire trade' and of Australian affairs, but 'Mr McDougall, in common with other officers, has been working more or less in a "water-tight compartment", and has communicated direct with the Prime Minister. The consequence is that it is most difficult to form a just estimate of the value of the office filled by him.'[17]

Coleman recommended the liaison functions and McDougall himself be brought more directly under control of the High Commissioner, so that the value of his services could be assessed. Stuart Smith and one typist should be transferred to general High Commission staff, and Fitzpatrick dispensed with. McDougall need no longer be paid by Development Branch and CSIR, reducing his remuneration by £1000; the loss of Fitzpatrick would save another £500. Adding insult to the substantial financial injury, Coleman recommended McDougall be designated 'Economic Officer'. By a process that is unclear, he had for some time by then assumed the more prestigious title 'Economic Adviser'.

---

16   Ibid., Daly to Scullin, 25 March; Moloney to Scullin, 9 April 1930.
17   Coleman, 'Report', pp. 10–11.

The 1930 Imperial Conference

McDougall wrote regularly to Scullin, hoping to educate the new Prime Minister on the case for empire cooperation and tactics for the Imperial Conference to be held in November 1930. He provided useful figures and arguments, memoranda on many subjects and gossip suggesting trends in British Government thinking.[18] As the conference approached, he briefed ministers and drafted speeches. Scullin's opening speech showed his influence. Like Bruce in 1923, Scullin stressed the importance of markets, but he also appealed for rationalisation: 'the better ordering of our production, both in agriculture and in industry and of our trade, has become a matter of vital urgency.' Australia would welcome conferences between industrialists to consider rationalisation, and he hoped Australian agriculture would be given some advantage 'over our foreign competitors'.[19] He expounded figures showing the importance to Britain of the Australian market, in comparison with the markets of Argentina and Denmark, and the value of Australian preference, and then declared, 'if British industry will co-operate with us in the development of our industries, we…will do everything in our power to help you secure the lion's share of our import trade'.[20]

McDougall gained personally from the conference: he impressed Scullin and other delegates with his range of contacts and hard work; he got to know Australian ministers and senior public servants Parker Moloney; E. Abbott, Deputy Comptroller-General of Customs; and Australian Trade Commissioner in Canada, L. R. Macgregor. Best of all was the opportunity to meet in person and cement his friendship with Rivett. Otherwise he was disappointed. The conference did not consider preference in detail, nor extension of the IEC towards the role of an imperial secretariat. These matters posing problems for all empire countries were deferred to a conference in Canada in 1932.

In London Rivett approved minor changes to CSIR's liaison arrangements, but insisted upon McDougall's value. Only a week earlier McDougall had gained RGC recommendation for a grant of £6000 for the Plant Industry Division. Furthermore:

> I have seen enough here to know that he has obtained a reputation in London on Imperial Economic Affairs which causes members of all parties, as well as important organisations, to consult him on many questions. He is also so highly regarded in Geneva and Rome as to be very frequently in request on economic and agricultural questions

---

18 See letters in NAA, CP103/12, bundle 19, and CP489/1, 430/AA/2.
19 NAA, CP498/1, 430/AA/9B, Stenographic Record, 1 October 1930, E 1st Meeting (1930), pp. 14–15.
20 Ibid., Stenographic Record, 8 October 1930, E 2nd Meeting (1930), pp. 16–20.

> in the international sphere. This is undoubtedly directly of value to Australia, and…has some real significance from the point of view of our International prestige. I do hope that whatever it may be necessary to do regarding Mr McDougall's financial position, there will be no reduction in his status likely to prejudice his work in these fields.[21]

This, and his own observation of McDougall's work at the conference, mollified Scullin. In a 'quite satisfactory' interview, he thanked McDougall 'quite prettily' for his assistance and apologised for the offensive wording: the intention had been economy, not reduction of status. His Commonwealth salary would be reduced, nevertheless, to £1000. It would cover liaison work, other than detailed tasks now allocated to Stuart Smith. McDougall agreed with Scullin's understanding 'that I had not been whole-time employed previously so…that arrangement would continue'.[22] He possibly thought this a reference to his work for the DFECB; he may have agreed because it gave him freedom to write paid articles. The arrangement was to cause him problems later in the 1930s, when Bruce, as High Commissioner, persuaded the Lyons Government to give permanency and superannuation rights to High Commission staff. Legislation to amend the *High Commissioner Act* was passed in 1937. Lyons refused to include McDougall, referring to his 'peculiar position' and to the 'fact that for some years he has not been regarded as [a] full time officer'. Lyons noted McDougall's 'objection to being regarded as an officer on the High Commissioner's staff' and the fact that he had been given 'the right of private practice'.[23]

A furious Bruce responded in seven pages, protesting at 'ignorance' of the position or deliberate misrepresentation, reiterating the events of 1930, and arguing that any right to private practice 'has never been exercised and never could be exercised', given the 'very full whole time job' McDougall did, involving much confidential government information. Bruce added pressure by sending copies of his letter to senior members of the Lyons Cabinet who knew McDougall's work at first hand: Earle Page, R. G. Menzies and R. G. Casey. Lyons rather weakly protested that McDougall had accepted an honorarium of 100 guineas from the Australian Wine Board, with approval of both Bruce and the Government, and suggested, with more justification, that he had wished to be treated as 'someone apart from the regular staff of the High Commission'. He gave in, nevertheless. McDougall was granted security of tenure, furlough and superannuation rights, but 'insuperable difficulties' prevented his being permitted to make lower superannuation payments—a concession given to others who had served at Australia House since the early 1920s. He was

---

21 NAA/CSIR, A10666, [2], Rivett to Scullin, 24 November 1930.
22 Ibid., McDougall to Rivett, 1 December 1930.
23 NAA, A461, G348/1/7, part 2, Bruce to Lyons, 1 September; Lyons to Bruce, 9 September 1937. Lyons claimed that McDougall's objection to being regarded as full-time had been expressed in a letter to Scullin of 26 November 1930, which has not been found.

deemed to have been serving 'de facto' from the date of Scullin's memorandum appointing him 'Economic Officer' in 1930. McDougall's first secure and regular appointment was gazetted on 9 December 1937; he was fifty-three years old.[24]

The loss that did occur in 1930 was more than financial. For some years, Joyce McDougall's 'nerves had been very jangled'; McDougall hoped that a holiday on her own in Italy early in 1926 might improve her health. But in 1929 he told Bruce that his family had spent some time living in Sicily, in an unsuccessful attempt to help Joyce, who had been 'far from well for the last couple of years'.[25] He and Joyce subsequently agreed that while she remained in Italy their children should be educated in boarding schools near Geneva, where he could visit them while on international business. McDougall's reduced salary was paid in a currency devalued as the Australian pound floated against the pound sterling and was then effectively pegged to it as Britain left the gold standard. He also suffered one of the percentage cuts imposed on public service salaries.[26] His resources in Swiss and Italian currencies were strained and it was agreed that the family would return to Adelaide early in 1932.[27] Joyce visited London at least once in the prewar period, but her condition became so difficult that McDougall felt unable to visit Australia himself. She subsequently moved to New South Wales, where she died in 1986. McDougall saw nothing more of his children until his daughter, Elisabeth, took a post in the British Foreign Office after World War II. He was never to know his five grandchildren, who grew up in Australia.[28]

Beaverbrook

McDougall had begun to write paid, anonymous articles for *The Times*.[29] It is likely that he was occasionally paid by other journals. His articles always advanced some part of his cause: it was thus a moot point whether he thought he was working for Australia or for money in writing them. The issue is clearer in the case of work he undertook for *Daily Express* proprietor and vigorous campaigner for free trade, Lord Beaverbrook, whose paper, as noted above, had welcomed so embarrassingly McDougall's EMB publication. Invited to lunch

---

24   Ibid., Bruce to Lyons, 13 October; Lyons to Bruce, 22 November; Official Secretary, Australia House, to Prime Minister's Department, 14 December 1937.
25   NLA, MS6890/1/9, letter to Norman, 3 January 1926; *LFSSA, 220, 221*, 25 March and 17 April 1929, pp. 772–3.
26   NAA/CSIR, A10666, [2], McDougall to Rivett, 10 February, 12 March and 22 July 1931; Rivett to Mulvaney, 26 August 1931; NLA, MS6890/3/5, copy of letter from Pearce to Scullin, 13 November 1931. See also Bruce's 1937 letter to Lyons, cited above.
27   NAA/CSIR, A10666, [2], McDougall to Rivett, 23 September 1931; A10666, [3], McDougall to Rivett, 28 January 1932; A9778, M14/32/2, Rivett to Richardson, 31 March 1932.
28   Information from McDougall family.
29   NLA, MS6890/2/3, note from Elspeth, 15 January 1931.

with Beaverbrook, McDougall hoped he had persuaded the press baron that Australia must continue some protection.[30] He did not mention to Bruce that he had agreed to write for Beaverbrook, at a rate of 10 guineas per thousand words. Notes from Beaverbrook and his secretary suggest that he wrote a small number of factual articles on empire trade, which Beaverbook apparently studied carefully. He proposed a new edition of *Sheltered Markets*, which he found 'very valuable from our standpoint', though 'much of it can be eliminated'. Nothing came of that idea, but, according to the correspondence, McDougall was paid £114/9/- in total. Apart from the supply of occasional facts, McDougall ceased to work for Beaverbrook after February 1930, possibly because of the embarrassment in Australia. In May, Beaverbrook wrote, 'I hope you are not afraid of my company these days', and, in praise McDougall might have not welcomed, 'the help you have given me has been of decisive importance in our campaign'.[31]

Organisation for Empire Cooperation

Empire cooperation was a popular idea in many quarters, and so were proposals for organisations to facilitate it. McDougall was tempted by several variations of that common theme, all of which proved abortive. He was torn, as he had been in 1924, between a wish to continue serving Australia and Bruce and the attraction of positions offering security and status. More than that, he was most attracted to positions offering intellectual freedom, beyond the confines of national policies and specific constituencies. He understood himself well enough to reject such boundaries.

In December 1929, he was approached about an unidentified position, 'which would enable me to keep in touch with the things in which I am so tremendously interested and would probably be very attractive from a financial point of view'. He decided then to remain at Australia House as long as the new government retained confidence in him; even part-time, his connection with the imperial bodies would enable him to 'influence Imperial Economic ideas'.[32] He was tempted early in 1930 by a short-lived offer of a position in the Conservative Party's Research Department. Soon after it was made, by outgoing Director, Lord Eustace Percy, Neville Chamberlain took over the organisation and wrote that he had decided against creating that position.[33]

Bruce was in London in 1930, and helped devise a scheme for a similar organisation funded by business. Walter Elliot was involved, as was Brendan

---

30   NAA, M111, 1929, McDougall to Bruce, 11 December 1929.
31   NLA, MS6890/3/1, Beaverbrook to McDougall, 25 January, 10 and 20 May 1930.
32   NAA/CSIR, A10666, [2], McDougall to Rivett, 23 December 1929.
33   NLA, MS6890/3/2, Chamberlain to McDougall, 29 March 1930.

Bracken, then a director of Moody's-Economist Services Limited, and Sir Robert Horne. Moody's was prepared to establish a department to collate information and statistics, not for party propaganda, but for 'serious economic work', and to pay McDougall £5000 per annum as its director. The plan hinged on Horne obtaining a guarantee of £10,000 for five years, and it is hardly surprising that it evaporated. McDougall tried possible sponsors. His cousin Sir Arthur Duckham, then about to become President of the Federation of British Industries, hoped politically influential industrialist Dudley Docker might help approach Sir Montague Norman, Governor of the Bank of England. McDougall thought Norman might be interested as he was said to believe a general tariff would be introduced and to be anxious that it not shelter inefficiency. McDougall also discussed the idea with Assistant Editor of *The Times*, R. M. Barrington-Ward.[34] Bruce wrote several times from Australia asking about its progress.[35]

In Australia Bruce helped Casey, then seeking to enter Australian politics, to canvass a somewhat similar scheme: a plan to provide a policy and organisational base for conservative politics, including a 'Bureau of Economic and Political Research'. A paper by Sydney businessman T. S. Gordon fleshed out the idea, well beyond priming politicians with facts, figures and views of government agencies, to that of a clearing house for economic and market information from Europe, Asia and other areas of interest to Australia. It could have seemed attractive to McDougall. Many possible directors were considered, but Casey regarded him as the 'plum' in his list; Bruce, en route to London, was asked to sound him out on the basis of an annual salary of £1500 for three years.[36] McDougall was doubtful. While 'it has its attractive aspects…[it] also seems from a personal point of view to contain a high element of risk'. He would become a party man; he doubted his personal appeal in Australia, especially as 'the intellectual level of the genus Politicus Australis is deplorably low'. As in 1924 his *métier* lay elsewhere:

> I have become a fairly specialised animal. I am highly developed about Empire trade, agricultural policy and tariffs. I have also devoted myself without stint to Empire economic co-operation here in London. I am regarded here as the one overseas person who can be intelligent on these subjects, therefore I feel that I can be of much greater service to Australia, the Empire and to England here than I could be in the Australian job.[37]

Once again McDougall was unwilling to be pigeonholed. He saw himself working for something bigger than Australia alone but it was not enough to say

---

34 National Library of Scotland [hereinafter NLS], Acc. 6721, Walter Elliot Papers, Box 3, Brendan Bracken to Elliot, 15 July 1930; McDougall to Elliot, 7 January 1931.
35 NLA, MS6890/3/3, Bruce to McDougall, 15 December 1930, 8 and 22 January, 11 February 1931.
36 NAAV, M1146, [128], cable from Casey to Bruce, 31 October 1931.
37 NAA/CSIR, A10666, [2], McDougall to Rivett, 4 November 1931.

he was working for the Empire. He had identified three broad sets of interests—'Australia, the Empire and England'—and believed it possible to serve all three simultaneously. He appreciated how complex the Empire had become, but was not yet able to admit that the interests of its several parts could be in conflict.

In 1931, as in 1924, he made the right decision. The research organisation idea lapsed—despite the interest of 'very powerful men' in industry and commerce—through 'indifference'.[38] Cost, difficulties of finding the right staff and doubts about how much the bureau might be used were all factors.

Broad-Based Economic Consultation

A joint memorandum prepared by leaders of the Federation of British Industries (FBI) and the Trades Union Congress (TUC) just before the 1930 Imperial Conference urged

> the necessity for adequate machinery for economic consultation between the various parts of the Commonwealth. Unless such machinery can be set up, a proper investigation of the various problems cannot be achieved and the Governments of the Empire will not have a sound conception of the economic considerations involved or the detailed knowledge to guide their policy. At the moment, better machinery exists, in the shape of the Economic Organization of the League of Nations, for the discussion of economic questions between this country and foreign countries than exists for the purpose of considering Commonwealth economic problems.

These words could well have been written by McDougall, and were at least inspired by him: he had long worked with the FBI. The joint memorandum proposed an 'investigatory' British Commonwealth trade conference of government, industry, agriculture, commerce, finance, shipping and labour to review 'every question affecting inter-Commonwealth trade', with a view to drawing up an agenda for the next Imperial Conference. Regular conferences, together with a Commonwealth economic secretariat, would give the Commonwealth adequate machinery for dealing with economic problems of vital importance.[39]

---

38 W. J. Hudson, *Casey*, Oxford University Press, Melbourne, 1986, p. 82. Bill Hudson drew my attention to Casey's correspondence about the idea, in NAAV, M1146, [128].

39 NAA/CSIR, A9778, M14/31/12. The memorandum is part of a much longer, undated and apparently incomplete document. It includes a report of a visit by Sir Arthur Duckham and shipbuilder Sir James Lithgow to Canada to discuss promoting reciprocal trade, and a report of a Joint Preparatory Committee (for the Imperial Conference) representing the FBI, the Association of British Chambers of Commerce and the UK Chamber of Shipping.

In January 1931, McDougall wrote a memorandum arguing that the likely British adoption of tariffs demanded nonpartisan, broad-based study of problems, provision of data and appropriate safeguards, in order to ensure the support of finance, industry, commerce and labour. He suggested formation of a small research organisation. In a second version, dated 2 February, McDougall suggested a body including representatives from all interested groups to tackle issues including tariffs, rationalisation of empire industry, and home and empire agricultural policy. Political representation 'would be best served by the inclusion of three independently minded men'. He meant that they should be open-minded on the tariff question, and nominated Elliot, Sir Oswald Mosely and Liberal Member of Parliament E. D. Simon. Sending the February version to Elliot, he wrote: 'I really feel that the way to tackle this problem is to enlarge the scope and importance of the idea [of a consultative group].'[40]

The idea for a consultative group was discussed with Elliot and others early that year. McDougall referred to it as 'my idea'. He reworked his memorandum of January and February more fully in June 1931. McDougall's approach was reminiscent of a declaration the previous December by leading Conservatives, including Walter Elliot, that the public lacked confidence in politicians. It also called for 'rigorous economy', scientific methods of production, 'a reasonable measure of protection', imperial economic reorganisation and use of Britain's large consumer market as a tradeable asset.[41] Beaverbrook had also made 'a consistent attack both on parliament as a talking shop and on the Executive as a captive of the interests'; he called for inputs from finance, industry and unions.[42] McDougall wrote:

> The record of the present Government and of the last Conservative administration, makes it clear that industry cannot rely upon political parties to carry out this work. Political parties are inevitably tied by immediate electoral considerations and the political atmosphere is especially unsuitable for economic sanity.
>
> The civil service is too wedded to existing economic practice and is insufficiently constructive for the initiation of new policy.
>
> The proposal of the F.B.I. that when a party pledged to fiscal change takes office, it should be urged to establish a Tariff Board is a recognition that reliance cannot be placed either on politicians or on the Civil Service

---

40 NLS, Acc. 6721, Box 3.
41 Ibid., reports of the declaration, published on 17 December 1930.
42 Andrew Fenton Cooper, *British Agricultural Policy, 1912–36: A Study in Conservative Politics*, Manchester University Press, Manchester, 1989, pp. 114–15.

as at present constituted. Such a Tariff Board would, however, be of necessity a part of the Government machine and could not be regarded as expressing the voice of industry, commerce and finance.

No machinery yet exists whereby industry as a whole, commerce, shipping and finance, can jointly consider economic problems in such a way as to secure definite action.

He went on to consider the possible aims and structure of his proposed body.[43] Disillusioned with politics and bureaucracy, McDougall had begun to think beyond existing structures. The idea of gathering expertise from many quarters would remain a feature of his strategies in the future. Pragmatic and flexible, he would always aim to achieve the broadest possible composition of networks.

McDougall wondered about continuing to represent Australia on imperial bodies while acting as part-time director of this secretariat; he thought the functions would work well together. Casey recalled: 'I had often talked to him about some such job…It is a most worthwhile Imperial task [which] might well become more important to the Empire generally than the job he is now doing.' Rivett agreed with him.[44]

By September 1931, however, nothing had developed and McDougall feared 'the financial crisis may render further progress difficult'. In December he was approached, tentatively, about a part-time position developing an empire policy for the FBI—apparently now acting alone—but he doubted it would materialise.[45] The final blow was probably the sudden death early in 1932 of his cousin Arthur Duckham, then FBI President and a driving force in the scheme. McDougall remained interested in the possibilities of the idea. Early in 1932, at the suggestion of Ormsby-Gore, he wrote to Neville Chamberlain, offering himself for membership of a tariff advisory committee, which he expected to be responsible for 'shaping the tariff policy'.[46]

A Ferment of Ideas

In the 1920s McDougall had cultivated 'intelligent', young reform-oriented Conservatives like Elliot and Harold Macmillan, who were interested in the imperial vision.[47] Like McDougall's, their ideas developed in new directions in

---

43   NAA/CSIR, A10666, [2], 'Proposal for a Consultative Group and Economic Secretariat', 9 June 1931, sent on 10 June to Rivett.
44   Ibid., Casey to Rivett, 28 July; Rivett to McDougall, 30 July 1931.
45   Ibid., McDougall to Rivett, 2 September and 12 November 1931.
46   NLA, MS6890/3/4, 19 February 1932.
47   *LFSSA*, 53, 10 February 1926, p. 158.

the ferment of the Depression years. Both were influenced to some extent by McDougall's work, and his ideas reflected some of theirs. All three believed in economic planning and sought to broaden inputs into that planning.

Elliot 'owed a special intellectual debt' to views discussed by an 'Imperial Study Circle', which included McDougall. Other colleagues or contacts of McDougall in the group included E. M. H. Lloyd, economist Sir Arthur Salter, Conservative politician Lord Eustace Percy and Sir Edward Davson, a member of the IEC. The group's thinking 'reflected...progressive economic opinion on intra-imperial trade issues'. Elliot believed the economic crisis was no aberration, but the result of 'technological change and growth of productive capacity leading to a permanent state of "glut"'. He had moved beyond imperial solutions to a view that the problem 'could only be solved by governmental organisation and management'. He sought 'a measure of centralisation of control' to bring 'the activities of separate industries...into harmony with the national interests'. In March 1930 he told the Royal Institute of International Affairs: 'if there is one lesson which the twentieth century is teaching us, it is the necessity for long-range planning.' He wanted 'the present economically "ignorant" legislature' replaced or combined with an industrial chamber.[48]

In his autobiography, Harold Macmillan records influence on his own thinking of three anonymous articles published in *The Times* in January 1932, under the general heading 'A True Tariff Policy'.[49] The articles were written by McDougall, who was pleased to learn from *The Times* political reporter that they were 'being quoted and their arguments used in talks taking place in Cabinet committees'. Some attributed them to Elliot; McDougall swore him to secrecy.[50] Subtitles suggest the thrust of their arguments: 'The Lever of Progress'; 'The Producer's Angle'; 'Efficiency as a Condition'. Discussion in the third article about 'the sort of tariff commission which should be set up' led Macmillan to develop the idea of a commission 'planning the growth of the nation's economic life', extending it ultimately to a 'sub-parliament', representing labour, management, producing and consuming industries, to consider, inter alia, imperial trade policy and financial and monetary problems. The idea resembled McDougall's suggestions for combining diverse interests in a deliberative body linked to an imperial secretariat, although it lacked the dimension of imperial organisation. Macmillan continued to apply this planning approach to a range of economic issues in the mid 1930s.[51]

---

48 Cooper, *British Agricultural Policy*, pp. 160–4.
49 *The Times*, 13, 14 and 15 January 1932.
50 NAA, CP103/12, bundle 21/12, McDougall to Bruce, 21 January 1932; NAA/CSIR, A10666, [3], McDougall to Rivett, 21 January 1932.
51 Harold Macmillan, *Winds of Change: 1914–1939*, Macmillan, London, 1966, pp. 355–8.

Preparing for Ottawa

In the late 1920s, British interest in both tariffs and empire grew. Disillusion with trade liberalism was driven by persistent unemployment at about 10 per cent, as British industry failed to regain its share of export markets in the face of strengthening European competition and a slow shift towards newer sectors of industry—electrical goods, chemicals and motorcars—which depended more on domestic and imperial markets and had less to fear from a preferential tariff. As world cereal prices weakened, agriculture also needed help. Imperial policies gained support from the political right and much of Labour. By 1929 most of industry supported protection; in 1930 prominent merchant bankers signed a pro-protection manifesto, and industry organisations campaigned for working-class support for a tariff.[52] This was the fertile ground in which McDougall laboured for what he believed to be the imperial cause.

The onset of depression increased protectionist pressure, but politics remained the stumbling block. The Labour Government elected in 1929 depended on Liberal support; key economic portfolios were filled by free-traders, including intransigent Philip Snowden as Chancellor of the Exchequer. Cabinet was divided.[53] The National Government formed in August 1931 bowed to economic and political pressure, enacting a 10 per cent revenue tariff on a wide range of goods. It was expected that a full imperial preference system would be created at an Imperial Economic Conference in Ottawa in 1932.

The British Government planned bilateral negotiations with the dominions, both at home and in London, as a preliminary to the Ottawa meetings. Bruce, who had been returned to Parliament after the fall of the Scullin Government late in 1931, was appointed Assistant Minister in the new Lyons Government and headed a Cabinet subcommittee in Canberra preparing for the conference; McDougall was placed in charge of negotiations in London. Talks beginning in February 1932 examined Australian and British requests for tariff concessions in detail. McDougall's method of dealing with each commodity was to submit the facts in the form of a memorandum, as a basis for discussion. His questions were referred back to the Board of Trade, thence if necessary to the FBI and the individual industry. The process was amicable. He thought British representatives appreciated the opportunity for such close discussion with one dominion, which helped them to 'clear their own minds and to see the whole situation in proper perspective'.[54]

---

52 Tim Rooth, *British Protectionism and the International Economy: Overseas Commercial Policy in the 1930s*, Cambridge University Press, Cambridge, 1992, pp. 38–48, 70.
53 Ibid., pp. 48–54.
54 NAA, CP103/12, bundle 21/10, McDougall to Gunn, 21 April 1931; bundle 21/13, McDougall to Gunn, 14 April.

McDougall was able to argue at a detailed level without losing sight of principles. His mastery of minutiae, of details of complex industries and of their places in the Australian economy, in the empire and in world trade, can be observed in his memoranda; in the records of meetings; in endless detail about chilled and frozen beef, pork and bacon, light and fortified wines, barley for malting or for stockfeed; in lists of percentages of imports and exports grouped and sorted a dozen ways; in explanations of where Britain obtained supplies of this food and that raw material, where they might be obtained if duties were imposed, where suppliers might sell if not to Britain, and what effect an extra penny a pound duty might have for British consumers. His power of persuasion rested in considerable measure upon his grasp of this material. He had help with its preparation, but his strength lay in his ability to take it as ammunition for argument. It was never better demonstrated than in these discussions.[55]

McDougall did not limit his efforts to officialdom. A *Times* leading article dwelt on the unprecedented opportunity at Ottawa for effective economic cooperation. There should not be sacrifices; each empire government should aim to encourage empire trade to the greatest extent compatible with its own economy. 'The most promising line of approach is through a broad development of the principle of complementary production', it declared, giving as an example one of McDougall's favourite cases: the Australian electrical industry. Sending a copy to Rivett, McDougall explained: 'I never write *Times* leaders, what I do is go and see the leader writer and discuss with him what he is going to write!!'[56]

Ottawa

Opening speeches at the Ottawa Conference dwelt on the example of cooperation about to be presented to the world. Its real business was hammered out in bilateral discussions leading to conclusion of 12 trade agreements, seven of them between the United Kingdom and a dominion. In the series of meetings between British ministers and their Canadian, Australian and New Zealand counterparts—essentially the centrepiece of the conference—high moral purpose degenerated to abrasive bargaining. W. K. Hancock reminds us that it was a gathering 'of anxious and suffering nations, desperately intent upon a task of economic salvage'. While the participants could be criticised for intellectual

---

55 Records of McDougall's discussions with British officials are in UKNA, BT 92/12, as are the reports of Britain's Senior Trade Commissioner, R. W. Dalton, of his discussions in Australia. McDougall's reports to Canberra are in NAA, CP103/12, various bundles. His memoranda are, as always, scattered through these and other sources.
56 'The Importance of Ottawa', *The Times*, 22 March 1932; NAA/CSIR, A10666, [3], McDougall to Rivett, 31 March 1932.

inconsistency and muddled thinking, 'it would be unjust and profoundly misleading not to keep constantly in mind the crisis atmosphere' in which they met to deal with 'unprecedented economic calamity'.[57]

The Australian delegation comprised two ministers: Bruce and H. S. Gullett, Minister for Trade and Customs. It had ten official advisers,[58] six official 'consultants'[59] and more unofficial advisers and lobbyists. Delegates had one hundred and sixty-nine documents to digest. No decision had been made beforehand about 'exactly what they would do, or how'. The British delegation, representing a non-party government, had 'agreed to disagree' on policy. I. M. Drummond points out that no British minister or civil servant had been involved in such trade negotiations with any country; they were called upon 'to construct seven major trade agreements in thirty-one days'. By the end they seem to have been pleased 'not so much with the agreements' *terms* as with the documents' *existence*'.[60]

Current economic conditions must have made bargaining more desperate than it might have been in kinder times. But the underlying problem was that the empire had outgrown the visionary idea of complementary trade. There were conflicting interests that no imperial preference scheme could resolve. McDougall's desire to see the Australian canned-fruit industry encouraged in schemes for rationalised, complementary production was thwarted in his preliminary discussions by British policy to develop horticultural and processing industries.[61] Canada and New Zealand were in conflict over butter: low wheat prices had stimulated Canadian dairy production, which then faced US tariff restrictions against cream. Canadian farmers agitated against imports of New Zealand butter, which enjoyed comparative advantage.[62] Britain was anxious not to offend Denmark, an essential supplier of dairy products, or Argentina, a major supplier of meat and recipient of British investment.[63]

Australia's primary concern to secure the British market for its meat was successful, but at a cost. Meat had not been included in Britain's 10 per cent

---

57 Hancock, *Survey*, pp. 215, 229.
58 E. Abbott and A. C. Moore, Department of Trade and Customs; L. E. Stevens and C. B. Carter, Department of Commerce; J. F. Murphy, Prime Minister's Department (Secretary to Delegation); A. E. V. Richardson, CSIR; E. C. Riddle and L. G. Melville, Governor and Economic Adviser, respectively, Commonwealth Bank of Australia; L. R. Macgregor, Australian Trade Commissioner in Canada; and McDougall. NAA, A1667/1, 430/B/18, 'Report of the Conference'.
59 W. C. Angliss, meat interests; S. McKay, Chambers of Manufactures; R. W. Knox, Chambers of Commerce; F. H. Tout, Graziers' Federal Council; H W. Osbourne, Dairy Produce Control Board; M. B. Duffy, adviser on labour questions. Ibid.
60 Ian M. Drummond, *Imperial Economic Policy 1917–1939: Studies in Expansion and Protection*, George Allen & Unwin, London, 1972, pp. 220–1, 284.
61 NAA, CP103/12, bundle 21/13, McDougall to Bruce, 16 June 1932; UKNA, BT11/92, McDougall to Wilson, 18 June, minutes of discussion with McDougall at MAF, 24 June 1932.
62 Hancock, *Survey*, p. 212.
63 UKNA, BT11/92, Stacy to Carlill, 24 May 1932; Cooper, *British Agricultural Policy*, p. 153.

general tariff imposed by the National Government. Chamberlain preferred a quota system, believing that the British livestock industry contributed to the 'general prosperity' of the nation; that Argentine exporters could dominate the market even with a 10 per cent tariff; and that frozen meat from southern dominions could not compete with the chilled Argentine product. Dominions 'wanted the foreigners to be pushed out of the U.K. market altogether', but were forced to consider a quota which, after hard bargaining, was extended to include restrictions on foreign imports of mutton and lamb. It was a 'makeshift' and unsatisfactory arrangement, which did not deal with the fundamental problem of falling prices.[64]

By 1932, dairying had become the most successful intensive farming industry in Australia; butter ranked third after wool and wheat in overall exports and had overtaken wheat in the UK market. A depression-induced surge in rural production was particularly marked in dairying. After Ottawa, in the unrestricted market of the United Kingdom, the percentage of butter imports from Australia doubled that of the late 1920s and prices fell from 109 shillings in December 1930 to 67 shillings in April 1933. A British Government proposal for restricting supplies was supported by both McDougall and Bruce, at least as a temporary measure. It was opposed in Australia, where the dairy industry wielded considerable political power. Bruce's arguments for voluntary regulation were supported by Lyons, but rejected by State premiers and ministers for agriculture. Australian meat and butter imports into Britain continued to increase until, under terms agreed at Ottawa, Britain proposed to introduce quotas for the second half of 1934. Despite Bruce's continued urging that cooperation was preferable to compulsion, Australian governments, State and federal, refused to cooperate—alone among the dominions. Bernard Attard comments that Australia expected too much of its relationship with Britain: 'it demanded a guaranteed proportion of the British market far greater than it had ever previously supplied or could hope to consistently in the future.'[65] Attard also concludes from this case that Bruce's influence 'over the formulation and implementation of commercial policy' by a government constrained by various political and economic conditions was more limited than has been argued by others, including John O'Brien, who has suggested that 'Bruce and McDougall had all the semblances of a government in exile'.[66]

The Ottawa Agreements could not create the cooperative, rationalised system McDougall had hoped for. I. M. Drummond has suggested that the agreements created ill feeling within and beyond the Empire. The meat quota agreements

---

64 Ibid., pp. 152–6.
65 Bernard Attard, 'The Limits of Influence: The Political Economy of Australian Commercial Policy after the Ottawa Conference', *Australian Historical Studies*, Vol. 29, no. 111, October 1998, pp. 325–43.
66 John O'Brien, 'Empire v. National Interests in Australian–British Relations During the 1930s', *Historical Studies* [Australia], Vol. 22, no. 89, 1987.

antagonised Argentina and did nothing to protect British producers.[67] O'Brien argues that Australian success was a hollow victory, leaving Britain determined not to let empire considerations outweigh national interests again, and Australia unduly complacent, with its prospects of success in subsequent negotiations with Britain damaged.[68] Hard bargaining characterised imperial trade relations in the 1930s. Kosmas Tsokhas suggests Australia did well despite having to concede some reduction in its own tariffs: Australian exports benefited from a depression-mandated devaluation and Australian exports to Britain returned to pre-1929 levels by 1936; British exports to Australia remained at only half the earlier level. Australia's Tariff Board interpreted Ottawa clauses, intended to ensure equal competition between British and Australian industry, as requiring 'a marginal advantage to the Australian manufacturer'. Australia used the 'diversion' of wool sales from Japan to bargain for expanded meat exports to Britain, at very little actual cost to the wool industry.[69]

Harmonising interests, even of countries owing allegiance to a common heritage and crown, proved difficult if not insuperable. Yet Francine McKenzie concludes that

> ...the importance of imperial preference lies not only in the domain of reason, or economic calculus, but also in the realm of emotion and political symbolism...imperial preference cannot be fully understood without considering its emotional force...Preferential tariffs touched upon issues of power, identity and alliance, subjects about which bureaucrats, politicians and ordinary citizens had strong, sometimes visceral, feelings.[70]

Complex attitudes to empire and nation persisted.

There is no doubt, however, that the United States viewed the Ottawa Agreements with implacable hostility.[71] The incoming Roosevelt Administration was viscerally anti-imperial. Anti-imperialism certainly fed on opposition to British colonial policies, and in the 1930s was linked to what was described as a US free-trade policy. But this was 'qualified free trade'. Tariffs, traditionally used to protect developing American industries and agriculture, were not objectionable; discrimination and tariff preference were.[72] Imperial preference was, in the view of Secretary of State, Cordell Hull, 'a grievous injury to US

---

67 Drummond, *Imperial Economic Policy*, pp. 264–6.
68 O'Brien, 'Empire v. National Interests', pp. 561–86.
69 Tsokhas, *Making a Nation State*, pp. 105–11.
70 McKenzie, *Redefining the Bonds of Commonwealth*, pp. 261, 265.
71 Elizabeth Borgwardt, *A New Deal for the World: America's Vision for Human Rights*, The Belknap Press of Harvard University Press, Cambridge, Mass., 2005, pp. 24–5.
72 W. M. Scammell, *The International Economy since 1945*, second edn, St Martin's Press, New York, 1983, pp. 11–12.

commerce'. Hull led a campaign to free up international trade and dismantle preference; in 1934 the Administration gained congressional authority to negotiate bilateral trade agreements; an agreement in 1938 with Britain brought 'a marked benefit to US agriculture'.[73] McKenzie identifies 'intertwined' motives underlying 'the passion and indignation' inspired by the imperial tariff in the United States. One was to eliminate empires from the international community and involved an element of altruism. But Americans also believed 'that by liberalizing the postwar international economy, American values of freer trade, unrestricted competition, and democracy would become universal and the US would be in a position to dominate the postwar world'.[74]

Cooperation and the Empire Marketing Board

The idea of a permanent body to coordinate empire economic policy had gained a foothold in London, partly as a result of EMB publicity and Beaverbrook's vigorous campaigning, but also as a counter to rising tariffs and economic nationalism in Europe and the United States. It was much less welcome in the dominions where, in varying degrees, opposition to potential imperial control was combined with reluctance to bear the cost of such a body. The economic crisis threatened all funding in Britain, and was particularly dangerous for the EMB, which was still the object of Treasury suspicion. The 1930 Imperial Conference had praised its work, but its friends and staff grew anxious as the crisis deepened.

British opinion had favoured broadening control and financing of the EMB since 1929. McDougall developed ambitious plans to remodel it as a truly imperial body. A committee, including McDougall, formed to consider relevant resolutions of the 1930 Imperial Conference, recommended extending membership to representatives of the overseas empire and reconstitution as a body incorporated under Royal Charter with a fixed annual income.[75] In the Research Grants Committee, McDougall argued for a policy of making more funds available for overseas research, and chaired a review allocation committee set up for that purpose.[76]

Insurance magnate Sir George May headed a committee of accountants, businessmen and labour representatives with the task of identifying economies in government. The majority report, published on 1 August 1931, recommended

---

73 Robert Skidelsky, *John Maynard Keynes. Volume Three: Fighting for Britain 1937–1946*, Macmillan, London, 2000, pp. 188–9.
74 McKenzie, *Redefining the Bonds of Commonwealth*, pp. 34–5.
75 NAA/CSIR, A9778, M14/31/5, Report of Committee, 21 May 1931.
76 NAA/CSIR, A10666, [2], McDougall to Scullin, 10 February 1931.

reductions worth £96 million, mainly from social services, unemployment insurance and government salaries. It also recommended abolition of the EMB. The Labour Cabinet split over its implementation, resulting in formation of a National Government on 25 August.

An anxious Rivett wrote to McDougall that the dominions should offer 'as a matter of good business…financial support to keep the Board going as an Imperial concern charged with the central oversight of the work of stimulating intra-Empire trade and of developing, as far as possible, team work in research'. McDougall 'stirred up two or three significant people to write to the press and… to Neville Chamberlain, Sir Robert Horne and to several industrial leaders'. He asked Walter Elliot to request that South African statesman Jan Smuts, then about to become President of the British Association for the Advancement of Science, should publicly acknowledge the EMB's importance in centralising scientific and economic research. After formation of the National Government, he correctly predicted that Cabinet, which included supporters like Chamberlain and J. H. Thomas, would maintain the board with a reduced expenditure for 1932–33. He discussed possible economies with Tallents and was the only overseas member of a subcommittee to deal with funding cuts—'a heavy and most invidious task'.[77] He persuaded the committee to request a lesser reduction and, given the need to reduce foreign imports, for £25 000 to be spent on a 'great national campaign to buy British goods from home and overseas'. The national campaign was launched in November. Four million posters displayed the 'Buy British' slogan from windows of government offices and public transport; letters 15 ft high faced Trafalgar Square. The press offered free advertisements and the Prince of Wales led a series of broadcasts.[78]

In preparation for Ottawa, McDougall had written memoranda giving detailed backgrounds of the various imperial bodies, and proposals to transform them into vehicles for imperial cooperation, as discussed in the two preceding years and refined in discussion with Tallents and Chadwick. He kept in touch with Bruce, who agreed dominions should share in EMB funding, albeit in token amounts during the financial crisis.[79] One suggestion was that the IEC and EMB be integrated, with the IEC being the 'economic consultative side' of an organisation including a secretariat for an imperial council of ministers meeting annually.[80] After discussion with Sir Geoffrey Whiskard of the Dominions Office and McDougall, Tallents drafted a plan to combine IEC, EMB and IAB into one body controlling all imperial economic activities. Besides machinery, he and McDougall considered tactics to persuade the more reluctant dominions at

---

77 NAA/CSIR, A10666, [2], Rivett to McDougall, 11 August; McDougall to Rivett, 19 and 20 August, 10 and 23 September 1931.
78 NAA/CSIR, A9778, M14/31/12, McDougall to Mulvaney and to Rivett, 19 November 1931.
79 NAA/CSIR, A10666, [2], McDougall to Rivett, 10 December 1931.
80 NAA, CP103/12, bundle 20, memorandum IEC No. 3, 'Imperial Economic Consultation II', 13 January 1932.

Ottawa.⁸¹ McDougall's final scheme abandoned the idea of a single body. Under the supervision and budgetary control of an imperial council, the EMB should serve as an umbrella organisation over the Imperial Bureaux.⁸² He chaired an 'animated discussion' with overseas EMB members on methods of finance; H. A. F. Lindsay, Government of India Trade Commissioner, suggested a levy on empire imports into Britain, leaving McDougall to draft a proposal.⁸³

The Cabinet Estimates Committee recommended a review of the EMB, a 'Colonial Development Fund' established in 1929 chiefly as a means of relieving unemployment, and the Advisory Council for Agricultural Research, to resolve overlaps and inefficiencies. It argued that Britain's adoption of a tariff in 1931 removed the *raison d'être* of the EMB. McDougall told Bruce he doubted any serious intention to abolish it. In the same letter, he enclosed a cutting from the *Manchester Guardian*, reporting schemes for a new organisation to replace the EMB and outlining one that was 'largely the work of Dominion representatives here' and was 'believed to have a big backing'. The item resulted, McDougall explained, from 'a certain leakage of the very confidential paper which I gave to you when you were in London…I am going to suggest that the paper should be circulated at an early date'.⁸⁴

The Skelton Committee

On the Ottawa Committee on Methods of Economic Co-operation, Canada's Minister for Mines, W. A. Gordon, stated bluntly that if the EMB were to continue, it should be a British, rather than an imperial, board. Canada had the staff and facilities to solve all its own problems. Bruce proposed, as a solution to the resulting *impasse*, that representatives of all empire governments consider the problem of cooperation and report in 1933. Dominions Secretary J. H. Thomas undertook to keep imperial machinery functioning until then.⁸⁵ McDougall was disappointed but hopeful. After a 'heart to heart' talk with Canadian Prime Minister, R. B. Bennett, Bruce believed there was a fair prospect of Canadian cooperation.⁸⁶ EMB staff were less sanguine. Elspeth Huxley wrote: 'I gather that the staff is fed up to a man…Personally, I feel that a speedier and more dignified end would have been cleaner than this year's reprieve…its real purpose in life is over.'⁸⁷

---

81 Ibid., 'Some Notes on Inter-Imperial Economic Machinery', drafts sent to Bruce on 8 and 23 March.
82 Ibid., 'The Future of Imperial Machinery for Economic Co-operation and Consultation', 3 May.
83 NAA, CP498/1, 430/AA/13, memorandum reporting the meeting on 30 May 1932.
84 NAA, CP103/12, bundle 20, McDougall to Bruce, 7 April 1932.
85 NAA/CSIR, A9778, M14/32/5, Richardson to Rivett, 5 September 1932.
86 NAA/CSIR, A10666, [3], McDougall to Rivett, 8 September 1932.
87 NLA, MS6890/2/4, Elspeth to McDougall, 23 August 1932.

McDougall did his best to ensure the 1933 committee reported favourably. He sought a memorandum of support from Rivett and prepared one himself. It shows him thinking still much as he had before Ottawa, with perhaps more attention to dominion sensibilities concerning status and representation.[88] He kept in touch with an interdepartmental committee preparing the British case and worked with British delegates Sir Horace Wilson and Sir Fabian Ware, heartily approving their decision to propose Canada's O. D. Skelton as chairman.[89] Skelton, Deputy Minister of the Department of External Affairs, 'felt strongly about Canadian nationalism and shared [former Prime Minister Mackenzie] King's suspicion of and opposition to foreign, and particularly imperial, entanglements'. During King's premiership, Skelton had been, unobtrusively, 'one of the half-dozen most powerful men in the country'. He remained in his position, becoming 'as indispensable to the new Prime Minister as he had been to the old'.[90] He had constantly opposed both the IEC and the EMB, on grounds of independence and national self-respect: 'Canada does not want to go "on the dole".'[91] When the committee met, McDougall worked hard to convince Skelton 'that I was just as keen as he was on maintaining the spirit of nationhood of the dominions'. In hearings he did what he could to draw out witnesses with 'helpful leading questions'.[92]

The Skelton Committee on Economic Co-operation sat in London, often three times a week, throughout February and March 1933, hearing evidence from imperial bodies and holding intensive discussions. Records of these show considerable unanimity. There was broad agreement on the value of empire cooperation but also, as McDougall put it at a late stage, nobody wanted 'an Imperial Economic Secretariat'; nor did they want 'a large common fund which could be drawn upon by any government for the support of its own scientific or economic services'.[93] This admission must have cost him some pain: he certainly had wanted the first and something approaching the second, but was forced to acknowledge the general mood.

Disagreement came when delegates turned to discuss what they did want. McDougall, at one extreme, argued for the value of centralised research and information dissemination, drawing on his long experience of working with the EMB. His formidable opponent was Colonel G. P. Vanier, Official Secretary at the Canadian High Commission. Vanier, a decorated war hero, was to continue a

---

88 NAA/CSIR, A10666, [3], McDougall to Rivett, 26 October, 17 and 23 November, 8 December 1932, 2 February 1933; memorandum, 'The Empire Committee', 30 January 1933.
89 Ibid, McDougall to Rivett, 9 February 1933.
90 Lester B. Pearson, *Memoirs 1897–1948: Through Diplomacy to Politics*, Victor Gollancz, London, 1973, pp. 71–2, 76.
91 Constantine, 'Anglo–Canadian Relations', pp. 362–5.
92 NAA/CSIR, A10666, [3], McDougall to Rivett, 16 February; A10666, [4], 16 and 22 March 1933.
93 NAC, RG25, External Affairs, vol. 1632, 731, FP(16–23), Meetings of Committee on Economic Consultation and Co-operation, Minutes of 5 April 1933 p. 2.

senior diplomatic career and to be Governor-General of Canada from 1959 until his death in 1967. He spoke persuasively of a changing relationship, significantly choosing the term 'Commonwealth':

> The Commonwealth has evolved...and its members have developed in natural and cultural resources, in national consciousness and in economic integration. The Commonwealth has become decentralised...
>
> We recognise today that the co-operation we seek to effect must take place in Canberra and Cape Town...as well as in London, and that there should be instruments or agencies of co-operation not only between New Zealand and the United Kingdom but also between New Zealand and Canada. The United Kingdom remains, and will long remain, easily first, in the range of its scientific and economic achievements, as well as in the numbers and quality of its workers in those fields; London will remain the most important focussing and contact point, but in their varying and modest ways, Dublin, St Johns, Salisbury, Delhi and Ottawa have come to occupy a place in the picture.

Vanier described scientific research as 'an essential phase of national activity... an indispensable and integral part of a nation's intellectual life'. Cooperation should not be achieved through a 'permanent central authority', but through periodical conferences and exchanges of workers and research programs. Coordination of statistical services and research into economic questions was essential, but a permanent central body would not deal competently with the changing range of economic issues and varying aspects of problems throughout the Commonwealth.

> If such an effort were made, it would involve navigating in the perilous waters of governmental policy, and probably prove more embarrassing than helpful...expert assistance must be available, on the spot, to aid each Government in the daily tasks of administration, particularly now that the economic position is changing from week to week and from day to day.

Once again, Vanier advocated *ad hoc* cooperation.[94]

The southern dominions objected that periodical conferences posed practical and financial difficulties. McDougall acknowledged that in many fields there were local problems, but there were also fundamental problems better tackled jointly: the pasture research being undertaken in several empire countries needed 'an intellectual general headquarters in the Commonwealth'; it already existed at Aberystwyth.[95]

---

94 Ibid., FP(1–15), minutes of 20 March 1933, pp. 20–8.
95 Ibid., minutes of 24 March, pp. 5–8.

Both Vanier and McDougall spoke at considerable length and with authority for their governments: Vanier because Skelton chaired the committee, McDougall because by then Bruce was in London as Resident Minister. Most overseas representatives were less certain of official opinion and stressed that their views were personal. These views ranged between the extremes: South Africa generally agreed with the Canadian position; New Zealand, representing 'one of the smaller dominions with perhaps not the same financial resources', was appreciative of the advantages provided by the EMB but took a middle view; Newfoundland, 'hitherto…largely ignorant of the advantages to be derived from co-operation', hoped they could continue.[96] Representatives of India, the United Kingdom and the colonies were close to McDougall's position, with some reservations.

There were differences of opinion about publicity and market promotion. All agreed that EMB work had brought a valuable change in public opinion about the empire; India and Southern Rhodesia wanted it to continue; Canada and New Zealand thought it should be left, in most cases, to individual countries. Both were conscious of their current conflict of interest over butter: publicity 'should not tread upon the dangerous ground where there were competing products from various parts of the empire'.[97]

Most delegates wanted some form of imperial organisation to continue. Canada opposed 'a central organisation with large funds, a roving commission and growing staffs', but conceded that one or two functions of the EMB should continue on a cooperative basis. New Zealand opposed creation of an 'Empire League of Nations', but supported 'some form of imperial economic council' to facilitate liaison between empire governments on proposals for action by specialised bodies.[98] All agreed that the IEC should continue, as should trade surveys and market intelligence. Impressive evidence had been given to the committee about EMB work in these fields: it was filling gaps in what was offered by the International Institute of Agriculture at Rome, and had obtained the first outside information about agricultural production in Russia, and in Argentina. E. M. H. Lloyd predicted an empire service could outdo that of the United States, which served only the needs of its own growers.[99]

All agreed on the importance of continuing the Imperial Agricultural Bureaux. McDougall's opinion, confirmed by the written record, was that Orr's evidence, combining 'a certainty of utterance with a modesty of demeanour', created 'a profound effect on the committee'. Sir Rowland Biffen's 'somewhat cynical

---

96 Ibid., R. S. Forsyth and W. C. Job, minutes of 20 March 1933, pp. 37–44.
97 Ibid., FP(16–23), Minutes of 27 March, p. 23, R. S. Forsyth.
98 Ibid., Minutes of 29 March, pp. 1–8.
99 Ibid., Volume 3433, File 1-1933-14, Stenographic Notes of Oral Evidence, 'Evidence of Sir Stephen Tallents, Mr Lloyd and Mr Hildred', 1 March 1933, p. 14, statement by Lloyd.

detachment made his advocacy of the Bureaux all the more impressive'.[100] Biffen admitted to scepticism when the idea of a Bureau of Plant Genetics was suggested late in 1928:

> I pointed out then that there were a couple of journals in existence, and that the whole of our literature on what was…a new subject was fairly well centralised…but now I am perfectly convinced that nobody can get on without this abstracting service, merely because of the enormous growth of this highly specialised subject…in the last two years we have had to look for papers in no less than 330…journals…the output of literature in genetics at present works out at something like thirty pages of print a day…the average worker is simply overwhelmed with material which he has to consult.[101]

When McDougall put forward his idea for a levy on exports, Canada objected that the income would be 'unstable' and fluctuating, with administrative difficulties. Moreover, it would not be fair: a Canadian levy would come mainly from wheat exports, which could not be helped by any empire body. Only India was prepared to support a levy outright.[102]

Tallents explained a change in the role of the EMB, stemming from the meetings of heads of empire scientific bodies at the 1930 Imperial Conference. Until then, the function of the RGC had been to assess applications for assistance, but

> we came to the conclusion that the right policy of the Board, now that it had found its feet, was, with the help of [scientific institutions in the empire], and in consultation with them, to work out a programme and to some extent take the initiative in consulting them as to what they regarded as the major problems…We have really passed from being merely a target for applications and giving advice to something more active.[103]

Canada was determined to avoid continuation of such an 'active' role by any imperial body. None should have any executive role or right to initiate work. McDougall proposed, as a compromise solution, a right to recommend *services*, as distinct from policies, on research and economic intelligence.[104] The committee's report was signed on 11 April on this basis.

---

100 NAA/CSIR, A10666, [3], McDougall to Rivett, 9 March 1933.
101 NAC, RG25, External Affairs, Volume 3433, File 1-1933-14, Stenographic Notes of Oral Evidence, evidence of Sir Rowland Biffen, 8 March 1933.
102 Ibid., Volume 1632, 731, FP(16–23), minutes of 29 March, pp. 14–21.
103 Ibid., Volume 3433, File 1-1933-14, Stenographic Notes of Oral Evidence, evidence of Sir Steven Tallents, 27 February 1933, p. 17.
104 Ibid., Volume 1632, 731, FP(16–23), minutes of 5 April, pp. 1–9.

McDougall—'dreadfully disappointed'—continued to hope. The outcome provided 'a basis upon which Empire Governments can, if they will, build up again a really satisfactory system of Imperial economic and research services'.[105] The report was signed on the understanding that empire ministers would discuss the findings when they attended the World Economic Conference in London that summer. They might 'take a more liberal view of the recommendations than it was possible to achieve on the Committee itself'. They did not. Bruce pressed for a special meeting on imperial economic problems and the Skelton report. Ministers gathered at 10 Downing Street, but Bruce failed to get any discussion of the report or of the EMB; they simply agreed to recommend the report to their governments. 'Everyone was determined to get away for their summer holidays; everyone was tired and the two-and-a-half hours—of which only one was spent on economic subjects—was all the time that Ministers were prepared to devote to the economic problems of the British Empire.' For once, McDougall's optimism failed him: 'No doubt after one has had a holiday…one will shake off the sense of discouragement.'[106]

The British Government accepted that 'the [Skelton] report represents the maximum on which unanimity was possible'. I. M. Drummond writes: 'Thus finally expired the dream of an imperial economic general staff—killed, and rightly, by a fact: that economic policy is political, not just technical.'[107] McDougall had perhaps failed fully to accept that decisions taken at a Cabinet table must take account of what seemed good policy and of what could be afforded, but also of what electorates would be willing to bear. It could be a difficult task for a single government. It verged on the impossible for representatives of several governments each, in 1932–33, with suffering constituencies and conflicting needs. McDougall's determination to maintain his own independence of action had freed him from worries about constituencies, budgets and pleasing ministers or department heads. The disadvantage was his reluctance to allow for factors constraining even the most powerful of men.

The negotiations at Ottawa showed clearly the flaw in the 'sheltered markets' theory. The ideas of the 'empire visionaries' could work only where economies were complementary. The economies of the British Empire might once have been so, but in the twentieth century they were changing rapidly; so were the national aspirations of empire governments. Any semblance of control by an imperial secretariat was simply unacceptable. Rivett's expression of national aspiration had a moral twist: 'That Great Britain should finance research in the Dominions was always, to my mind, rather a scandal and we need not regret

---

105   NAA/CSIR, A9778, M14/33/3, McDougall to Rivett, 12 April 1933.
106   Ibid., A10666, [4], McDougall to Rivett, 25 May and 3 August 1933.
107   Drummond, *Imperial Economic Policy*, pp. 227–8.

the passing of that little bit of charity; rather may one be ashamed that the Dominions failed to shoulder the load when their responsibility was pointed out to them.'[108]

Yet the empire governments were faced with the heavy responsibilities of extreme economic pressures, and it was clear that efforts to devise empire solutions based on research, tariffs, quotas or anything else could not solve their chief problem: falling prices of staple commodities on a world scale. By 1933 McDougall was familiar with international efforts to solve that problem, and was ready to look for solutions beyond the empire.

---

108   NAA/CSIR, A9778, M14/33/9, Rivett to Sir Charles Martin, 10 November 1933.

Part B: The League of Nations

Prologue

Voluntary cooperation between independent nation-states to create a broad-based international organisation only became possible when the ultimate consequences of national rivalries had been demonstrated in the slaughter of 1914–18. All three principals in this study participated in that conflict, which must have influenced their thinking. On the Western Front, Orr served with distinction as a medical officer in the trenches and took part in military action in battles including the Somme and Passchendaele.[1] In 1939–45 his only son was to die serving in the Royal Air Force. In 1916 McDougall was appointed to the AIF Cycle Corps, intended for messenger duties and reconnaissance after shelling. The latter proved impracticable in the appalling conditions; after a period largely involving traffic control, his war service was spent just behind the front lines as a quartermaster. He reached the Western Front just before the Battle of the Somme. There, no-one escaped privation, fear or horror. He wrote of watching the 'terrible blind groping' of German artillery, 'feeling with his shells for our batteries…as I wasn't on duty at the time when his search came in my direction I made tracks for shelter and then away'. 'I shall be very glad for a chance to sleep without my boots and clothes', he wrote to his mother, and, in a less-restrained moment, contrasted the pleasant garden surrounds of their headquarters out of the line with the 'dugouts, shells and the unspeakable smell of the partially buried', adding, 'where we were was quite hot enough for my liking but it wasn't a patch on the front line'. He did think it necessary to write a letter for his wife in the event of his death.[2]

Bruce later revealed something of the horror he experienced at the Dardanelles. In 1921, while in England on PLB business, he was appointed Australian delegate to the Second Assembly of the infant League of Nations. There he spoke passionately in favour of arms reduction:

> I want you to realize what it does mean to the soldier…to see the whole of his company, or even the whole of his battalion, wiped out practically to a man, for no result except that it is all part of the great strategy of war to make men attempt the impossible…If you had seen men mutilated and dying without the possibility of being helped, if you had ever heard the cry of a wounded man out between the lines with no possibility of assistance being given him, and with a likelihood that he may lie dying there for days…then I venture to say that you would look on this question with a different eye.[3]

---

1 John Boyd Orr, *As I Recall*, Doubleday, New York, 1967, pp. 68–78.
2 NLA, MS6890/1/6, letters to Kit, 26 July and 18 August; to mother, 6, 13 and 17 August 1916.
3 Lee, *Stanley Melbourne Bruce*, p. 20.

By the time Bruce attended his next League Assembly, in 1932, he had developed reservations about the League's ability to provide for collective security. As Prime Minister, like his predecessor, Hughes, he had not been prepared to 'surrender any part of [Australian] autonomy to an international organisation', and was particularly wary of threats to Australian tariff policy in League attacks on trade barriers. But he was firmly in favour of extending its role in social, economic and technical areas, where he believed it could be effective. He 'genuinely set value on the League as a forum for discussion between states, as a conciliator, as a trend-setter in humanitarian activity and...as an educator of opinion'.[4] He was to take a very active role in the League throughout the decade and hoped its potential for social and economic reform might go some way towards defusing international tensions.

In what was 'almost...an afterthought', Article 23 of the League Covenant had authorised the League to deal with 'the international aspects of social, economic and humanitarian progress'. The League developed a 'double aspect...on the one hand the centre for world-wide co-operation in purposes which all could support, on the other a political agency closely bound up with a treaty which was far from being universally approved'.[5] To meet the requirements of Article 23, bodies were established within the League to deal with economics and finance, communications and transit, health, traffic in drugs, traffic in women, child welfare, intellectual cooperation and refugees. In the 1930s the League's social and economic institutions became 'concerned more and more intimately with the ordinary problems of the life of individuals as well as of nations', representing 'in the aggregate an immense contribution to human welfare and a necessary element in the complex life of the modern world'.[6]

By 1939 more than 60 per cent of the League's budget was spent on economic and humanitarian work.[7] The greatest public impact was perhaps made by the League's work on nutrition: it published a highly regarded scientific report on nutrition in the early 1930s, and the report of the Mixed Committee on Nutrition, resulting from the Bruce–McDougall initiative of 1935, was a bestseller amongst League publications. This success demonstrates the strength of the League as a clearing house for information and statistics, as a meeting place for experts and as a focus of international cooperation and recommendations on aspects of public policy. Experts gathered at the League, whether as individuals or as representatives of institutions, drew on a variety of professional and occupational

---

4   W. J. Hudson, *Australia and the League of Nations*, Sydney University Press, Sydney, 1980, pp. 5–6.
5   F. P. Walters, 'The League of Nations', in *The Evolution of International Organizations*, ed. Evan Luard, Frederick A. Praeger, New York, 1966, pp. 28–31.
6   F. P. Walters, *A History of the League of Nations*, Royal Institute of International Affairs, London, 1952, pp. 175–6. Walters gives as examples work on standardising guidance lights for shipping, road signs and medical practice.
7   League of Nations, A.23.1939, Report of the Special Committee of the League of Nations, 'The Development of International Co-Operation in Economic and Social Affairs', Geneva, 1939, p. 7.

backgrounds and worldwide links to bureaucracies, political structures and academic communities. Citizens of non-member states participated in the technical work. The technical organs of the League formed a network of connections with administrative departments of member and non-member countries in fields such as health and social welfare; 'the world's best experts' were prepared to serve on them 'not only for the sake of the work itself but still more in the conviction that thereby they were helping the cause of peace and international co-operation'.[8] The League Health Organization has been described as 'a co-ordinating body—a sort of executive committee—for a worldwide biomedical/public health episteme that recently had acquired confidence in its ability to alleviate human suffering by reducing, if not eliminating, disease'. It enjoyed 'a symbiotic relationship' with the Rockefeller Foundation, which funded projects, including a bureau on epidemic diseases in Singapore. US expertise contributed to much of this work, the partnership enhancing 'not only [the organisation's] effectiveness, but also its legitimacy'.[9]

Important as this humanitarian work was, it was never likely to prevent war. In the 'first "realist" monograph on international relations in the twentieth century', E. H. Carr proposed that international relations are primarily based on power—military, economic and political—and hard bargaining between conflicting interests.[10] He shocked many believers in the classical harmony of interests, McDougall among them, though not the more politically aware Bruce.[11] McDougall retained much of what Carr called 'the optimism of the nineteenth century...based on the triple conviction that the pursuit of good was a matter of right reasoning, that the spread of knowledge would soon make it possible for everyone to reason rightly...and that anyone who reasoned rightly...would necessarily act rightly'.[12] This was the essential basis of the Bruce–McDougall nutrition approach. The League of Nations resolution resulting from their initiative sought amelioration for international trade problems without resort to international compulsion. The measures it suggested lay almost entirely in the realm of national policy: no international body was to set prices, export quotas or limits to production. International responsibility was to collate and provide information, and to encourage. National bodies would collect information, educate to create demand for foods believed to promote health and devise creative policies to encourage cheap and adequate supplies of those foods. The process would be driven by an educated market.

---

8   Walters, *A History of the League of Nations*, p. 176.
9   Martin David Dubin, 'The League of Nations Health Organisation', in *International Health Organisations and Movements 1918–1939*, ed. Paul Weindling, Cambridge University Press, Cambridge, 1995, pp. 56, 63–4, 72–3
10 E. H. Carr, *The Twenty Years' Crisis 1919–1939: An Introduction to the Study of International Relations*, second edn, Macmillan, London, 1946; Jonathan Haslam, 'Carr, Edward Hallett (1892–1982)', *ODNB*, 2004, <http://www.oxforddnb.com.virtual.anu.edu.au/view/article/30902> [accessed 29 April 2005].
11 Stirling, *Lord Bruce, the London Years*, p. 140.
12 Carr, *Twenty Years' Crisis*, pp. 24–5.

In the short term, the idea failed: 'it required *fundamental* changes in economic policy, in the role of the state in the economy and in the very structure of economic activity in individual countries. None of this...came to pass.' In the longer term, and largely through the education on which McDougall relied, change did occur. Bringing public attention to the problem of inadequate nutrition

> provided a valuable corrective to the confusion of thought which tended to turn shibboleths of 'finance', 'economic laws', and 'free trade', balanced budgets or gold standards, into ultimate criteria of economic policy, and pointed to a saner approach to economic problems.[13]
>
> [It was] a 'damning commentary' that the nutrition approach sought to emphasize living standards. Such an emphasis *should* have been a truism.[14]

But it was necessary. Madeleine Mayhew has pointed to the policy conflict in Britain in the 1930s between nutrition scientists—notably John Boyd Orr—and the bureaucracy, centred on the question of the relationship between inadequate nutrition and income, a conflict between science and economics.[15] Paul Weindling describes the nutrition research of the League Health Organization as 'the product of scientific experts frustrated with the fundamental irrationality of the prevailing social order'. The League's work on nutrition 'exemplifies how scientists were keen to extend their expertise in support of radical reforms... British nutritionists could criticise the British government by invoking the new standards and perspectives on nutrition endorsed by [the League], but which they themselves had formulated'.[16]

An immediate result of the League resolution on nutrition was the formation of national nutrition committees in some 40 countries, including Australia. Early surveying of the state of nutrition by the Australian Advisory Council on Nutrition was marred by amateurism, ignorance of overseas sampling techniques and by a decision to avoid the contentious relationship of income to nutrition. Findings were 'vague and ambiguous'. But the work was continued and steps were taken towards national food standards: 'nutrition was now widely regarded as a national responsibility.'[17] It seems very likely that the action of League

---

13   Sean Turnell, 'F. L. McDougall: Eminence Grise of Australian Economic Diplomacy', *Australian Economic History Review*, Vol. 40, no. 1, 2000, pp. 67–8.
14   Ibid. Turnell is paraphrasing a view expressed by A. G. B. Fisher in 'Economic Appeasement as a Means to Political Understanding and Peace', *Survey of International Affairs 1937*, Vol. 1, pp. 56–108.
15   Madeleine Mayhew, 'The 1930s Nutrition Controversy', *Journal of Contemporary History*, Vol. 23, no. 3, 1988.
16   Paul Weindling, 'Social Medicine at the League of Nations Health Organisation and the International Labour Organisation Compared', in Paul Weindling, ed., *International Health Organisations and Movements*, p. 144.
17   James Gillespie, 'The "Marriage of Agriculture and Health" in Australia: The Advisory Council on Nutrition and Nutrition Policy in the 1930s', in *Migration to Mining: Proceedings of the Biannual Conference of the Australian Society of the History of Medicine*, ed. Brian Reid, Northern Territory University, Darwin, 1998, pp. 6–14.

experts and others who promoted these bodies in many countries accelerated a process of education in nutrition that has become integral to public and professional thinking about health.

This section deals with the nutrition initiative and with the League's sponsorship of conferences attempting to solve other international problems. One was the plight of the international wheat market in the early 1930s, where the only acceptable remedy seemed to be limitation of production. Its failure spurred McDougall, encouraged by Orr and supported by Bruce, to devise his 'marriage of health and agriculture'. The proposal received overwhelming support in Geneva. Following that success, Bruce and McDougall devised plans to extend the approach of increasing consumption to a political solution for world tension and to a scheme for improving the effectiveness of the League's social and economic activities. The first idea generated little enthusiasm; the second was warmly received, but simply ran out of time.

5. The Wheat Crisis of the 1930s

Summary

By the late 1920s the international wheat trade was in crisis. As production from the New World increased, traditional importers in Europe encouraged domestic production and placed high tariffs on imports. Prices for European consumers rose, export prices fell and growers and governments suffered. McDougall had been involved in international discussion of agricultural problems as a member of the League of Nations Economic Consultative Committee, and one of a group of 'experts' chosen to advise the League on agricultural problems. He became familiar with statistics produced by the International Institute of Agriculture in Rome and developed a broad understanding of worldwide agricultural production. He protested against Eurocentric views and saw himself as spokesman for the world beyond Europe. He began to argue for a more rationalised system of agricultural production and for enabling greater consumption of nutritious foods by raising standards of living. Not yet ready to abandon imperial cooperation, McDougall wrestled with the difficulty of reconciling this international outlook with British aims to protect domestic wheat production while avoiding conflict with the dominions. He suggested special treatment for British wheat.

As the crisis in exporting countries worsened, the only solutions raised involved restriction of production. In 1933 a special conference on wheat was held in conjunction with a League-sponsored Monetary and Economic Conference. Bruce, then Resident Minister in London, led the Australian delegation, mediating between pressure from the conference to institute crop and export restriction and reluctant Australian governments. An agreement reached after some days of tension established a Wheat Advisory Committee, on which McDougall represented Australia. Cooperation with restrictions proved impossible to enforce; the wheat crisis was only solved by drought in North America.

The Crisis

> Wheat…is a commodity of the greatest economic importance, and yet in certain countries…it is a crop that exercises a psychological influence even greater than is warranted by its almost unique position in world trade.[1]

---

1 F. L. McDougall, 'The International Wheat Situation', *International Affairs*, Vol. 10, no. 4, 1931, p. 524. Address given to Royal Institute of International Affairs at Chatham House, 7 May 1931.

At the time McDougall wrote these words, the value of world wheat production as a whole was exceeded only by that of rice. Its value as an internationally traded commodity was higher than all commodities except raw cotton. Yet only 18 per cent of all wheat produced in the world entered international trade.[2] Many countries took extraordinary measures to safeguard domestic production of this staple food. By the end of the 1920s domestic prices were rising in those countries and the wheat export market was facing a crisis. The British Empire produced more than half the world's export surplus; it was not a problem amenable to solution by an imperial tariff policy.

Four wheat exporters—Canada, Argentina, Australia and the United States—produced some 90 per cent of total world wheat exports in the 1920s. Their rapid growth as wheat exporters since the late nineteenth century had followed similar patterns; Canada led with exports of wheat and flour worth $10.9 million in 1901, $33.5 million in 1921 and $495 million in 1929.[3] Expansion in all four depended on similar political and technical factors: land settlement and migration policies favouring agriculture; development of wheat varieties suitable for climates more extreme than those of Europe; expansion into lands suitable for the use of agricultural machinery; provision of cheap transport by railway and steamship; and increased demand from Europe as a result of World War I. Wheat-growing constituencies wielded considerable political power, except in Argentina, where cattle-owning landlords were dominant and wheat was grown mostly by non-national, immigrant, short-term tenant farmers.[4] There were other differences: the United States, with a large home market and complex economy, depended less on primary exports; the other three needed income from those exports to finance further development and imports. Wheat was the chief export of Canada and Argentina, both exporting about two-thirds of their crop; the pastoral industry predominated in Australia. Canada and the United States, with well-developed storage systems, were not under pressure to export all their surplus. In Canada, Federal Government control, established during World War I, was replaced in the early 1920s by prairie wheat pools, based on cooperative marketing principles and responsible for marketing more than half of prairie wheat.[5] In the interwar years Canada supplied more than one-third of world wheat exports and Argentina about one-fifth. US wheat exports fluctuated from almost one-quarter in the 1920s to less than one-tenth in the dust-bowl years of the mid 1930s; for some of those years the United States was a net importer. Australia's share rose to almost one-fifth in those years—from a more normal contribution of just more than one-tenth. See Tables 2 and 3.

---

2   Ibid., pp. 524–5.
3   William E. Morriss, *Chosen Instrument: A History of the Canadian Wheat Board: The McIvor Years*, Reidmore Books/Canadian Wheat Board, Edmonton, 1987, p. 10.
4   Carl E. Solberg, *The Prairies and the Pampas: Agrarian Policy in Canada and Argentina, 1880–1930*, Stanford University Press, Stanford, Calif., 1987, pp. 226, 229.
5   Morriss, *Chosen Instrument*, p. 14.

Table 2 Wheat Exports of the 'Big Four' as a Percentage of the World Wheat Trade: Five-Year Averages, 1922–37.

	Canada	Argentina	United States	Australia
1922–27	36.9	17.5	23.2	11.3
1927–32	34.6	20.4	17.7	13.7
1932–37	37.8	24.1	9.0	19.1

Table 3 Wheat Exports of the 'Big Four' in Millions of Bushels: Five-Year Averages, 1922–37.

	Canada	Argentina	United States	Australia
1922–27	286.8	135.9	180.9	87.9
1927–32	277.8	163.3	144.1	110.1
1932–37	214.5	138.6	1.9	109.8

Sources: Tables from Carl E. Solberg, *The Prairies and the Pampas: Agrarian Policy in Canada and Argentina, 1880–1930*, Board of Trustees of the Leland Stanford Jr University, 1987. All rights reserved. Used with permission of Stanford University Press, <http://222.sup.org>

Six European countries had traditionally supplied the European market. Russia (as McDougall and many contemporaries still called the Soviet Union) had been the world's chief wheat exporter until 1914. Edmund de Waal records that his ancestors, the Ephrussi family, had once been based in Odessa, where in the nineteenth century they cornered the market in the Ukraine—the largest grain producer in the world—shipping it across the Black Sea and up the Danube River through Europe. Members of the family settled in Vienna and Paris where they were known as *les Rois du Blé*. They diversified into banking and infrastructure, maintaining an opulent lifestyle comparable with that of the Rothschilds.[6] The other five European exporters were Bulgaria, Hungary, Romania, Yugoslavia and Poland.

Russian wheat did not re-emerge on the market until 1926; for three years after that it represented up to 1.3 per cent of world exports, but the proportion rose suddenly in 1930–31 to 7 per cent. Official statistics—not necessarily reliable—suggested production had increased from 1929–30 by an amount equal to Canada's annual total. McDougall believed that Russia was under the same pressures to sell wheat as other exporters. The Five-Year Plan required money from exports for machinery and raw materials. Although there was scope for greater internal consumption, it seemed likely that Russian wheat exports

---

6 Edmund de Waal, *The Hare with Amber Eyes: A Hidden Inheritance*, Vintage Books, London, 2011, pp 21–37.

would continue to increase.[7] The rise in Soviet exports was short-lived: forced collectivisation, terror, repression, near civil war and famine removed Soviet wheat from the competition, though not from the fears of outsiders.[8]

The wheat problem continued to worsen nevertheless. A widespread bumper crop in 1928 exceeded world demand by an estimated 100 million bushels. The resulting carryover formed the basis of a surplus that would undermine trade well into the 1930s.[9] Exporters and many importers continued to encourage production. The other five European exporters had adopted postwar 'agrarian reform' policies, breaking up large estates to establish peasant holdings with higher costs of production. Their governments sought to maintain a reasonable living standard for peasant farmers, leading to proposals for inter-European tariff preference.[10] One study concluded that 'farming in Europe…is a tradition, a way of life, a civilization, and any attempt to regard it merely from the cash aspect must fail'.[11] McDougall wrote of the 'cult of wheat', in which the peasant was recognised as a stable, conservative element of society:

> No one can begin to grasp the world wheat situation unless he clearly recognises that European countries will do everything to maintain the peasant, and that wheat together with other cereals in Eastern Europe and sugar beet are the mainstay of the peasant.
>
> The continental European attitude towards the oversea wheatgrowing countries is that they are upstarts, whose farmers, aided by the biologist and the agricultural engineer, threaten the foundation of society.[12]

Wheat production was supported, for similar reasons, in European importing countries. Tariffs against cheap overseas wheat appeared in the 1890s, their purpose to protect local producers from 'the ruin of the culture of the soil' and to encourage 'a wise and harmonious balance' between industry and agriculture.[13]

In the late 1920s European exporters and importers alike encouraged consolidation of acreages with government-assisted finance and debt reduction, and technologies to increase yields. French tariffs rose rapidly from the mid 1920s, supported by milling quotas and price and import controls. By 1929 France had an exportable surplus. Neither Italy nor Germany reached export

---

7   McDougall, 'The International Wheat Situation', pp. 525, 531–2.
8   Lizzie Collingham, *The Taste of War: World War Two and the Battle for Food*, Allen Lane, London, 2011, pp. 220–1.
9   Robert Holland, 'Imperial Collaboration and Great Depression: Britain, Canada and the World Wheat Crisis, 1929–35', in *Theory and Practice in the History of European Expansion Overseas: Essays in Honour of Ronald Robinson*, eds Andrew Porter and Robert Holland, Frank Cass, London, 1988, p. 110.
10 McDougall, 'The International Wheat Situation', pp. 525–7.
11 Royal Institute of International Affairs, quoted in Wilfred Malenbaum, *The World Wheat Economy 1885–1939*, Harvard University Press, Cambridge, Mass., 1953, pp. 32–3.
12 McDougall, 'The International Wheat Situation', p. 527.
13 Malenbaum, *World Wheat Economy*, p. 158.

status, but tariffs in both also rose steeply. Mussolini's 'Battle of the Wheat' began in 1925 with research, agricultural extension work and land reclamation as well as milling quotas, financial assistance and price and distribution controls. There were similar measures in Germany and in other European countries. Inevitably, stocks accumulated in North America and world prices fell. As tariffs rose even higher, consumers in European countries paid almost double the world price and were forced to reduce purchases of more costly foodstuffs. Growers in the overseas exporting countries suffered reduced incomes; their purchases of other commodities declined.

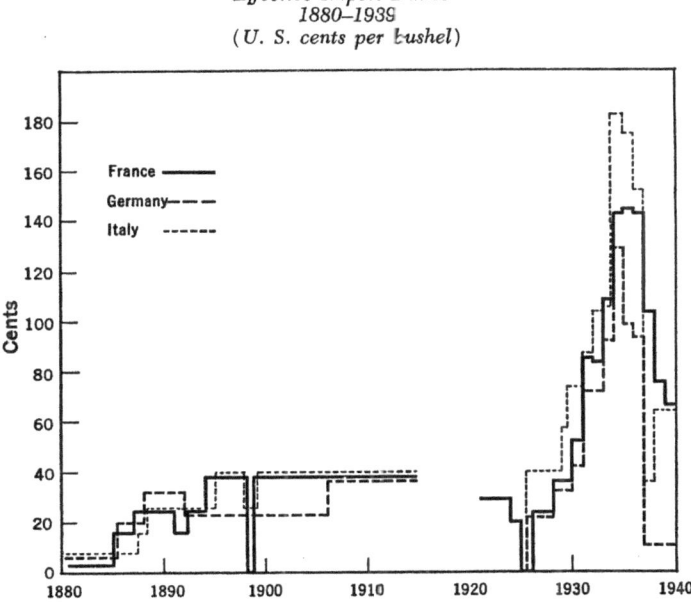

Figure 11 Tariff Levels on Wheat in France, Germany and Italy.

Source: Reprinted by permission of the publisher from: Wilfred Malenbaum, *The World Wheat Economy, 1885–1939*, Harvard University Press, Cambridge, Mass., p. 162. Copyright © 1953 by the President and Fellows of Harvard College. Copyright © renewed 1981 by Wilfred Malenbaum.

International Consultation

McDougall's experience of international consultation had begun late in 1928 when he was appointed to the Economic Consultative Committee (ECC) established by the International Economic Conference called by the League of Nations in 1927. At first he saw himself as spokesman for the developing industrial economies of the British dominions, countering the attempts of established European powers to use the League of Nations to lower trade barriers. He also sought to have the work of both the League and the International Institute of Agriculture serve the needs of his imperial vision, in ways less threatening to dominions' commercial policies. The crisis in agricultural prices soon led him to focus on the difficulties of 'overseas' producers of cereal crops on a large scale and on the problem of low wheat prices crippling the international economy as a whole.

Through the ECC, McDougall had encountered concerns about the International Institute of Agriculture (IIA), which had been established in 1905 under the auspices of the Italian Government to collect agricultural statistics and technical and economic information. McDougall found IIA publications unsatisfactory and was interested in discussions about relations between the Institute and the League, which had produced a more impressive range of statistics. He was invited to join a new IIA committee on agricultural economics, and persuaded Bruce that the League must concern itself with agriculture and therefore work to reform the Institute.[14]

He was also a member of the ECC's agriculture subcommittee, which, in 1929, recommended setting up a body of agricultural experts to advise the League on questions of commercial and economic policy affecting agriculture. The recommendation was opposed by Italy; McDougall's compromise proposal, that the advisory body be considered a temporary expedient, was unacceptable to agricultural countries like France. In subsequent discussion between the Institute and the League, it was agreed to establish a list of experts familiar with general problems of the agricultural economy. The experts could be involved 'according to the nature of each problem to be studied' in consultations between Geneva and Rome'.[15] The list, comprising experts connected either with the League or with the IIA, included McDougall.

Thus, in January 1930 he was part of a group summoned by the League to consider general agricultural problems, their allocation between the League and the Institute and, in particular, the problem of cereals. The group spent much time analysing the complicated causes of the agricultural depression, but a draft report did not win unanimous approval. McDougall, ever the eager draftsman, spent an evening rewriting passages, with the intention of submitting it to the group the next day. A misunderstanding led to his redraft being sent to all members, who took it as a substitute for the disputed draft. Some were happy

---

14 *LFSSA*, *215*, 27 February 1929, pp. 749–51; NAA, M111, Bruce to McDougall, 30 April 1929.
15 NAA, A11583, Imperial Conference 1930, 'EE(30)' Series—Memoranda—Papers Nos EE(30)1–EE(30)64, Imperial Conference Paper EE(30)6.

to accept it, others wanted changes, Britain and Czechoslovakia objected.[16] McDougall was doubtless embarrassed and probably learned something of the extra difficulties of operating in the international sphere.

His amended draft is worth consideration nevertheless, as it shows important new ideas emerging and older ones developing. He approved emphasis in the original on a principle enunciated in 1927—'the interdependence of all branches of economic activity'—and he agreed that 'agricultural prosperity is a first condition for industrial development and profitable trade between the nations'. The original draft listed problems of 'immediate urgency'. He accepted the first, point (a), that increasing production brought about by the application of science to agriculture had the potential to create severe international competition, with high prices for consumers but low returns for producers. Point (c) urged assistance by further study of agricultural credit, including producers' and consumers' cooperatives. McDougall added to that list government financial assistance, means to regulate price fluctuations, and reductions of labour costs through mechanisation.

His treatment of point (b) is significant. The original dealt with the threat to traditional markets from 'the uncontrolled overproduction of…certain countries peculiarly adapted to the cultivation of certain crops'—that is, the burgeoning cereal exports of the New World. McDougall's version reads, in part: 'the experts are convinced that steps must be taken to secure *a more rational system of production throughout the world* and also to ensure the spread of the idea of orderly marketing into the arena of international trade' (emphasis added). A first step must be 'a better supply of information…prepared and published by an international authority' and based upon 'full and reliable' national statistics 'as far as possible…compiled upon a comparable basis'. He listed desirable categories of figures beyond those already provided: the trend of demand, quantities of stocks held, tendencies in certain countries to substitute some products for others; early information on prospective crop yields, statistics of animal stocks and information as to average costs of production. The last would be challenging, but would enable administrators to 'form an estimate of the comparative advantage enjoyed by their country' in regard to specific crops. To this section McDougall added a point new for him:

> …the agricultural experts believe that one most important method of improving the position of agriculture throughout the world would be through the gradual raising of the standard of living of the peoples

---

[16] Archives of the League of Nations, Geneva [hereinafter LN], 1928–32, 10D 17138/12676. Note in particular, McDougall to Stoppani, 5 February, and reply, 11 February 1930.

of Asia and Africa. This would lead to a larger demand on the more nutritious and palatable foodstuffs and would directly or indirectly assist agriculture in all countries.[17]

This document shows that by early 1930, after a brief apprenticeship in international organisations, McDougall had achieved a considerable understanding of the scope and complexity of the problems of international agriculture. In terms of the development of his thinking, the essentials of his most important idea were already there. The 'marriage of health and agriculture' would depend upon both increasing consumption of 'more nutritious and palatable foods' and rationalising agriculture.

Later in 1930 the 'experts' were asked to supply an account of some 10 or 20 pages on various aspects of the 'situation in their respective countries', with a view to possible publication. McDougall protested, asserting the importance of a universal perspective, questioning the value of 'a series of uncorrelated statements' by a panel comprising 17 European representatives and five overseas states, without important agricultural producers like the British and Dutch colonial areas, China, Brazil, New Zealand and South Africa. It was essential to have 'the fullest information upon agricultural conditions and prospects in Russia in view of the increasingly significant and novel part which Russia is likely to play in the world situation of the near future'. The committee could be open to a charge of giving 'but superficial consideration to a most fundamental problem in economics and of taking an extremely narrow and localised view of what is a world international problem'.[18]

McDougall objected to a statement in a later draft report to the League Council that 'the majority of experts believe that restrictions of sown areas will be imposed by hard realities in every country'. He argued that in France, Italy and Germany 'wheat is now sheltered behind extremely formidable tariffs [so that] the hard realities of the situation will not have any substantial effect upon the individual wheat grower' who in France would receive twice the price received by his counterparts in Canada, Argentina or Australia. He wanted the paragraph altered to point out that 'hard realities' would cause restriction of production in 'the very countries where the cost of production is lowest'.[19] His protests apparently brought about a new draft, which, he was told, was 'in no small degree the result of your inspiration'.[20]

---

17 Ibid. The printed draft, with McDougall's handwritten additions and alterations, is dated 8 January 1930.
18 LN, 1928–32, 10D 16685/12676, McDougall to Stoppani, 28 November 1930.
19 Ibid., 10D 26670/2016, McDougall to Secretary, LEC, 13 and 15 March 1931.
20 Ibid., 10D 30344/22556, draft sent to McDougall for comment on 1 August 1931.

McDougall had been the spokesman for the overseas empire on imperial bodies, now he saw himself as, and was becoming in effect, the spokesman at the League for the world beyond Europe:

> At present I am really worried over the tendency of the League of Nations to concentrate its economic work on purely European problems and I think I should be particularly well qualified to see that the Second Committee of the Assembly took due note of the growth of this undesirable tendency.[21]

British Agricultural Policy

In Britain, meanwhile, agricultural policy had become a matter for intense political discussion by 1929. In opposition, the Conservative Research Department under Neville Chamberlain tackled agricultural policy, attempting 'to shape the still rather vague concept of Imperial rationalisation into a definitive planned form'. Wheat production at home could be rationalised by subsidies to efficient British producers. The import market might be regulated by a quota system, which would have the advantages of providing a lever in trade negotiations with other countries, of avoiding food taxes and of allowing wheat imports from Argentina, thought necessary to avoid retaliatory measures against British investments there.[22] Chamberlain's policy was supported by many business and empire lobbyists, though not by Beaverbrook, who had transformed his empire free trade campaign into one of support for British agriculture and cheap food.[23] In negotiations surrounding the 1930 Imperial Conference, it became clear that British millers would support a quota scheme only if they were offered protection against imported flour. Agreement to such protection would risk the Conservatives being drawn 'into conflict with importers, consumers and the Dominions'.[24]

The Labour Government settled, uneasily, on support for state trading in grains through import boards and bulk purchase.[25] Late in 1930 the Conservatives supported in principle the Labour Agricultural Marketing Bill for elected commodity boards with the power to buy and sell at fixed prices. The Conservatives also embarked on a new strategy of gaining industry support by offering measures of agricultural protection, including safeguarding against unfair competition and long-term contracts, as a *quid pro quo* for organisation.[26]

---

21 NAA/CSIR, A9778, M14/31/5, McDougall to Rivett, 4 June 1931.
22 Cooper, *British Agricultural Policy*, pp. 94–9.
23 Ibid., pp. 100–4, 113–24.
24 Ibid., pp. 106–8.
25 Ibid., p. 103.
26 Ibid., pp. 127–39.

The new National (all-party) Government formed in August 1931 'sanctioned a relatively high degree of state involvement in agricultural policies'. Government was involved in agricultural rationalisation through 'independent' boards with discretionary powers, marking 'a turning point in British farming'.[27]

McDougall had sufficient contacts to keep him informed of these political manoeuvres; he may well have contributed to them in private discussion. It is not surprising, therefore, that in the early 1930s he extended his earlier ideas on rationalisation of imperial industry and science to the complex and controversial problem of agriculture, aiming, like Chamberlain, to achieve 'maximum efficiency…having regard to the natural advantages of constituent parts for supply of particular commodities'. His memoranda written in advance of the imperial conferences of 1930 and 1932 took into account 'two vital factors in British prosperity': cheap food and raw materials, and a prosperous agriculture in both Britain and the wider empire.[28] In 'An Empire Agricultural Policy', McDougall advocated a variety of 'instruments of rationalisation': preferential tariffs, bulk purchase, import controls, anti-dumping legislation and 'indirect measures'—credit facilities, research, market intelligence, consumer education and 'discouragement of uneconomic agricultural development'. Selection of method would be based on the needs of particular industries established in studies conducted before implementation. Efficiency should be a condition of all assistance. Wheat was an exception, though, and the principle of natural advantage should not apply. Although British wheat cost much more to produce than Canadian or Australian, it should be supported for 'reasons of national psychology and agricultural employment', perhaps by a guaranteed price or by a compulsory minimum percentage of English wheat for milling. As total empire wheat production was three times the British requirement, the solution was not a tariff, but it could be 'inter-imperial marketing co-operation'.[29]

Agricultural rationalisation along these lines would stimulate rural industries overseas, encourage animal industries in Britain and protect against dumping of agricultural goods on the British market. Empire farmers would gain greater purchasing power, enabling them to buy more British manufactures. McDougall expected Britain to use its dominance as a market for primary products to bargain for a predominant share of dominions' industrial imports. He did not think that dominions' industrial development would necessarily reduce their demand for British manufactures. He expected any tariff-related increases in food costs in Britain to be negligible in comparison with their benefits to British industry and employment.[30]

---

27 Ibid., pp. 143–5.
28 NLA, MS6890/4/2, 'An Empire Agricultural Policy', 3 April 1930, p. 1.
29 Ibid., pp. 10–23.
30 Ibid., pp. 11, 15, 27–9.

'An Empire Agricultural Policy' spelled out a program for the 1930 Imperial Conference to initiate. It represented the 'sheltered markets' idea in its final form, concentrated and rationalised, taking account of all that McDougall had learned and experienced since 1925, of politics, world markets, specific commodities and of patterns of supply and demand. The memorandum was filled with facts and figures. But on the hard questions of conflicting imperial interests it offered little.

McDougall wrote 'Home and Empire Agriculture' late in 1931 for Sir John Gilmour, Minister of Agriculture in the new National Government. Its argument was similar to the one he had made in 1930: there was little competition between home and empire; policies benefiting one would help the other; and an empire agricultural policy should not increase the costs of British manufacturing. Most of the memorandum discussed details of the application of a tariff policy to particular agricultural commodities. He noted at the end that he had given little attention to Indian or colonial agriculture; these could be helped by applied research, market intelligence, agricultural education and effective transport.[31] A second paper written for Gilmour considered means of combating the 'bogey of over-production' in a time of depressed commodity prices and 'revolutionary' changes in agricultural production. McDougall predicted increased demand, once confidence was restored, for animal products, fruits, vegetables and raw materials. He suggested efforts to increase milk consumption, in the interests of public health, 'as has been overwhelmingly demonstrated by the large-scale experimental feeding of Scottish school children'. The decline in demand was not caused by oversupply, he argued, but by financial maladjustment.[32]

These memoranda were carefully considered and well received in the Ministry of Agriculture and Fisheries (MAF). Admired for the extent and thoroughness of their coverage, they seemed sympathetic to the British point of view. H. L. French concluded, 'if all the representatives from the various Dominions were as reasonable as Mr McDougall and equally capable of understanding the problems of home as distinct from Dominion agriculture, the task of reaching a satisfactory agreement at Ottawa would be greatly reduced'. H. E. Dale acknowledged, despite his differences on some points, that the memoranda 'are…thoroughly worth reading…they disclose a very reasonable attitude of mind on the part of Mr McDougall'.[33] They also show his ideas developing towards the 'marriage of health and agriculture'.

---

31   UKNA, MAF 40/17, 'Home and Empire Agriculture', 14 December, with covering letter to Gilmour, 15 December 1931.
32   Ibid., 'Home and Empire Agricultural Policy II', 19 January 1932.
33   Ibid., Minute, H. L. French, 4 March; H. E. Dale, 6 April 1932.

Overseas Exporters

Governments in overseas exporting countries were attempting to deal with the wheat problem. In Australia there had been three successive poor seasons by 1930, as overseas markets slowed and prices fell. Much arable land was in fallow, the balance of payments was in deficit, the flow of capital had stopped, and interest and debt repayments were at risk. Federal and State governments resorted to the traditional remedy for low commodity prices: urging farmers to 'Grow More Wheat'. Farmers responded eagerly, encouraged by a federal Bill introduced in April 1930 to guarantee 4 shillings (48 pence) per bushel for wheat delivered to railway sidings, and to establish federal and State marketing boards and a compulsory wheat pool. Many borrowed to plant more. Wheat acreage increased that sowing season by 21 per cent, yielding a record 213.6 million bushels—an increase of 30 per cent.[34]

The proposed price guarantee evaporated in political wrangling. State Governments in Western and South Australia opposed a provision that States share in the liability, the Federal Opposition objected to the implied socialisation and the Senate rejected the Wheat Bill. State legislation for compulsory wheat pools in Victoria and New South Wales failed to gain the consent of sufficient growers. As world wheat prices continued to fall, the Scullin Government rejected a suggested flour tax because it would increase the price of bread. A Wheat Advance Bill hastily passed in December 1930 guaranteed 2/6d per bushel (30 pence) for the 1930 crop, without compulsory pools or State liability. The Commonwealth Bank refused to advance the funds, arguing it could not be guaranteed against loss. A new Wheat Bill in March 1931—proposing a £3.5 million bounty on the 1930 crop with another £2.5 million in aid to farmers, to be financed by fiduciary notes redeemable by a later loan—was rejected by the Senate, as was a Wheat Marketing Bill providing compulsory pooling and higher-priced flour. In September 1931, attempts to compensate farmers for losses from the 'Grow More Wheat' campaign were officially abandoned. Farmers received approximately 20 pence per bushel, instead of the 48 they had expected on planting. Many (at one estimate, 20 000) were forced off their land; country businesses were bankrupted; it was 'one of the greatest disasters in Australian economic history'.[35]

Australia was not alone. Robert Holland suggests that the wheat problem in Canada altered the imperial relationship. Canadian growers had borrowed heavily to fund the expansion and mechanisation underlying wartime increases in production, encouraged by federal control of prices and marketing through

---

34 Edgars Dunsdorfs, *The Australian Wheat Growing Industry 1788–1948*, Melbourne University Press, Melbourne, 1956, pp. 267–9.
35 Ibid., pp. 270–5.

a Wheat Board. When prices fell in 1920, the board was dismantled and many were ruined. The outcome was establishment of producers' wheat pools in the mid 1920s. The concept, writes Holland, was 'developed in the hothouse world of American agrarian radicalism'; operation of the pools in Canada threatened 'the fabric of imperial collaboration between Liverpool and Medicine Hat, apparently seamless for so long'. The pools aimed to keep prices high by limiting supplies. *Wheat*, written by agricultural economists W. W. Swanson and P. C. Armstrong, and published in Toronto in 1930, supported the pools. The authors predicted that organisation and storage infrastructure would eventually develop in Argentina and Australia, to redress what was perceived as an imbalance of power between importers and widely dispersed sellers. Buyers, meanwhile, relied on the mounting costs of maintaining wheat surpluses to work in their favour. British millers, led by James V. Rank, began to combine and rationalise while conducting 'a bitter publicity drive against "the monopolistic" pools'.[36]

Canadian wheat pools hedged their advances to growers through bank loans. Rapid price decline in 1930 during 'the great Soviet wheat dump of 1930–31' led to fears that banks would demand liquidation of stocks in the open market. Prairie premiers, 'all heading farmer-dominated parties', guaranteed the funds, on the pools' undertaking to repay any government losses. Prices continued to fall; the Provinces sought federal guarantees, first refused by the new Bennett Government, but later agreed without parliamentary approval. Attempts to reform the pools failed: in August 1931 all three provincial pools withdrew from the central selling agency, owing their governments a total of $24.3 million. By November 1932, 75 million bushels of wheat were in store but federally supported measures to withhold wheat from sale failed to hold the price above 50c per bushel; bank loans taken out by the pools in 1929 had posited a price of $2. By 1933, the gross value of agricultural production in the prairie Provinces had fallen from approximately $1 billion dollars in 1926–27 to $163 million. The west became an area of net emigration: 'Pitiful caravans of destitute families, carrying the remnants of their belongings, trekked away from shattered dreams.'[37] In the United States, large unsold stocks were held by the Federal Farm Board, an agency of a government facing a record budgetary deficit.[38]

Holland suggests that failure to gain agreement on assistance for wheat at the 1930 Imperial Conference, which considered and dismissed the idea of an imperial tariff, destroyed the last hope for the pools in Canada, and that, 'through prairie spectacles, British capitalists—with the connivance of a Labour government—had sought an unholy alliance between Moscow and Buenos Aires

---

36 Holland, 'Imperial Collaboration', pp. 109–13.
37 Ibid.; Morriss, *Chosen Instrument*, pp. 34, 38–43.
38 McDougall, 'The International Wheat Situation', p. 531.

to smash the price level on which their community aspirations were based'. A complacent British view that 'the small prairie producer would have to accept an East European standard of living to survive' was shaken in 1931 by fears that agricultural nations might default. Default by Australia would threaten the strength of the pound. Consideration was therefore given to proposals for a wheat quota and to some weakening of attachment to the 'cheap food' policy.[39] Millers continued their opposition, but gave reluctant consent to offers of tariff preference for wheat and limitation of Russian supplies (both 'practically worthless') at the Ottawa Conference. Canada's Prime Minister, R. B. Bennett, rejected other British suggestions for a wheat quota, or preference, and resented attempts to protect British-milled flour. It became apparent that Britain could offer no real help, and the possibility of 'some great Imperial trade agreement' vanished. A disillusioned Bennett, convinced that 'the British were unprepared to enter into some equitable partnership', declared that he must seek 'the alternative economic strategy open to Canada—some accommodation with the United States. As he made plain to the U.K. delegation…"The Americans could not have treated us worse"'. A small preference was agreed, but 'was excoriated by rural representatives in the Dominion Parliament as a gross irrelevance'. It proved ineffective: dominions' share of British wheat imports actually fell in the following year.[40]

The Imperial Economic Committee, its terms of reference enlarged by the 1930 Imperial Conference, determined to tackle the question as an imperial problem. Its preliminary discussions show that McDougall had given much thought to the question: he noted the large amount of information available, but warned there could be no intra-imperial solution. The IEC should consider it seriously, nevertheless, as 'there is a most urgent need for the producers themselves to have a clear appreciation of the world position', particularly the alarming possibility of the Soviet Union returning to its prewar level of output. He suggested encouragement of mixed farming to avoid 'the extreme economic risk incurred in any country by too much reliance on wheat'.[41]

Soon afterwards McDougall's article 'The Dominance of Wheat', published in *The Times* on 25 February 1931 and ascribed to 'a correspondent', demonstrated his new worldwide perspective and recognition of a problem of great complexity: low prices; European production encouraged by artificially high tariff levels; technical and biological advances likely to nullify any restriction of crop acreage; subdivision of estates and political conservatism in traditional European exporters; a relatively inelastic market and declining consumption in Europe;

---

39 Holland, 'Imperial Collaboration', pp. 114–15.
40 Ibid., p. 117; Cooper, *British Agricultural Policy*, pp.148–51.
41 UKNA, DO35/201/5, 8115/19, 'Summary of discussion on foodstuffs remaining for inquiry under Sub-Head 1 of the New Terms of Reference', Meeting of the Imperial Economic Committee, 2 December 1930, IEC/109 (12.12.30).

and Canada's exportable surplus doubling the total European market. Again he suggested that diversification offered solutions such as the Danish conversion of cheap dumped cereals into stockfeed for more profitable animal products, and the more elastic markets for dairy products, meats, fruit and vegetables, textile fibres and oilseeds. As always, study of world production and consumption figures was an essential prerequisite to the framing of solutions. With some prescience, he predicted problems for international bodies attempting to regulate production or to control national policies.

International Solutions

Between 1930 and 1933, twenty international conferences dealt wholly or in part with the problem of wheat prices. Some involved regional groupings. European growers laid claim to their traditional markets, urging an inter-European preferential system.[42]

Forty-seven countries, many with weighty delegations, were represented at the International Wheat Conference convened by the IIA in Rome in March 1931. Canada sent its London High Commissioner, three other delegates and two expert industry advisers. McDougall was on his own, but his energy and familiarity with the issue enabled him to make important contributions. Delegates considered two solutions raised at earlier European conferences: restriction of production and intra-European preferences. A consensus view was that European nations could not reduce acreage, but that overseas exporters would be forced to do so if wheat prices continued to be unprofitable.[43] McDougall described the debate thus:

> Those who advocated compulsory restrictions were unable to answer any one of three very simple questions. First, would the advocate's own country practice restriction? Secondly, if so, how would it be enforced in a peasant agriculture? Thirdly, what assurance could be given of Russian participation? The mere asking of these questions rendered any further discussion impossible.[44]

His own speech—pointing out that overseas exporters were the chief sufferers in the crisis, and proposing concentration on orderly marketing rather than restriction—was supported by Canada and aroused considerable interest. It led to private meetings, in conjunction with Canada, with representatives of Argentina, the USSR and finally with European exporters.[45] He reiterated

---

42 Malenbaum, *World Wheat Economy*, pp. 198, 200–1.
43 Ibid., p. 202.
44 McDougall, 'The International Wheat Situation', p. 529.
45 NAA, A981, Conference 366, McDougall to Scullin, 31 March 1931.

Australian objections to international bodies promoting sectional interests, and he succeeded, with India's support, in having the conference recommend leaving negotiation of intra-European preference to regular diplomatic channels.[46]

McDougall thought his greatest achievement at this meeting was persuading Russian delegates to cooperate. They apparently made the first approach, then McDougall and the Canadians spent several hours in discussion with them over two days. McDougall attributed the success to his 'complete…even a brutal frankness', admitting the British Empire could not solve the problem alone, stating that international cooperation was essential and suggesting that the USSR must make clear whether it would serve its economic ends by cooperating for higher prices or its political ends by selling its own wheat cheaply and deepening the depression. The Russians admitted the Soviet Government would welcome a plan for raising prices; for the next few years, money would be more important than fostering revolution. McDougall hinted that if it were possible to blame the USSR for failure of the talks, a boycott already imposed by France and Belgium might spread, but he assured them he was convinced of a genuine desire for cooperation.[47] Russian opposition was crucial in defeating European preference and quotas at the conference.[48]

While conference resolutions recognised the difficulties of reaching agreement on production control and did recommend crop diversification, the more important recommendations were to create an international body to determine quotas, fix prices or both, and to make arrangements between wheat exporters for orderly marketing of unsold stocks. For that purpose, a meeting of exporters was convened by Canada in London in May 1931.

McDougall expected that the May conference would be confined to short-term solutions for the immediate problems of unsold stocks and the 1931 harvest.[49] Even this goal proved unattainable. Although the United States and the USSR both attended, the conference only achieved establishment of a committee to develop proposals for a clearing house of market and other information. McDougall chaired a committee that favoured establishment of an international wheat organisation and recommended orderly marketing by means of a quota. But the United States would not support quotas, except on the condition of reduced production. It has been suggested that other exporters supported quotas only because they were confident of US refusal, and that there were inconsistencies and fallacies in their views.[50] Although the Canadians believed Britain would propose a quota at Ottawa, McDougall thought British opinion

---

46   Ibid., CP498/1, 430/AA/3.
47   Ibid.; NAA, A981, Conference 366, 'Notes on discussions with the Russian Delegation', 8 April 1931.
48   Malenbaum, *World Wheat Economy*, pp. 202–3.
49   McDougall, 'The International Wheat Situation', p. 531.
50   Malenbaum, *World Wheat Economy*, pp. 203–4.

was moving away from quotas because they could not raise prices. Australia certainly needed higher prices more than disposal of surplus stock, and the only solution lay in international action.[51]

The standing committee established by the May conference met in July, and, at US insistence, confined discussion to establishing a 'Clearing House of Information' to serve wheat-exporting countries. McDougall admitted that these discussions had not made 'any contribution of substantial importance' to growers' problems.[52] What was to have been the 'International Wheat Information Service' gained only three potential participants: Hungary, Romania and India.

The Wheat Conference of 1933

A full-scale Monetary and Economic Conference was held under League auspices (but not confined to League members) in London in July 1933. Its Preparatory Commission recommended 'special attention' to 'the production and export of wheat' and discussion of means to liberalise tariff barriers. Formally requesting that wheat be placed on the agenda, Argentina submitted that the world wheat surplus had doubled since 1927. Import barriers would not ease until prices rose, consumption could not increase while widespread unemployment persisted and the policy of accumulating stocks to raise prices had failed. The only solutions remaining were production and export controls. Although exporters had failed to agree in 1931, the worsening situation since then might have changed attitudes.[53]

The four principal exporters met in Geneva for preliminary discussions in May 1933. McDougall was instructed from Canberra that he should 'not be sympathetic' to the idea of crop limitation in the absence of any 'offsetting advantage not now apparent'. It would, inter alia, force Australia to surrender market advantage in Britain and in Asia; prevent expansion and development, which were 'the most economic means of ending unemployment'; threaten Australia's ability to meet external liabilities and to make use of lands already developed and thus lose the benefits of overseas borrowings; and it would advantage the holders of large North American stocks. But as declaration of 'a very definite attitude' at the outset could be 'injudicious', McDougall should await development of other representatives' views.[54]

---

51 NAA, A981, Conference 368, McDougall to Scullin, 28 May 1931.
52 Ibid., Conference 371, McDougall to Scullin, 15 July 1931.
53 Ibid., Conference 117, part 1, copies of letter from Agriculture Minister Antiono de Tomaso to Argentina's delegate Raoul Prebisch, 25 October 1932; submission to the Conference Secretariat on 11 December; letter from J. H. Thomas to Bruce, 5 January 1933, requesting Australian views.
54 Ibid., Conference 117, part 2, Lyons to Bruce, 3 May 1933.

The delegates were inclined to blame European agrarian protection for the problem, but agreed that emergency action to raise prices must be taken; one method could be voluntary limitation of acreage. Export quotas, disposal of surplus stocks and tariff modification were also discussed; a standing committee to monitor the situation was recommended; no agreement was reached on the desirable extent of limitation.[55] Australian State premiers, who were responsible for agriculture, roundly rejected any restriction scheme.[56] Bruce warned Prime Minister Joseph Lyons to keep an open mind: the new US President, Franklin D. Roosevelt, was known to support restriction of production and wheat producers' agreement might well encourage US support for other economic measures. League officials believed European tariff mitigation depended on reciprocal action by overseas exporters to limit outputs.[57]

The world conference began amid pessimism, which was to prove justified. Concurrent with it, but formally outside its scope, the four principal wheat exporters met in a series of discussions that came to assume pivotal importance for the success of the conference itself. The United States' Henry Morgenthau, then Head of the US Federal Farm Board and a close friend of Roosevelt, chaired discussions; Bruce represented Australia; Bennett led the Canadian delegation, and was eager for a solution to the wheat problem. It has been suggested that he favoured acreage reduction and that the absence of imperial aid forced Canada into seeking 'a slice of the New Deal's inflationary action', working, albeit in a subordinate position, with the Americans.[58]

Looming over discussions was the threat of the huge North American surplus and the possibility that US wheat might be dumped onto the international market. On 15 June, Bruce repeated an earlier request for instructions. Acting Prime Minister, J. G. Latham, replied that Australia was reluctant to surrender a tactical advantage in bargaining for removal of trade barriers against meat and dairy produce; there were practical difficulties in compulsory restriction and State governments opposed the idea.[59] Latham's position was supported by representations from wheat industry organisations and by a confidential cable from Toronto advising that Bennett's statements in London were simply politicking; informed opinion in Canada considered limitation impractical.[60]

By 21 June the United States had declared itself willing and indeed anxious to restrict acreage and limit exports in the coming season. Canada was at

---

55 Ibid., copy of Agreed Statement, submitted by McDougall to Bruce, 19 May 1933.
56 Ibid., Economic 21, letters from Premiers of South Australia, Victoria and New South Wales, 24, 26 and 27 May 1933.
57 Ibid., Conference 117, part 2, Bruce to Lyons, 30 May 1933.
58 Morriss, *Chosen Instrument*, p. 47; Holland, 'Imperial Collaboration', pp. 115–18.
59 NAA, A981, Conference 117, part 2, Bruce to Lyons, 15 June; Latham to Bruce, 16 June.
60 Ibid., Economic 21, Australian Trade Commissioner in Canada to Commonwealth Government, 21 June 1933.

first inclined to make any limitation conditional upon relaxation of import restrictions and tariffs, and, like Australia, was hampered by the difficulties of a divided jurisdiction. Fear of US dumping, Russian exports and a premature announcement of Provincial support helped force Canadian agreement.[61] Argentina was prepared to cooperate. Bruce cabled again. Wheat was becoming the pivotal point of the monetary conference and was governing the US attitude on wider issues. Australia could be in an embarrassing position if cooperation was not given soon. Without agreement, the United States might well flood the market, particularly Australia's new markets in the Far East. The alternative of export limitation was now being discussed as well.[62] Australian refusal might be held responsible for destroying any hope of international action to deal with the economic crisis as a whole. A record of discussion, dated 22 August, indicates that on 16 June Bruce explained that acreage restriction, if opposed by State Governments, would be unenforceable in Australia. Morgenthau then pointed out that the US Farm Relief Act would permit dumping of the US surplus overseas. At meetings on 19 and 21 June, it became clear that the United States and Canada were willing to restrict acreage, while the Argentine representative 'stated his conviction' that his country would accede to a general agreement. On 21 June, Bruce said he would be prepared to make two suggestions to Canberra: acreage restriction, which, he repeated, was 'difficult if not impossible'; and limitation of exports. The latter seems to have been a new suggestion introduced into the discussion by Bruce.[63]

Australian policy had to be reversed, and that quickly, by a federal government without power to ensure States' compliance. Latham was in Melbourne, Lyons in far north Queensland, ministers, State premiers and industry leaders scattered. Cables flew across the country. For a week the London talks hung in the balance. Lyons was persuaded not to consult industry organisations for lack of time, but he wanted the States to be consulted.[64] Premiers were to meet in Melbourne on Saturday, 1 July. Bruce cabled details about alternative schemes for crop limitation and export restriction and a further warning that Australian refusal would prevent negotiations with European importers. If there were any possibility of either scheme being introduced by Australia, an indication of tentative acceptance should be given immediately.[65] Lyons gave general agreement by telegram 'because impossible for us to be responsible for

---

61 *Documents on Canadian External Relations* [hereinafter *DCER*], Vol. 5, 1931–1935, ed. Alex I. Inglis, Department of External Affairs, Ottawa, 1973. Documents 696 and 697. Secretary of State for External Affairs [hereinafter SSEA], Ottawa, to High Commissioner, London, 27 May; SSEA to Canadian Minister in Washington, 1 June 1933, pp. 564–6.
62 NAA, A981, Conference 117, part 2, Bruce to Lyons, 21 June.
63 Ibid., 'Discussions on Wheat in London, May–August 1933'.
64 NAA, A981, Economic 21, Lyons to Latham, 23 June.
65 Ibid., Conference 117, part 2, Bruce to Lyons, 24 June.

breakdown', leaving details to Latham and other ministers.[66] By 27 June, Bruce felt compelled to make some statement, but his explanation of the difficulties of State and producer cooperation and of Australia's determination not to participate without concessions from European importers was countered by US and Canadian insistence that Europeans would not act without agreement by the overseas exporters. Bruce cabled again: the only way to prevent Australia being blamed for the deadlock was to make a statement accepting in principle some form of restriction.[67]

Ramsay MacDonald, Chairman of the Monetary and Economic Conference, cabled Lyons, pointing to the 'striking unanimity among countries who have hitherto held divergent views [and the] unique opportunity for agreed and effective action'.[68] MacDonald's cable was put before premiers who had already been advised that Australia was 'practically forced in her own interests to collaborate'. The Australian delegate could not 'continue to stand aloof while others work out a plan which might be less favourable to Australia'.[69] Messages had already been sent to Bruce and MacDonald accepting restriction in principle.[70] In London, on 30 June, the basis of a scheme to reduce wheat production by 15 per cent of the average crop for the years 1931–33 was agreed. Canada and the United States would restrict acreage; Argentina and Australia would accept an export quota. On 1 July, Australian premiers reluctantly agreed to limit wheat exports to 142 million bushels from the 1933–34 harvest and to 113 million (plus any unexpended quota from the previous year) in 1934–35.[71] The decision was thought so sensitive that Latham took the unusual course of telegraphing newspaper editors, inviting 'sympathetic consideration' of the view 'in the highest quarters' that agreement was essential to the success of the London conference. Leading articles seemed to show the message had served its purpose.[72]

The four exporters' agreement in principle was followed by informal, but intense and difficult discussions throughout July and August between the four and other exporters and importers. Danubian countries and the USSR agreed to limitation in principle; importers agreed not to encourage production and recognised that higher prices should mean lower tariffs. Although Bruce had assumed principal representation in these discussions, he relied heavily on

---

66 Ibid., Economic 21, Lyons to Latham, 26 June.
67 Ibid., Conference 117, part 2, Bruce to Commonwealth Government, 27 June.
68 Ibid., MacDonald to Lyons, 28 June.
69 Ibid., Economic 21, letter to all State Premiers from Latham, 26 June.
70 Ibid., Conference 117, part 2, Latham to Bruce and MacDonald, 29 June.
71 Ibid., Lyons to Bruce, 1 July.
72 Ibid., Economic 21, personal and confidential telegram, sent to editors through Commonwealth Investigation Branch, 2 July; Latham to NSW Premier, B. S. Stevens, who suggested the idea, 7 July 1933.

McDougall: 'if anything comes of the wheat proposal it will be very largely due to McDougall', wrote Orr in September. McDougall's secretary sent apology after apology to Rivett for McDougall's failure to attend to correspondence.[73]

The Wheat Advisory Committee

Twenty-two nations reached agreement late in August 1933 but there was to be no end to the travail. The agreement established a Wheat Advisory Committee, representing all principal exporters and importers, to monitor world supply. The committee had a delicate and demanding task. Quotas were based on crop estimates; exporters eyed each other with suspicion. Vagaries of weather meant unpredictability. The committee was hampered by lack of knowledge: crop estimates varied even where information was freely available, but there was none from the traditional bogey, the USSR. Nor was there information about US stocks in private hands, yet these could be released to the market in response to any slight price rise. Seasonal differences meant that planting in one hemisphere could not be calculated until accurate crop forecasts were available for the other.

McDougall represented Australia. He played a leading and characteristic role: asking questions, summarising, steering discussion, suggesting procedures, oiling the wheels.[74] The work began in an inauspicious atmosphere, with Canadian and US claims that Australian crop estimates were exaggerated, and Australian fears that subsidised US wheat was about to flood promising new markets in the Far East.[75] Producer confidence waned as prices continued to fall.[76]

The committee nevertheless assumed ambitions well beyond mere 'monitoring'. McDougall was member of one subcommittee to consider a minimum price scheme and of another to investigate ways of increasing consumption.[77] A draft scheme, largely drawn up by McDougall early in 1934, provided for scales of minimum prices, maintenance of domestic prices at levels to discourage low-priced exports and limits on export subsidies.[78] In commending the proposal to governments, the committee described the idea as 'indispensable for the restoration of confidence' necessary for any effective adjustment of supply

---

73   NAA/CSIR, A9778, M14/33/9, Orr to Rivett, 4 September; M14/33/3 and M14/33/9, M. Divine to Rivett, 22 and 29 June, 6 and 20 July, 24 August 1933.
74   NAA, A981, Conference 117, part 3, Minutes of the Committee for 27–28 November 1933.
75   Ibid., Conference 117, part 2. See in particular communications from Commonwealth Government to Bruce, 25 September, 11 and 19 October; Bruce to Lyons, 13 October; Bruce to US Ambassador Bingham, 20 October; US Embassy, London, to Bruce, 17 November; Bruce to Lyons, 23 November 1933.
76   Ibid., Conference 117, part 3, report by Andrew Cairns, minutes, pp. 27–8.
77   *DCER*, Vol. 5, Document 716, High Commissioner, London, to SSEA, Ottawa, 29 November 1933, pp. 578–9.
78   NAA, A981, Economic 22, McDougall to Lyons, 15 April 1934.

to demand.⁷⁹ The scheme as presented was viewed in the British Ministry of Agriculture as 'naïve to the point of absurdity'; the committee in fact seemed to be considering 'export monopolies under Government control' in each exporting country.⁸⁰ One official quoted a recent report that Australian State aid for wheat farmers was effectively 'keeping in production land incapable of producing payable wheat crops even in normal times', and added that minimum prices would do nothing for the problem of oversupply.⁸¹ President of the Board of Trade, Walter Runciman, and Minister of Agriculture, Walter Elliot, thought the proposals unworkable. Reluctant to offend the dominions and hopeful that 'Argentina will not play', they adopted a policy of 'wait and see'.⁸²

Argentina obliged, by requesting an extra 40 million bushels of quota. There were suspicions Argentine wheat was being dumped on world markets and that farmers were encouraged to plant more.⁸³ With inadequate storage facilities, Argentina believed it was justified, because neither the United States nor Canada had reduced sowing to the agreed level. Fearing an unauthorised 40 million bushels would create chaos, other exporters agreed to meet the request. McDougall believed Argentina might be persuaded by pressure from other signatories, particularly Britain. In Bruce's absence, he urged this course in Whitehall, and he persuaded Lyons to send a cable stressing the threat of disaster, not just to the agreement but also to 'the whole idea of international collaboration'. The Canadian High Commissioner acted similarly. Urged by Canberra and Ottawa, the British Government made diplomatic representations in Buenos Aires in cooperation with the United States.⁸⁴ In the background McDougall helped ensure US cooperation, which depended upon British action, by informing J. V. A. MacMurray, who was responsible for wheat negotiations in the absence of US Ambassador Bingham, as soon as British intentions were confirmed.⁸⁵ Acting as spokesman for the United States and Canada, McDougall also sought further help from British ministers to prevent collapse of the agreement. Further diplomatic moves failed to dissuade Argentina, which refused any commitment 'inconsistent with their placing upon the world's markets their entire stocks of wheat by the end of December 1934'.⁸⁶

In the event harvests in all except Argentina were well below expected levels, reduced by drought in North America and by poor weather and rising wool

---

79 Ibid., section of Committee's report to Governments, 16 April, attached to letter from McDougall to Lyons, 18 April 1934.
80 UKNA, MAF 40/144, memorandum by E. M. H. Lloyd, 1 March 1934.
81 Ibid., note by C. Houghton, 2 March 1934, quoting E. T. Crutchley, UK Trade Commissioner in Australia.
82 Ibid., D. Fergusson to J. R. C. Helmore, 4 May; Elliot to D. E. Vandepeer, 5 May; Vandepeer to Helmore, 7 May 1934.
83 NAA, A981, Economic 22, McDougall to Lyons, 9 and 14 April 1934.
84 Ibid., McDougall to Commonwealth Government, 26 April; Commonwealth Government to Dominions Secretary, No. 28, 30 April; Dominions Secretary to Australia and Canada, 3 May 1934.
85 Ibid., McDougall to Lyons, 3 May.
86 USNA, 561.311, F1 Advisory Committee/339, Andrew Cairns to McDougall, 25 May 1934.

prices in Australia. The Wheat Agreement failed to accomplish its main purpose of raising prices, which had changed very little by late 1935. The large carryovers that had been the root cause of the problem were eventually reduced not by actions internationally agreed but by weather conditions (see Table 4). Though some reduction in sown areas did occur through the 1930s, it is doubtful whether any of the four exporters would have complied fully had nature not taken a hand. No agreement was reached on export or acreage controls in discussions in March 1935 and the agreement was formally terminated in May, although the Advisory Committee continued as a consultative and liaison group until World War II.[87]

Table 4 Export Quotas Allocated under the Wheat Agreement 1933: Actual Exports Shown in Parentheses (in million bushels).

	1933–34		1934–35	
United States	47	(29.1)	84	(–3.9)
Canada	200	(194.4)	263	(164.9)
Argentina	110	(147.1)	154	(181.5)
Australia	105	(86.1)	150	(109.1)
Total exports: four exporters	462	(456.7)	651	(451.6)

Source: Edgars Dunsdorfs, *The Australian Wheat Growing Industry 1788–1948*, Melbourne University Press, Melbourne, 1956, p. 308.

Conclusion

McDougall had used every means he could to form international concern with the problem of wheat, and of agriculture generally, into a rational, informed and balanced debate, encompassing all aspects of the problem and leading to cooperative solutions. As at Ottawa, sectional interests prevailed. The failure of the wheat negotiations, and the failure of the general Economic Conference of 1933, left both Bruce and McDougall disheartened and disillusioned. Bruce later spoke of the 'disastrous and negative' policy of restricting production, 'a policy that I described at the time as one of desperation and danger'.[88]

McDougall had once called himself an 'incorrigible optimist'. He was to prove in a remarkably short time the truth of that assertion. As the following chapter shows, his 'inventive mind' was able to create hope out of despair.

---

87 Malenbaum, *World Wheat Economy*, pp. 208–11.
88 Cecil Edwards, *Bruce of Melbourne: Man of Two Worlds*, Heinemann, London, 1965, p. 240.

6. 'The Marriage of Health and Agriculture'

Summary

The year 1935 marked the pivotal point in McDougall's thinking. This chapter begins with an account of the development of scientific knowledge of human nutritional needs in the early twentieth century, particularly in understanding the importance of vitamin-rich 'protective foods' such as dairy products, vegetables and fruit. Surveys following establishment of dietary standards in the 1920s showed that substantial proportions of populations, even in advanced countries, could not afford a diet adequate for health. Concerned doctors and scientists, including John Boyd Orr, sought action but were met with resistance from the British Government, which feared the cost. Measures to increase milk consumption were taken in Britain, but on grounds of assistance to the agricultural economy.

In 1934 McDougall began to make the connection between poor nutrition and restrictive agricultural policies such as extreme protection. His suggestion for a campaign by scientists for adequate diets was enthusiastically supported by Orr, and by Bruce. McDougall completed his seminal memorandum, 'The Agriculture and the Health Problems', in early January 1935. This memorandum analysed the causes of agricultural problems, argued the benefits of improved nutrition and called for a reorientation of agricultural policy, meaning that industrial countries should concentrate on producing more of the protective foods, benefiting both consumers and producers. The 'nutrition initiative' was taken up by the International Labour Office at Geneva and the League of Nations; both passed resolutions calling for further investigation and action. The League established a 'Mixed Committee' of lay and specialist members, including McDougall, to report on nutrition to the 1936 Assembly and produced a four-volume report on the subject. Although immediate changes in government policies were negligible, nutrition became a subject of wider public debate and education.

Starvation in the Midst of Plenty

Nineteenth-century understanding of human nutrition was rudimentary. By that century's end, orthodox theory proposed five food groups: protein for growth, fats and carbohydrates for energy, salts for bones, blood and thyroid function, and water. Scientific studies were chiefly concerned with food as fuel and requirements in terms of quantity. Despite centuries-old experience that scurvy could be prevented by citrus fruit, and more recent experiments showing cod-liver oil prevented rickets, the connection between components of diet and health had scarcely been made.[1] In the early twentieth century, understanding of the complex nature of foodstuffs developed. Amino acids were identified and shown to be present in differing proportions and types in various proteins. In 1901, after prolonged research in the Dutch East Indies, it was shown that brown rice would protect against beri-beri. Scientists began to suspect the presence of mysterious substances helping to prevent some diseases. In 1912 Cambridge chemical physiologist F. Gowland Hopkins published his conclusion that 'minimal quantitative factors' were likely to play a part in prevention of scurvy, while Polish chemist Casimir Funk, then at London's Lister Institute, predicted many diseases would prove to be due to absence from diets of 'special substances...which we call vitamines'.[2] Vitamin C was identified in the following decade, and its effectiveness was proved in the aftermath of World War I when Dr Henriette Chick found that babies in food-starved Vienna did not develop scurvy when given orange juice.[3] In the 1920s, vitamins A and D were identified. Pellagra was shown to be linked to diets based largely on maize, although the precise deficiency, vitamin B_3 (niacin), was not yet understood.[4]

Dietary standards incorporating these early discoveries were devised in the 1920s. With a standard established, it was possible to calculate the cost of an adequate diet. Such exercises invariably showed that cost to be beyond the reach of the very poor, while other studies continued to demonstrate the benefits of improved diets. Nutrition policy clashed with economic and commercial policies. Tension developed between measures to raise agricultural prices and efforts to improve health.

In 1928 the French Government asked the League of Nations Health Organization to include nutrition in its program, and studies were undertaken in several countries, including one on the effect of the economic crisis on health. Conferences in Rome and Berlin considered both the effects of the crisis and the fundamental problem of establishing an adequate diet. In 1932 Dr George McGonigle, Medical Officer in Stockton-on-Tees, showed that tenants who were moved into improved housing and obliged to pay higher rents reduced

---

1 J. C. Drummond and Anne Wilbraham, *The Englishman's Food: A History of Five Centuries of the English Diet*, Jonathon Cape, London, 1939, pp. 429–30.
2 Ibid., pp. 498–508; Barbara Griggs, *The Food Factor: An Account of the Nutrition Revolution*, Penguin, London, 1986, pp. 36–8.
3 Ibid., pp. 80–4; Drummond and Wilbraham, *The Englishman's Food*, pp. 519, 533–4.
4 Griggs, *The Food Factor*, pp. 41–50.

their expenditure on food and suffered a higher incidence of tuberculosis.[5] In 1934 the League commissioned its officers Drs E. Burnet and W. R. Aykroyd to investigate nutrition policies in Britain, France, the United States, Denmark, Sweden, Norway and the USSR. The Burnet–Aykroyd Report of June 1935 acknowledged the relationship between nutrition and the economy, as well as the need to educate both the medical profession and the public on principles of good nutrition.[6]

Edward Mellanby, best known for his work demonstrating vitamin D deficiency as the chief cause of rickets, published works on nutrition, chaired international conferences and was a member of many other committees in Britain, including the Medical Research Council, of which he was Secretary from 1933 to 1949.[7] From 1927 Mellanby had led fruitless efforts to put the new knowledge of nutrition into practice. The Ministry of Health delayed appointment of a committee for the purpose and then stacked it with holders of opposing views to avoid the ministry having to 'come down on the side of those with a positive policy'. After Mrs Annie Weaving starved herself to death in order to feed her seven children, the *Week-End Review* commissioned the 'Hungry England' inquiry early in 1933. The Health Ministry committee accepted the inquiry's findings but could not agree on action. The Nutrition Committee of the British Medical Association (BMA) published its own report in November 1933, which estimated higher needs, at a cost of 5/11d for a working man, compared with 5 shillings for the 'Hungry England' estimate. The ministry, subjected to widespread press criticism, condemned the BMA report as 'a Labour Party tract' and a 'stunt' by McGonigle. It feared being involved in 'a far-reaching economic issue, which is most important to avoid—an issue which might easily affect wages, cost of food, doles etc'. A joint committee of BMA and ministry nutrition committees agreed on a compromise sliding scale of minimum needs; the ministry committee chairman resigned on the issue and the ministry continued to deny any connection between low incomes and mortality.[8] The ministry response demonstrates the difficulty of persuading governments to tackle the question of cost inherent in the problem of nutrition.

Industry promotion could persuade consumers of the nutritive value of particular foods, increasing sales and farm incomes. In the United States, the dairy industry promoted the nutritional value of milk. Biochemist Elmer McCollum, who had identified vitamin A, wrote magazine articles at the request of the industry,

---

[5] 'Poverty, Nutrition and the Public Health', published in *Proceedings of the Royal Society of Medicine*. See Susan McLaurin, 'McGonigle, George Cuthbert Mura (1889–1939)', *ODNB*, 2004, <http://www.oxforddnb.com.virtual.anu.edu.au/view/article/60875> [accessed 20 March 2005].
[6] LN, 'Nutrition and Public Health', *Quarterly Bulletin of the Health Organization*, Vol. IV, no. 2, June 1935. Also published separately as Offprint No. 2 of the *Quarterly Bulletin*.
[7] B. S. Platt, 'Mellanby, Sir Edward (1884–1955)', rev. Michael Bevan, *ODNB*, 2004, <http://www.oxforddnb.com.virtual.anu.edu.au/view/article/34980> [accessed 22 November 2005].
[8] Mayhew, 'The 1930s Nutrition Controversy', pp. 447–52.

promoting the value of milk as 'the greatest of all protective foods'. McCollum joined the US Advisory Committee on Alimentation, established in 1918, and contributed popular articles on nutrition to *McCall's Magazine* for more than 20 years. US milk consumption doubled between 1918 and 1928, as did demand for lettuces; orange consumption tripled.[9]

Milk led the way in demonstrating the importance of nutrition and in state-sponsored schemes for greater consumption in Britain. During the 1920s a number of experiments, including EMB-funded work in Scotland, appeared to show improved growth rates in children. The intention, according to Orr's own account, was to 'show the nutritive value of milk and increase the sale of the more profitable liquid milk'.[10] The results influenced nutrition pioneer Gowland Hopkins and George Newman, Chief Medical Officer of both the Ministry of Health and the Board of Education. In the 1930s more surveys were supported by the Milk Marketing Board and were complemented by animal experiments—some conducted at the Rowett Institute. 'The connecting theme in much of this milk-feeding experimentation was the influence of John Boyd Orr.' Orr became convinced 'that an increased consumption of milk was important, especially for children'. Peter J. Atkins suggests that the research agenda may have been influenced by 'Orr's wish to use the results as a political lever. The underlying science was at times questionable but the publicity was positive and Orr's political and networking skills are not in doubt.' By 1933 more than one million British children received school milk, either as charitable feeding under the *Education Acts* or from voluntary milk clubs, increasing, from 1927, with encouragement from the National Milk Publicity Council. The milk club scheme provided a new market for some 9 million gallons per annum. It was taken over by the Milk Marketing Board in 1934 and funded by the Ministry of Agriculture for £1 million over two years. Half the elementary school population participated by 1939.[11]

The purpose, however, was to meet the economic needs of agriculture, rather than the health needs of the population. Milk and dairy products were second only to 'fatstock' in the gross agricultural product of England and Wales; the milk sector alone represented 27 per cent of the total. Some three-quarters of members of the National Farmers' Union were milk producers to some extent; with the 'generally distressed nature of agriculture' in the 1920s and 1930s, farmers shifted towards milk production. Walter Elliot, in opposition in 1929, managed passage of a Private Member's Bill that became the *Education (Scotland)*

---

9   Griggs, *The Food Factor*, pp. 76–7. See also E. Melanie DuPuis, *Nature's Perfect Food: How Milk Became America's Drink*, New York University Press, New York, 2002.
10 Orr, *As I Recall*, p. 114.
11 Peter J. Atkins, 'Fattening Children or Fattening Farmers? School Milk in Britain, 1921–1941', *Economic History Review*, Vol. LVIII, no. 1, 2005. Atkins notes that a 1987 thesis by E. C. Petty questions the reliability and methodology of the surveys, which were also controversial at the time.

Act of 1930, enabling Scottish authorities to provide subsidised or free school milk: 'His main argument was one of efficiency: pupils would get more from their education and taxpayers would have a better return for their investment in education.'[12] As Minister of Agriculture from September 1932, Elliot introduced a second *Agricultural Marketing Act* and oversaw establishment of the Milk Marketing Board in September 1933. The *Marketing Acts* gave boards of producers 'powers of compulsion in the creation of a monopoly', in accordance with Elliot's view on the necessity for planning, since the 'fundamental failure' of agrarian capitalism must be countered by 'restructuring along corporatist lines with guidance and support from the State'.[13] By then Elliot had moved away from the imperial solution, opposing dominion preference and pressing for greater self-sufficiency in dairy products. But the Ottawa decisions 'derailed' that policy. Preference for imports of empire dairy products meant a milk surplus, which, without those imports, 'would have found a profitable outlet in manufacturing'. In 1934, therefore, Elliot introduced a *Milk Act* to 'compensate farmers and manufacturers for the sacrifice of their interests at Ottawa'. As part of that legislation, he proposed a 'Milk in Schools' scheme, inspired, like his 1930 Act, by Orr's experiments. He won Cabinet support, however, 'with an agricultural and economic rather than a nutritional or public health rhetoric'.[14]

Although the *Milk Act* subsidy for butter and cheese manufacture was three times the budget of school milk, Treasury feared the school measure might create a precedent for welfare food and clothing. While social reformers lobbied departments of health, agriculture and education in favour of poverty alleviation and provision of a basic diet, demanding wider provision of free milk (then available only on certification by the local medical officer), the Board of Education continued to claim that child malnutrition had been exaggerated. The 'progressive dairy legislation' of 1934 was possible 'only because it appealed to the self-interest of the farming and milk trade lobbies'. Weight is added to this contention by the failure of attempts to have only pasteurised milk supplied and to make available more supplies of free or cheaper milk. Such measures were 'perceived as threatening the prosperity of dairy farmers and milk traders'.[15]

The primary aim of the 1933 *Agricultural Marketing Act* was to establish 'an equilibrium of price levels'. Market-supply provisions of the Act promised 'to effectively control the "glut" of production which was perceived as swamping agricultural prices'. There was some backlash against Elliot's orderly marketing

---

12 Ibid.
13 Ibid. Atkins is here drawing upon Fenton Cooper, *British Agricultural Policy 1912–36: A Study in Conservative Politics*, Manchester University Press, Manchester, 1989, Chapter IX: 'Walter Elliot and the Corporatist Challenge', pp. 160–80.
14 Ibid.
15 Ibid.

approach, on the grounds of rising prices for domestic staples.[16] Orr records his own opposition to the idea of producers' boards, since they would increase the cost of essential foods. As a member of a committee appointed to reorganise the Milk Marketing Board when it was threatened with bankruptcy because of oversupply of liquid milk, he pleaded, alone and in vain, for a plan to increase milk production. Orr was happier with later outcomes: funding to publicise the benefits of milk consumption, expansion of the school milk scheme and, in wartime, milk supplies distributed on the basis of need.[17]

In the mid 1930s, Orr worked on broad population surveys of diet and income, covering 1152 families divided into six income groups. Results were published in his book *Food, Health and Income*.[18] The work was supported by Elliot and through him by the Agricultural Marketing Boards.[19] When the survey results were in final stages of preparation for publication, however, 'a very senior civil servant' ordered civil servants to cease working on them. Orr and McGonigle were threatened with deregistration if they went ahead with a broadcast on the survey. McGonigle, still a practising doctor, withdrew; Orr broadcast alone. Fearing repercussions for civil servants, including staff at the Rowett Institute, Orr determined to publish under his own name, though much of the work had been done by others. In 1935, lest publication be somehow prevented, he delivered a public lecture, to which his friend, journalist Ritchie Calder, invited a large press contingent. 'Sensational reports' of the findings followed. Harold Macmillan agreed to publish; despite official displeasure, *Food, Health and Income* first appeared in 1936 and went through three editions. Orr recalled:

> The Establishment put up the strongest possible resistance to informing the public of what the true position was regarding under-nourishment among their fellow-citizens…The thought of mothers and children suffering malnutrition because they were too poor to afford the more expensive health foods was intolerable. At that time these foods were so abundant that the government was taking measures to reduce production so as to increase retail prices. I had never lost my hatred and anger against unnecessary poverty. Now, as a scientist, I had the chance of giving expression to that anger.[20]

---

16 Cooper, *British Agricultural Policy*, pp. 167–71, 174–5.
17 Orr, *As I Recall*, pp. 112–13.
18 John Boyd Orr, *Food, Health and Income: A Report on a Survey of Adequacy of Diet in Relation to Income*, second edn, Macmillan, London, 1937.
19 Orr, *As I Recall*, p. 115.
20 Ibid., pp. 115–18.

Countering 'intolerable pessimism'

For McDougall and Bruce, the failures in 1933 to solve the international monetary crisis, and more particularly the wheat crisis, were to prove a profound disappointment but, ultimately, a turning point. It took little more than a year for McDougall to devise what they could believe to be a way out of the morass.

Writing in *The Times*, McDougall condemned—in words Bruce had used at the Economic Conference, but which could easily have been his own—the 'doctrine of intolerable pessimism' regarding restriction of production as a method of restoring prosperity to a poverty-stricken world.[21] Groping for an answer to 'intransigent economic nationalism', he suggested an expanded imperial grouping with tariff protection, based on complementary trade. It could include close economic partners like Denmark and Argentina, and politically associated units like Egypt and Iraq. Free trade would not do, because it could not take account of national aspirations: France must grow wheat, Australia must have factories. He acknowledged difficulties in applying an Ottawa-type efficiency requirement to agriculture, but maintained that world trade would be strangled without some limitation on agricultural development in industrial nations.[22] 'McDougall must be a rather disillusioned man nowadays, but he obviously intends to fight to the last', commented Rivett.[23]

The summer months of 1934 were spent in tough negotiations under the International Wheat Agreement and with the British Government on the Ottawa meat quota. McDougall's partnership with Bruce strengthened and settled into the mode of operation they would follow for the next decade: McDougall writing memoranda and suggesting courses of action, each then lobbying at appropriate levels for agreed purposes. Bruce had 'further improved in the width of his outlook and the sanity of his point of view'.

> He and I are doing our utmost in all sorts of ways to convince the British Ministers and other influential persons of the extremely grave danger of the undue expansion of British agriculture to the detriment of Great Britain's imports from the dominions. I think we are having some success.

They were urging appointment of a committee to study the problem of agricultural costs being higher in Britain than in the dominions, its report to be considered by all empire governments.[24] At the same time, McDougall recorded

---

21 'Aid for British Agriculture', *The Times*, 13 March 1934—the second of a series of three, published anonymously on consecutive days, under the general title 'Factory and Farm'; copies sent to Rivett in NAA/CSIR, A10666, [4].
22 Third article, 'Planning for Reciprocity', *The Times*, 14 March. It was supported by a leading article of the same date.
23 NAA/CSIR, A9778, M14/34/3, Rivett to Casey, 24 March 1934.
24 Ibid., M14/34/8, McDougall to Rivett, 19 July 1934.

disappointment with Walter Elliot's alignment with British producers' interests. He had urged him to aim to be 'a good Minister rather than...a superlatively good Minister of Agriculture'. He thought Elliot took the point, but 'I have not seen any sign of a change of heart'. Elliot had, nevertheless, been 'quite enthusiastic' about *The Times* articles, which McDougall showed him before publication since they implied 'serious criticism' of his policy.[25]

Much of 1934 thus constituted a period of uneasy transition for McDougall, mopping up after Ottawa and admitting the existence of worldwide problems needing a global solution. Although his work on the Wheat Agreement had required him to think and act well beyond the British imperial framework, his thinking was still limited to a considerable extent by the empire idea. The period was also marked by a new personal relationship, possibly begun in Rome, where McDougall had attended his first meeting of the governing body of the International Institute of Agriculture in April. In May, Elspeth Huxley wrote delightedly that the friendship with 'the Italian Bavarian ski champion...sounds excellent'. In August, just before attending the League Assembly, McDougall spent two weeks in a French mountain village near Geneva. Elspeth hoped 'the mountain air is toning up the tissues, and the wide views soothing the spirit, and the companionship satisfying both the soul and the body'. She was later delighted to learn that 'the walking tour was a success...It sounds fun and you deserve a bit of affection...Hurray for the signora!' She continued to respond gleefully from travels in the United States as the relationship with 'Beattie' progressed.[26] The correspondence provides no further information about the lady, and the relationship seems to have lasted only a year or two.

McDougall wrote to Elspeth in a depressed mood from Geneva, prompting her to reply: 'you took a very defeatist point of view about the achievements of the past five years and evidently the atmosphere of Geneva is damping the optimist which is such an essential and characteristic part of your makeup.'[27] But perhaps the 'satisfaction of soul and body' proved a catalyst to creative thinking. In October, happily reunited with 'Beattie' in Rome, McDougall first publicly made the connection between nutrition and agricultural policy. At the IIA, he welcomed a British speech calling for increased consumption, but added that decreasing consumption was a consequence of prices raised by extreme protection. At the same time, national conscience in many countries was being aroused by new understanding of the importance of diet to human welfare:

> ...it was to be hoped that the nations would soon realise how much it was to their own advantage to enable their people to obtain adequate supplies

---

25 Ibid., A10666, [4], McDougall to Rivett, 13 March 1934.
26 NLA, MS6890/2/5, 3 May, 23 August, 22 and 30 September 1934.
27 Ibid., 22 September 1934.

of such valuable human foods as dairy products, meat and fruit…A wide application of the doctrine of autar[k]y means the persistence of poverty in the midst of plenty with disastrous consequences to the whole world.[28]

He had suggested in 1932 that decline in demand for foods such as fruit and vegetables was not a result of glut, but of 'financial maladjustment'. Earlier he had also suggested encouraging greater consumption of liquid milk, for health reasons. The leap in 1934 was to connect poor diets with restrictive agricultural policies. Although the connection was made, the only remedies suggested in that speech were avoidance of extreme protection and reduction of internal costs. The IIA could help by studying factors involved in production costs; these costs would include protection.

Figure 12 F. L. McDougall at the Institute of Agriculture, Rome. McDougall is seated in the centre of the back row.

Source: E. McDougall.

Once back in London, McDougall decided that a direct attack on high agrarian protection could not succeed in the short term and he moved with speed to devise a campaign to tackle the problem in another way: 'Possibly a fruitful method

---

28 NAA/CSIR, A9778, M14/34/12, 'Summary of Speech made by Mr F. L. McDougall in the general discussion which arose on Item 8 of the Agenda', sent to Rivett, 8 November 1934.

would be a rally of scientific and medical opinion in favour of a reasonable standard of nutrition.'[29] Bruce was persuaded. Orr was already 'giving as much publicity as possible to the urgent need for a food policy based on health needs' in speeches, writings and broadcasts.[30] On 2 November, McDougall consulted him about

> a point of view which, I think, should be pushed forward in every possible way, both nationally and internationally. In brief, I think the time has come when, in the interest of world recovery and the prosperity of the peoples of the British Empire, it is necessary to secure the greatest possible notice for modern ideas about nutrition.

Orr should launch the campaign with 'a letter of great weight' to *The Times*, warning of statisticians' predictions of a stationary population in Britain unless infant and child mortality rates were reduced, and stressing the beneficial effect on those rates of adequate diet. The objective would be to increase demand, first in Britain and then in other countries, for milk and other dairy products, fruit and probably meat.[31] The suggested diet, and its emphasis on dairy foods in particular, reflects the Eurocentric nature of most early research in nutrition. I am grateful to Barry Higman for pointing out that promotion in pioneering nutrition campaigns of a diet based on European preferences has had significant consequences for world food production.

Orr replied eagerly on 6 November. A similar idea had been 'fermenting in my innards...during the last month or two'. Had it not been for his friendship with Elliot, he would have taken 'a running kick' at the milk-marketing scheme: 'as near national lunacy as makes no difference.' There was 'a great expandable market for dairy products, fruit and meat': production was not yet at a level equivalent to the amounts needed for the general health of the community. 'The present schemes tend far too much to maintain the *status quo* on the unproved assumption that there is a glut.' He suggested an imperial scheme encouraging consumption of liquid milk 'until it reached a pint per head per day, which would double the output of dairy farming in this country', yet still allow New Zealand and Australia to send butter until 'every household in the country is using butter and has stopped the use of margarine', which was then cheap, of poor quality and thought to lack the nutritional benefits of butter.[32]

McDougall was already mapping out a broader campaign

---

29   NAA/CSIR, A9778, M14/34/12, McDougall to Rivett, 8 November 1934.
30   Orr, *As I Recall*, p. 115.
31   Copy sent to Rivett in NAA/CSIR, A9778, M14/34/12.
32   Ibid.

to try to get scientific and medical opinion to express views, to which wide publicity could be given, and then to try and get the Health Section of the League of Nations working and to get the International Institute of Agriculture at Rome to obtain and publish comparative figures on the consumption per caput of milk, dairy products, fruit and meat, in Western Europe, United States, Great Britain, Scandinavia and the Dominions.[33]

He attended a meeting of the Wheat Advisory Committee in Budapest from mid November until early December, but had a rough draft of a memorandum by 17 December. A second draft followed on 9 January 1935, a third on 24 January.[34] Copies were distributed to contacts from the Wheat Advisory Committee: a letter to Norman Robertson of the External Affairs Department in Ottawa suggested showing it to O. D. Skelton. Another was sent to Mordecai Ezekiel of the US Agricultural Adjustment Administration, from whom Elspeth Huxley had learned something of malnutrition figures in the United States, with a request that it be shown to Agriculture Secretary, Henry Wallace, as a personal, not an official Australian, paper.[35]

'The Agricultural and the Health Problems'

The final form of the memorandum runs to fourteen and a half double-spaced, typed pages.[36] It has a strong claim to being the most significant memorandum McDougall ever wrote. Yet there are some oddities. The title, with its two definite articles, reads awkwardly in English, perhaps less so in the League's other official language, French. Titles of earlier drafts varied, some having no articles; the final could be an infelicitous result of re-translation. But there seems little doubt that McDougall himself intended the final form. He was a meticulous draftsman and editor: League files contain many examples of his minor corrections. He was unlikely to have let any version pass without scrutiny. When the memorandum was published by FAO in 1956, McDougall, although officially retired, still spent time in his office and was involved at least to some extent in the publication.[37] It must be concluded that the two articles are there for emphasis: there are two problems and McDougall is proposing a solution for both of them. There is evidence, however, that McDougall was not entirely happy with the title. A letter to his mother, dated simply 7 May, but presumably

---

33   Ibid., Orr to McDougall, 6 November; McDougall to Rivett, 8 November 1934.
34   NAA/CSIR, A9778/4, M1 N43, related correspondence.
35   NAC, MG30 E163, Vol. 4, File 7(.-4), Mgl–Nar, Wheat Advisory Committee, Wheat Export 1938, both letters dated 26 February 1935.
36   NLA, MS6890/4/4 and MS6890/4/6. The version published in more lavish layout by FAO is 17 pages long.
37   Ibid., MS6890/5/4, *The McDougall Memoranda*, FAO, Rome, 1956.

written in 1935, describes a discussion between McDougall, Earl De la Warr, Parliamentary Secretary at the Ministry of Agriculture, and David Lubbock, Orr's son-in law and assistant on the nutrition campaign, concerning 'a good title or phrase for our campaign'. He wrote: '"Agriculture and Health" all right for the initiated, "Increased consumption" sounds rather poor. The best phrase we could think of was "More Abundant Life". I don't wholly like it. Can you give me any suggestions?'[38]

McDougall's memoranda were working documents and generally pedestrian in style. Arresting statements were reserved for speeches. The opening sentence of this memorandum is both awkward and dreary: 'Although the diagnosticians of the continuance of the depression vary in the emphasis they place upon the causes retarding world recovery, there would be general agreement today that the problems facing world agriculture are major factors.' From then on, happily, the memorandum becomes clear and persuasive. It includes some of the best exposition to be found in McDougall's work—for example: 'It would argue a bankruptcy of statesmanship if it should prove impossible to bring together a great unsatisfied need for highly nutritious food and the immense potential production of modern agriculture.'[39]

Considering the breadth of its subject matter, the memorandum is remarkably brief and simple. In 16 points, it summarises arguments McDougall had been making, in some cases for many years. It begins with the causes of agricultural problems: science-based increases in production levels, high protectionism and increasing US reliance on exports for economic recovery. The placement of the agricultural problems first, as in the title, is significant: this is the primary anxiety of statesmen everywhere.

Nutrition, however, takes up nine pages—the greater part of the text. McDougall makes it clear that 'the very important question' of nutrition in Asian countries is not dealt with; responsibility for a solution lies in the first place with 'the more advanced countries' where quantities of food consumed are generally adequate, but where substantial sections of populations—perhaps 10 million in the United Kingdom—suffer nutritional deficiencies of

> first and foremost milk, then fresh fruit and vegetables, dairy products including eggs and, for large masses of the poorer sections…meat. These commodities are exactly those which are causing the greatest concern to the Governments of the more advanced countries, whether normally importing or exporting agricultural products.[40]

---

38   NLA, MS6890/1/10.
39   *The McDougall Memoranda*, p. 6. Page numbers here and following refer to the printed version, not an earlier typed copy.
40   Ibid., pp. 5–7.

McDougall's suggestions for tackling the problem rest largely on education and efforts to make the protective foods more affordable for the poor: 'If a start was made in the United Kingdom and the United States of America through the mobilisation of scientific and medical opinion and the securing of adequate publicity, the results might be expected rapidly to spread to other countries.' An education effort could be assisted by international bodies—the League, the International Labor Organization (ILO) and the IIA—providing useful facts and figures. He suggests 'special prominence' be given at the 1935 League General Assembly, to be followed by an international conference on agriculture and health, in which 'the United States might be expected to play an important part'.[41]

He devotes two pages to the benefits of improved nutrition, which include better health and improved birthrates. 'The immense potential demand' of the US population could avoid the possibility of the world's greatest creditor nation expanding her own agricultural exports. Improved nutrition as administration policy would mean no exports of meat or dairy products and more grain required for stockfeed.[42] The advantage of his proposal for industrial countries is 'an opportunity for a reorientation of agricultural policy; they could continue to protect efficient production but increased demand would allow of a simultaneous increase of import trade with beneficial effects upon their industrial export industries'. For agricultural countries, the policy means more export opportunities.[43]

Propaganda alone will not solve the problem; the poor need help to afford the right foods; state intervention may be necessary. McDougall suggests that President Roosevelt's championing of 'the cause of social justice' has eased the way, as has the likely realisation that the alternative is 'not a swing back to the right, but wild schemes of confiscation from the extreme left'. This fear has brought conservative leaders in other countries to discussion of ways to secure 'a more equitable distribution of wealth'.[44]

Methods of state intervention are discussed briefly. 'Social provision', food as relief, is applied as free distribution in some countries, more to relieve pressing economic problems than for health. McDougall suggests universal provision of nutritious school lunches as a method less likely to lower the morale of recipients. Free education is widely accepted, but 'education given to undernourished children is frequently wasted'. The cost would be largely offset by savings in medical costs and of subsidies presently made to agriculture.[45]

---

41   Ibid., pp. 7–8.
42   Ibid., pp. 8–10.
43   Ibid., p. 17.
44   Ibid., pp. 10–11.
45   Ibid., pp. 12–13.

Distribution costs are high in urban areas of Britain and the United States, but lower in countries where there are retail markets without extra charges for delivery, packaging and credit. McDougall wonders about replacing milk deliveries with central milk pumps, rather like petrol stations with facilities added to sterilise containers. Distribution of essential commodities like milk, and preliminary processing like flour milling, might be treated as public utilities rather than profit-making enterprises—this from a member of a prominent milling family![46] He demonstrates the high profitability of such enterprises in a note showing the rise in share prices of four British companies: United Dairy, Tate & Lyall, Spillers and Ranks, which have at least doubled since 1928; some have trebled or quadrupled.[47]

Finally the effect of high protection policies is shown: butter prices in Paris are more than three times those in Britain. While conceding that agricultural policies in Europe have to maintain a high population of peasant farmers, McDougall calls for a simultaneous commitment to farmers' efficiency: 'An aroused public conscience might well insist that the price the farmer must pay for protection is reasonably high efficiency and prices must not be unduly restrictive on the consumption of the poorer classes.'[48]

'Mr McDougall's action in the Assembly'

In March 1935, McDougall travelled to Geneva and Rome. Orr sent ahead of him an introduction to W. R. Aykroyd, joint author of the League Health Committee Report. H. R. Cummings of the League organisation in Britain also wrote to Aykroyd, foreshadowing both Australian and British action on the issue.[49] The ground was being carefully prepared. Aykroyd read the memorandum, noting that his own report had marshalled evidence for McDougall's proposition that subnormal nutrition was the rule, rather than the exception, in even the best-fed countries. 'On purely health grounds [the Health Committee] must be in sympathy with Mr Bruce and his colleagues.' But more than sympathy was needed: 'By taking a definite standpoint as regards diet [health organisations] can indirectly influence economic policy.'[50]

Copies of the memorandum spread. Dr Ludwik Rajchmann, Director of the League Health Secretariat, requested a dozen more copies, and found Grace

---

46 Ibid., pp. 13–14.
47 This note is not included in FAO's publication *The McDougall Memoranda*. A version in NLA (MS6890/4/4) does include it; that in NLA (MS6890/4/6) does not.
48 NLA, MS6890/5/4, printed version, pp. 14–16.
49 LN, 1933–46, 8A, 17356/2133, Orr to Aykroyd, 11 March; 1933–46, 8A, 17008/8855, Cummings to Aykroyd, 8 March 1935.
50 Ibid., note by Aykroyd, 15 March 1935.

Abbott, head of the US ILO delegation, most interested.[51] A draft ILO resolution was in preparation by 9 April; cooperation between both organisations was rapid and effective. League officials approved the ILO draft, adding a clause affirming cooperation with the League Economic Organization and the IIA. Correspondence referred to 'Mr McDougall's action in the Assembly' or 'l'initiative McDougall'.[52]

At the Nineteenth ILO Conference in June, the report of the Director-General referred to problems of nutrition; statements were made by Miss Abbott and by Australian and New Zealand representatives. Australia's Sir Frederick Stewart moved a resolution calling for continued investigation into the relationship between nutrition and productive capacity, and for cooperation, particularly on the social aspects of nutrition, with the League and the International Institute of Agriculture. It was 'warmly supported'.[53]

Australia, with much to gain, supported the move at every level. The Prime Minister and his wife visited Europe briefly that summer. In Geneva on 23 June, Lyons toured the new League building, then nearing completion, visited the ILO and lunched with senior officials. In Rome on 28 June, he had 'a very nice chat' through an interpreter in an audience with the Pope, but 'found it difficult to discuss anything fully'. With the Cardinal Secretary of State, he discussed 'the matter referred to in a memorandum re agriculture and production—the need for greater consumption of good food'.[54] The memorandum was presumably McDougall's. Lyons explained to the Cardinal his hope that the Pope might issue a pronouncement

> drawing attention to the present mad policy of trade barriers [as a result of which] large numbers of people were suffering from the effects of insufficient and unsuitable food [although] there was actually overproduction in the world of all that they required. It was useless for any politician to draw attention to this, as he would be said to have an axe to grind, but if the Holy Father would speak to the world those in power might grapple with the evil.

The Cardinal Secretary requested a memorandum for consideration of the question.[55] The memorandum was subsequently presented. McDougall was at

---

51 LN, 1933–46, 8A, 2133/2133, Rajchmann to McDougall, 11 June 1935.
52 Ibid., 17649/2133.
53 *Commonwealth Parliamentary Papers*: 1934–35–36–37, Vol. II, no. 216, 'ILO: Report of Australian Delegation to 19th Session 1935'.
54 NAA, A461/10, 748/1/361, part 2, extract of an undated private note from F. J. McKenna of the Prime Minister's Department to Australia's External Affairs Liaison Officer in London, Keith Officer. Confidential copy sent to W. R. Hodgson in Canberra.
55 Australian Joint Copying Project, CO35/853, reel 5334, report from British Legation to the Holy See to Sir Samuel Hoare, 29 June 1935. According to this report, Lyons did seek to raise the issue with the Pope, who also requested a memorandum.

first 'a little horrified' to learn it had been issued to the press, but later amused by the 'vivid, not to say sentimental' tone of a leading article on the subject in the London Catholic weekly, the *Table*, and, in general, delighted by 'the interest in the Agriculture and Health movement in this country'. Rivett responded that the idea was 'getting some push from Canberra' and that the Director-General of Health had been instructed to pay close attention to nutrition.[56]

McDougall would have worked closely with Bruce on his speeches for the League Assembly; it was their habit to work even as they travelled on the overnight train from Paris to Geneva.[57] Inclusion of nutrition on the agenda of the Second Committee (Technical Subjects) was co-sponsored by Australia with Argentina, Austria, Britain, Canada, Chile, Denmark, France, Italy, New Zealand, Poland and Sweden.[58]

Figure 13 Assembly of the League of Nations, 1934.

Source: E. McDougall.

On 11 September, Bruce introduced it in Plenary Session, ignoring accepted practice that introductory speeches be general in character. He suggested

---

56 NAA/CSIR, A10666, 30[4], McDougall to Rivett, 4 and 11 October; Rivett to McDougall, 31 October 1935.
57 Interview, Professor J. G. Starke, April 1988.
58 LN, *Official Journal*, 1935, Records of 16th Assembly, Plenary Meetings, Third Plenary Meeting, 11 September, p. 42.

first that the dominating anxiety of the Abyssinian crisis—then threatening the principle of collective action to maintain peace and already referred to the League Council—should be avoided lest the problem be exacerbated. Instead, in a speech far more compelling than McDougall's memorandum, he recalled that the League's responsibilities extended to economic, humanitarian and social issues, and that the question of nutrition had been placed on the agenda of the Second Committee. National health, he said, was the basis of the wellbeing of every nation. Sanitation and infectious disease control had already been the subjects of government regulation to the incalculable benefit of populations. Now science, as shown in the League's own Burnet–Aykroyd Report, provided new means of controlling disease and misery with adequate nutrition. The problem was to make the more expensive protective foods widely available. Yet ministers of agriculture were wrestling with problems of apparent overproduction:

> Increased yields are being regretted and abundance is often officially deplored. At the same time Ministers of Health and their official and medical advisers are realising, more and more, that public health demands an increased consumption of many of the very products about which the Departments of Agriculture are so unhappy.
>
> Millions of pounds are being spent annually in subsidies, bonuses and other forms of assistance to agriculture. Side by side with that expenditure millions of pounds are being devoted to combating disease. Is it not possible to marry health and agriculture and, by so doing, make a great step in the improvement of national health and, at the same time, an appreciable contribution to the solution of the agricultural problem.[59]

Introducing a draft resolution in Second Committee, Bruce argued the suggestion was neither idealistic nor impracticable. Economic, political and social sciences had lagged behind the physical and other sciences. Even in the United States it was admitted that six million children were malnourished. A first objective might be elimination of malnutrition in 'the richest and most developed countries', with improvement for the 'teeming millions' in undeveloped countries 'a long-range objective'. Action should not wait until the economic situation improved; progress must be gradual but should start immediately. He listed some practical measures and 'submitted that action to bring about the greater consumption of health-giving foodstuffs would not only be beneficial to public health and efficiency but would bring about greater demand, larger production, increased and freer trade, and better financial conditions'.[60] Bruce would later emphasise

---

59 Ibid., Fourth Plenary Meeting, 11 September, pp. 51–3. Bruce's speech was also published in *The McDougall Memoranda*, pp. 18–24.
60 Ibid., Records of 16th Assembly, Minutes of the Second Committee, 19 September, pp. 15–17.

that the nutrition case was not based on 'high moral principles, but purely on economics', quoting 'an American' who told Orr that 'this idea of the marriage of food and agriculture is the starting-handle of the whole economic recovery of the world'. Bruce added: 'That point has to be stressed, because it's drifted away from hard economic fact and is now mainly governed by humanitarian considerations.'[61]

Delegates appeared to turn with relief from a problem they seemed unable to solve, in Abyssinia, to action they could take. Seventeen delegations joined in, spreading the committee debate over three days, 'to the surprise even of Geneva idealists'.[62] Britain's Earl de la Warr spoke of 'a challenge to statesmanship no one could afford to ignore': while Britain could not retreat from 'the policy of planning the market' undertaken in recent years, 'such planning should be for the purpose of expansion rather than restriction'.[63] Denmark's Christiani endorsed Bruce's 'masterly speech'.[64] Gautier of France 'agreed entirely', spoke of the need to improve the purchasing power of the agriculturalist, and urged the importance of education, 'since mankind was confronted with conditions of existence which were entirely new and called for education of a new kind. Hundreds of millions of persons were living in a constant state of semi-starvation.'[65] Italy's Fera admitted that his country had adopted measures of restriction, but was 'prepared to co-operate most loyally' with the governments that shared her determination to remove 'the obstacles standing in the way of the natural interchange of goods'.[66]

Some delegates offered examples of measures already taken in their own countries in provision of milk and other foods to children and the needy; all supported the need for further inquiry. The Second Committee appointed a drafting committee, including McDougall, to prepare the final text of a resolution, which it approved on 25 September. On 27 September the Assembly resolved that governments should examine the practical means of securing better nutrition and requested the Council

> (1) To invite the Health Organisation of the League of Nations to continue and extend its work on nutrition in relation to public health;
>
> (2) To instruct the technical organisations of the League of Nations, in consultation with the International Labour Office and the International

---

61 Edwards, *Bruce of Melbourne*, p. 416.
62 Hudson, *Australia and the League of Nations*, p. 177.
63 LN, *Official Journal*, 1935, Minutes of the Second Committee, 19 September, p. 17.
64 Ibid., 20 September, p. 18.
65 Ibid., p. 23.
66 Ibid., 21 September, p. 25.

Institute of Agriculture, to collect, summarise and publish information on the measures taken in all countries for securing improved nutrition and,

(3) To appoint a Committee, including agricultural, economic and health experts, instructed to submit a general report on the whole question, in its health and economic aspects, to the next Assembly, after taking into consideration, *inter alia*, the progress of the work carried out in accordance with paragraphs (1) and (2) above.[67]

The Mixed Committee

Action was rapid. The Health Committee appointed the Technical Commission, which met in November and produced a *Report on the Physiological Bases of Nutrition*.[68] That same month the League Secretary-General requested member governments to submit information on action taken to improve nutrition. The League Council, of which Bruce was a member, acted immediately, rather than delay action until its next session in January, to establish a 'Mixed Committee' of specialist and lay members to report to the 1936 Assembly.[69] McDougall reported encouraging press attention in Britain: there had been supportive articles in *The Times*, the *Economist*, the *Statist*, the *Spectator* and *Nature*.[70] He wrote to every expert and interested person he could think of, seeking letters to the press worldwide.

Both McDougall and Bruce worked hard to establish the Mixed Committee. McDougall suggested to League officials that decisions about membership should pay some regard to geographical representation; it should include 'the best available men' from the United States, the overseas dominions, France, Scandinavia, Central Europe, the Mediterranean and South America. There should be representatives of the technical committee on nutrition, the ILO and a British chairman. Some members should have wide knowledge of world agriculture, others should understand commercial policy, yet others should be economists.[71] By 18 December the membership was largely in place, composed as McDougall had suggested, and including McDougall himself.

---

67 Ibid., Plenary Meetings, 27 September, p. 125. The text of the resolution was published in *The McDougall Memoranda*, pp. 24–5.
68 LN, Series of League of Nations Publications, III. Health 1935 III. 6. A final version was published in June 1936, as P.1936 II.B.4.
69 Ibid., *Official Journal*, 89th Session of the Council, 28 November 1935, item 3650.
70 Ibid., 1933–46, 8A, 50/20905/19868, McDougall to Stoppani, 9 November 1935. Articles had been published in the *Economist* on 31 August, 14 and 28 September; the *Statist* on 5 October; and in *Nature* on 19 and 26 October 1935.
71 Ibid., 1933–46, 8A, 50/20095/20095, McDougall to A. Loveday, 10 October 1935.

Bruce was responsible for finding the British chairman. He sounded out Austen Chamberlain. Winston Churchill was interested and would have taken the role, had he not been preoccupied with defence problems. He did offer to make 'a big speech on the subject' at any appropriate occasion. Bruce really wanted Lord Astor, but had to wait until 18 January to see him in person. Astor, Chairman of the Milk-in-Schools Advisory Committee and former Parliamentary Secretary to both the Ministry of Food and the Ministry of Agriculture, readily agreed.[72]

The Mixed Committee was expected to report quickly. Members with disparate backgrounds would have little information beyond their own specialties and the purely scientific work of the health experts. Responses to the questionnaire to governments could not be expected before the committee began meeting in February 1936. Therefore it lay with the few persons who had some concrete ideas on the subject, foremost among them McDougall (*'vous figurez au premier rang'*) and the League Secretariat, to research and direct the committee along productive lines. The head of the League's Economic Relations Section, Paul Stoppani, suggested McDougall, *'qui êtes dans une certaine mesure le créateur de cette enterprise'*, discuss a plan of action with the Secretariat.[73] McDougall took with him to Geneva a draft of a preparatory document. Discussion of the document on 9 December centred on ways of avoiding minor factual difficulties and impolitic references and of giving the document more impact. By 14 December a shorter and simpler document had been prepared, in line with the ideas of the Secretary-General on the question. A further revision was taken to London by Godfrey Lloyd of the League Economic Section later in December.[74]

The League published a four-volume report on nutrition in 1936. Volume II was the *Report on the Physiological Bases of Nutrition*; Volume III, *Nutrition in Various Countries*; and Volume IV, *Statistics of Food Production, Consumption and Prices*.[75] Volume I was an interim report[76] of the Mixed Committee, which had decided at its first meeting that 'it would be quite impossible for it to cover the immense field which its terms of reference required' in time to present a comprehensive report that year. In 98 pages it did cover knowledge of nutrition and its relation to the health of various groups in populations, a survey by the ILO of workers' health and social and labour legislation, and the relationship between nutrition, agriculture and the economics of farming. Its 15 recommendations included: continued study of and education on the principles of nutrition; cooperation and exchange of information; consideration of means of meeting the nutritional

---

72 Ibid., 1933–46, 8A, 50/20905/19868, McDougall to Godfrey Lloyd, 14 December 1935.
73 Ibid., Stoppani to McDougall, 28 November 1935.
74 Ibid., notes of meeting on 9 December; Draft with annotations dated 14 December; letter from Stoppani to McDougall, 14 December 1935.
75 Ibid., 1936, II.B. 4–6.
76 *The Problem of Nutrition. Volume 1. Interim Report of the Mixed Committee on the Problem of Nutrition*, Series of League of Nations Publications, E. Economic and Financial, 1936, II.B.3.

needs of various sections of communities, especially the young and those on lower incomes; steps to improve and cheapen marketing and distribution of the protective foods; consideration of any need to modify general economic and commercial policies, including assistance to reorientation of agricultural production; coordination of the work of various nutritional authorities; and continuation and improvement of collection of national statistics on supply, consumption and adequacy of diets. It is a familiar list. McDougall had done his preparatory work well.

A final report of the Mixed Committee was never produced: after 1936 international events turned minds everywhere to more pressing problems. But the *Interim Report* achieved much: it was a mine of information and became a bestseller amongst League publications. It had put nutrition on the international agenda; national nutrition committees were established in many countries in response to the League initiative. Awareness of the science and importance of nutrition helped ensure that mistakes made in 1914–18 were not made in 1939–45 in countries that were able to avoid them; many British people, in particular, were healthier than they had been in peacetime.

The memorandum from which the *Interim Report* developed was a unique and inspired contribution to thinking on international commodity problems. It applied consumptionism to a problem of human health that could not be solved by any of the traditional policies. On the passing of the Assembly Resolution, Bruce, McDougall and de la Warr sent a telegram to Orr: 'Brother Orr, we have this day lighted such a candle, by God's grace, in Geneva, as we trust shall never be put out.'[77] The resolution represented a tiny speck of light in a darkening international world.

---

77 This version of their paraphrase of Bishop Hugh Latimer's advice to Nicholas Ridley at the stake is quoted in Hudson, *Australia and the League of Nations*, p. 177. Other versions of the story vary slightly.

7. A Vision for the League of Nations

Summary

Following the success of the nutrition initiative, both Bruce and McDougall held significant positions within the League of Nations, Bruce as President of the Council in 1936 and chairing committees seeking to restructure the economic and social activities of the League; McDougall as a member of the League Economic Committee. Both men sought to use the idea of raising living standards to achieve two intertwining goals: persuading the leaders and peoples of the fascist powers to adopt policies of international cooperation, generally known as 'economic appeasement'; and enabling the League to become a more effective agency of social reform.

This chapter deals with these complex issues as events unfolded. It begins with an account of their new roles, particularly of McDougall's new contacts with economists, and his use of the concept of consumptionism. Accounts follow of the threat posed by German autarky and rearmament; the development of McDougall's early ideas to alleviate it; British and US policies on economic appeasement; and the writing of the 'Hall Memorandum' reflecting wider discussion of McDougall's suggestions. A summary of McDougall's major memorandum 'Economic Appeasement' is preceded by discussion of the possibility that he coined the term. Although his memorandum received a cool response in Whitehall, the Imperial Conference in 1937 supported economic appeasement in principle. As war became inevitable, McDougall, Orr and Professor Noel Hall sought to persuade financial and industrial interests of the need for a broader policy of increasing prosperity and for improving national defence and health through increased consumption.

McDougall's hopes of promoting positive action through the League Economic Committee to remove trade barriers and pursue economic appeasement were disappointed, but Bruce successfully proposed a series of inquiries into means of raising standards of living, mitigation of depressions and agricultural credits. A report by Hall on the standards-of-living proposal detailed problems to be investigated. Three committees chaired by Bruce resulted in the 'Bruce Report', stressing links between social and economic problems and proposing a more effective structure to deal with them. War prevented implementation of both the Hall and the Bruce reports, although recommendations of the latter were embodied in the Economic and Social Council of the United Nations.

A Failing League

In 1936 the League of Nations began its move to permanent premises. An international design competition had attracted 377 entries and a committee of winning architects supervised work on a vast, white *Palais des Nations*, sprawling through Ariana Park on the north-western outskirts of Geneva. The long, irregular Secretariat building extended to a wing housing the Council; in front of a high Assembly building, the 'Court of Honour' portico led to terraces facing Lake Geneva and the Alps of Savoy; a domed library beyond was built with a $2 million grant from the Rockefeller Foundation. The international bureaucracy was provided with hundreds of meeting rooms and offices, a restaurant, post office and bank. Elegant formal spaces were fitted in marble embellished with brass. The Assembly Room accommodated some 2000 persons, including two tiers of onlookers, and was thus more suited to set speeches than to serious debate. The galleried Council Chamber—dwarfed by gigantic murals celebrating, in stark monochrome, the banishment of war—was criticised as 'over large, pretentious, even theatrical'. It conveyed the dignity of the world's highest presiding body at the expense of any understanding of the practical needs of productive negotiation. Members sat in a shallow crescent on a small stage and 'nothing less suitable to quick and spontaneous discussion could possibly be imagined'.[1]

Whatever the deficiencies of the building, they were less remarkable than the irony of its timing. By 1936 the mortal weakness of the League was clear. Despite some small successes, the decade already seemed a sorry tale of inability to deal with international crises. Mediation and condemnation failed to prevent Japan's annexation of Manchuria and departure from the League in 1931. The Disarmament Conference in 1932 achieved nothing, except Germany's departure from the League. In contravention of Versailles, the *Luftwaffe* was re-established in 1934, the German Navy was enlarged, and conscription introduced in 1935. Abyssinia's appeal against Italy in 1935 served merely to demonstrate lack of will and agreement between key members: British-backed economic sanctions against Italy were invoked in part, but France foiled the inclusion of vital oil sanctions. As argument continued, Germany reoccupied the Rhineland in March 1936 and Italy annexed Abyssinia in May. In view of this *fait accompli*, such sanctions as had been put in place against Italy were lifted in July.

For Bruce and McDougall, nevertheless, the last years of peace involved a desperate attempt to enhance the potential of the League for social, economic and even political change. The nutrition initiative was but their first major action in what has been called the 'Renaissance of the Economic and Social

---

[1] Walters, *A History of the League of Nations*, p. 699; *The Times*, 'New Home for the League', 17 February 1936.

Agencies'. In the years 1935 to 1939, work in broader fields had begun, and formal conferences and agreements were abandoned in favour of less-rigid, problem-solving processes of information sharing, concentrated no longer upon actions of governments but on 'the cares and interests of the individual and his family'.[2] This process was used, encouraged and often led by Bruce and McDougall. After 1935, both men sought to use League resources to apply the fundamental idea of the nutrition approach—that of increasing consumption—to wider problems of standards of living and prevention of war.

Both were positioned to invest their skills in the future of the League. Bruce was President of the League Council in 1936. That presidency 'seemed to mark a watershed in his life'. He identified with the economic and social purposes of the League and enjoyed the atmosphere of an international association, its demands and its contacts.[3] Since it had been thought that his high commissionership could conclude as early as 1936, Bruce might well have been thinking of his success in mobilising the League towards nutrition as pointing the way to a new and greater career.

The League Economic Committee

In January 1937 the League Council appointed McDougall to a three-year term on the Economic Committee, one of two standing committees of the Economic and Financial Organization, reporting to the Second Committee (Technical), which in turn reported to the Assembly. Most members of the committees were respected economists, appointed as individual experts, not as representatives of their governments. The Economic and Financial Committees both worked through expert committees and individual specialists to investigate problems of economic aspects of international relations; their members spoke with 'a certain freedom, and yet in most cases with intimate knowledge of the views of their respective governments'.[4]

Members of the Economic Committee tended to be officials and had the confidence of governments and also of business, which subscribed 'scores of millions of pounds' to loans authorised by the Financial Committee.[5] In 1937, the 15 members of the Economic Committee were predominantly European, but included O. Morato of Uruguay, Y. Shudo of Japan and Dr Henry F. Grady, former diplomat and academic, member of the US Tariff Commission and committed free-trader, then working with Secretary of State, Cordell Hull, to lower world

---

2   Walters, *A History of the League of Nations,* pp. 749–50.
3   I. M. Cumpston, *Lord Bruce of Melbourne,* p. 143.
4   Walters, *A History of the League of Nations,* pp. 177–8.
5   Ibid.

trade barriers. Notable Europeans included H. M. Hirschfeld, a Dutch civil servant later, controversially, administrator of finance, food and transport in the occupied Netherlands, Belgian sociologist Fernand van Langenhove, and R. Ryti, President of the Bank of Finland. A further group of 18 'corresponding members' included a sprinkling of diplomats, but also Professor A. Flores de Lemus, a distinguished right-wing Spanish economist, Dr M. H. de Kock, Deputy Governor of the South African Bank, John Leydon, the Secretary of the Irish Department of Industry and Commerce, and Indian politician Ramaswami Mudaliar, who would later take significant roles in the UN Economic and Social Council and in establishment of the World Health Organization.[6] The committee's Chairman was Sir Frederick Leith-Ross, Chief Economic Adviser to the British Government.

Membership of the Economic Committee gave McDougall access to the services of the Economic Relations Section of the League bureaucracy.[7] The appointment may have recognised McDougall's contribution to the work on nutrition as a member of the Mixed Committee and as a representative and often alternative spokesman for Bruce. It was undoubtedly flattering for a self-taught amateur, but there is no evidence of his feeling overwhelmed. He did take advantage of its facilities and contacts to undertake wider and more theoretical study of economics than he had done previously. He certainly read J. M. Keynes's *General Theory*, published in February 1936; *Consumers, Credit and Unemployment* (1938) by early Keynesian J. E. Meade, a member of the League staff from 1937;[8] work by D. H. Robertson, then an adviser to the League's Financial Section;[9] and by Swedish economist, politician and later Nobel Prize winner Gunnar Myrdal. Sean Turnell neatly describes McDougall as 'an inveterate consumer of ideas'. His advocacy in the late 1930s incorporated new insights, like nutrition, old favourites such as rationalisation and the fruits of his wider reading and recent contacts: Meade's 'consumer credits' to underpin consumption; Robertson's supposition that peace required emphasis on consumer, at the expense of capital, goods; and Keynes's suggestion for income redistribution to increase effective demand, and his concept of a 'spending multiplier'.[10]

McDougall met both Meade and Robertson while they worked with the League and corresponded for a time with pioneering Dutch econometrician Jan Tinbergen, who three decades later was to share the first Nobel Prize for Economics. From his own correspondence with Meade, Turnell notes

---

6 LN, *Monthly Summary of the League of Nations*.
7 Ibid., 1933–40, 10A, 27571/357, Stoppani to McDougall, 30 January and 3 February 1937.
8 Susan Howson, 'Meade, James Edward (1907–1995)', *ODNB*, 2004, <http://www.oxforddnb.com.virtual.anu.edu.au/view/article/60333> [accessed 21 November 2004].
9 Gordon Fletcher, 'Robertson, Sir Dennis Holme (1890–1963)', *ODNB*, 2004, <http://www.oxforddnb.com.virtual.anu.edu.au/view/article/35776> [accessed 21 November 2004].
10 Sean Turnell, '*Éminence Grise*: The Political Economy of F. L. McDougall', *Macquarie Economics Research Papers*, 13, 1998, p. 16.

that McDougall's circle of acquaintance amongst 'mostly young economists' destined to 'rise to spectacular heights in later years' extended well beyond the committee; they included Gottfried Haberler, Tjalling Koopmans, Ragnar Nurske, J. M. Fleming, J. B. Condliffe, J. J. Polak and Folke Hilgerdt. It was perhaps a time when economic theory, as distinct from policy, was more widely discussed than usual; a decade, says Turnell, 'most productive for political economy'.[11] In the 1920s, McDougall's thinking had been stimulated by contacts with young scientists and by innovators in the worlds of business and politics. Now he eagerly soaked up ideas complementary to the nutrition approach from the world of the social sciences.

He participated in inquiries initiated by the committee, including one on standards of living and another on prevention and mitigation of economic depressions. The contacts and challenges provided by this membership led him to espouse a more general aim of increasing consumption. In 1938 McDougall wrote that 'since 1935, Mr Bruce and I have jointly been responsible for a series of initiatives in the international field, all aimed at demonstrating the importance of increased consumption'.[12] The comment demonstrates the change in emphasis taking place in his writing and thinking. The idea of 'nutrition' had been subsumed into his broader concept of 'consumption'.

To a professional economist 'under-consumption' means the idea that consumption expenditure can be 'insufficient to absorb the total output of an economy at prices consistent with normal profits'.[13] McDougall used the words 'consumption' and 'under-consumption' frequently, and referred to the work of leading economists as he did so, but he used them as a layman, in a more literal sense. A 26-page memorandum spelled out at length the importance of adequate, varied diet for health, and the effect of purchases of cheap manufactured foods rather than fresh foods in reducing farm prices and incomes, so that farmers bought fewer manufactured goods and the cycle of depression continued.[14] McDougall added: 'the ignorance which is no doubt a contributory factor [to high rates of illness and mortality in poor districts] may be described as an under-consumption of education.' A world depression means 'the under-consumption of some thirty millions of employed and partially employed workers; under-consumption which is especially severe in respect to food deficiency'.[15]

Fears of renewed depression persisted through the 1930s and means of prevention and mitigation preoccupied economists. The influence of economists

---

11 Ibid.
12 CSIR/NAA, A10666, [5], McDougall to Rivett, 23 August 1938.
13 Turnell, 'F. L. McDougall: *Éminence Grise* of Australian Economic Diplomacy', 2000, p. 63.
14 NLA, MS6890/4/5, 'Economic Depressions and the Standard of Life', 24 June 1938. Attached is a letter with brief comments by Tinbergen.
15 Ibid., pp. 2, 11.

McDougall knew and read is obvious in his memoranda. One suggested the Economic Organization undertake work on 'increased consumption': 'Mr D. H. Robertson's plea for a "re-butterment campaign" has encouraged me to propose the inclusion of these considerations in the work of the Delegation.'[16] McDougall conceded that radical social policies during a depression might damage business confidence and prolong depression, but cited Keynes as an authority for his argument that stimulation of consumption was as valid a method of increasing economic activity as stimulation of investment, quoting in full a statement by Keynes sanctioning both.[17] He argued in favour of deficit financing in a depression to create employment, while acknowledging such policies must be 'undertaken in ways that will not prejudice the recovery of the normal level of investment'. He supported reducing 'excessive commodity stocks' as 'a valuable contribution'. Difficulties in devising methods of 'social distribution' would decrease with greater awareness of 'the importance of nutrition in social welfare and of the relation of efficient distribution of foodstuffs to agricultural and world prosperity'. Keynes had recently noted 'the part that control of stocks may play in smoothing out the trade cycle'.[18] Turnell writes that while 'McDougall eschewed linking his ideas with any particular economic theory…[in a 1938 memorandum,] he 'self-consciously identified them as being consistent with the underconsumptionist tradition'.[19]

Germany: 'That powerful and dangerous country'

The American loan funds upon which German postwar recovery depended vanished in the 1930s depression. The resulting economic chaos paved the way for political dominance of the Nazi Party and helped set much of the Nazi agenda. High tariff barriers confronting foreign foodstuffs hardened into a doctrine of autarky—economic self-sufficiency aiming at as little international trade as possible. What could not be produced economically would be produced with the assistance of protection—massive protection if necessary. What could not be produced at all would be replaced with substitutes or done without. With autarky came a controlled economy geared to large-scale rearmament. German rearmament heightened the political crisis, and the need to match it in the democracies created further economic strain. German autarky threatened the British Empire's triangular pattern of trade by reducing sales to Germany

---

16 FAO, RG3.1, Series D3, 'Trade Depressions and the Standard of Living', 14 June 1938. I am grateful to Sean Turnell for generously making his notes of this FAO series available to me.
17 Turnell notes on his copy that the quotation was taken from J. M. Keynes, *The General Theory of Employment, Interest and Money*, Macmillan, London, 1936, p. 325.
18 FAO, RG3.1, Series D3, 'Note on Consumption Policies in relation to the Trade Cycle', 25 August 1938.
19 Turnell, 'F. L. McDougall', 2000, p. 63.

of goods from many parts of the empire: industrial raw materials, wheat, wool, sugar and other colonial products. This real or potential loss of markets added to anxiety in London: British export sales to the empire depended upon the empire's ability to pay. Recovery from depression was fragile; by 1937 there was evidence of a further slump. In Britain and its empire, the need to bring Germany back into the international fold seemed as pressing on economic grounds as it was politically.

McDougall's determination to tackle the problem of Germany began early in 1936. In January he read *Inside Europe*, a widely popular book by US journalist John Gunther. So, presumably, did the Bruces: amongst books donated by Bruce to the National Library of Australia is a copy of it bearing the handwritten inscription 'Ethel Bruce'.[20] Gunther examined the personal history and sources of power of political leaders and institutions in most countries of Europe. In 1936, before 'warts-and-all *exposés*' had become commonplace, the book was unusual both in style and in content. It demolished a comforting theory that Germany lacked the resources to make war. It also dealt briefly with problems of the German economy.[21] Germany should be allowed to become 'strong but not too strong' and manifest injustices of Versailles should be redressed. Germany, a 'have-not' nation—that is, lacking colonies, as former German colonies had been allocated to 'victor' nations under the Versailles Treaty—should be allowed to regain self-respect. Yet Gunther also warned that German rearmament was being funded by borrowing from sources overseas, including the Bank of England.[22] Perhaps what most alarmed McDougall and other readers was the book's destruction of the mystique of national leaders. Suddenly the destinies of ordinary men and women lay in the hands of fallible mortals who, more often than not, were fatally flawed or psychologically damaged. Gunther's book may be one more reason 1936 seems to mark the beginning of general apprehension of impending war.

In 1935 some 12 million Britons had demonstrated against war in the 'Peace Ballot'. McDougall was influenced by this awakened 'moral sense' and by public disillusion with the League. His optimism, his belief in the efficacy of reason and persuasion, would not permit him simply to wring his hands. Recommending Gunther's book to Rivett, he urged 'we should give very serious thought to the European situation…[which] is full of danger and I believe it has got to be looked at not only from the foreign policy point of view but also from a wide economic standpoint'. He listed three options facing the British Empire in both defence and trade. 'Splendid isolation' was dangerous, because it invited attack; cooperation with France to use the League as an instrument of collective security

---

20 John Gunther, *Inside Europe*, H. Hamilton, London, 1936.
21 Ibid., pp. 98–102.
22 Ibid., p. 111.

would be ineffective while the League was crippled by resentment of the Treaty of Versailles, of which the League Covenant formed part, and by the disabilities of 'have-not' nations. Alliance with the United States seemed the ideal, but was ruled out by US isolationism. McDougall therefore proposed a fourth option: to persuade Germany to 'abandon the aggressive spirit' by offering her a general treaty of cooperation. In return for largely unrestricted trade with the British Empire, and subject to safeguards including adherence by all parties to ILO conventions, Germany would return to the League, undertaking to cooperate with the British Empire to secure progressive disarmament, abandonment of military aviation, international control of civil aviation and economic equality for her racial minorities.[23]

In the first half of 1936, while hope remained that League sanctions against Italy might succeed, McDougall wrote a series of papers linking economic policy with the political situation. A 26-page memorandum, written in February, expanded on the German treaty idea and an 'open-door' trade policy for colonies, meaning that colonial powers would not claim preferential advantages, thereby allowing all nations to trade with colonies on equal terms.[24] The overriding aim was to avert war:

> Fear of war and the necessities of defence reinforce the tendencies towards economic nationalism.
>
> These fears and difficulties lead to a dread of over-production at a time when under-consumption is the most real factor…It is suggested that the world's economic difficulties can best be lightened through improvement in political relations and through attempts to bring about increased consumption.
>
> The long continuance of stark poverty in what is potentially an exceedingly rich world must lead either, to revolutionary outbreaks or to what is more probable, embarrassed Governments attempting to escape from their internal difficulties through external adventures.[25]

There were economic advantages of a treaty and of liberalised trade generally, both to the 'have-not' nations and to the rest of the world. McDougall now dismissed the imperial solution as a temporary expedient. The Ottawa Agreements had 'helped the Empire out of the worst of the depression' and remained of some benefit, but imperial economic cooperation should not go so far as to make the Empire a closed economic system, with resulting dangers

---

23 NAA/CSIR, A10666, [5], McDougall to Rivett, 23 January 1936.
24 Ibid., 'The British Empire, World Peace and Prosperity', 7 February 1936.
25 Ibid., p. 5.

of 'war and…reduced prosperity'. The agreements had nevertheless provided 'material for concessions to other countries and a bargaining power which would have been absent had Great Britain retained her free trade policy'.[26]

McDougall acknowledged the persuasive case of the 'have-not' powers for treaty revision and the 'moral isolation' of Britain and France. Although German demands for access to colonies might be satisfied with nothing less, he did not advocate restitution. Instead, Britain might regain moral leadership by offering, at an international forum on the economic relations of colonies, a complete 'open door' overseen by the League of Nations. He noted that such a policy was already observed in the Belgian Congo and Dutch East Indies. It would be necessary to consider whether Britain should do so without similar action by France and Portugal. The British Empire should push the League to deal with issues of social justice, such as nutrition and ILO conventions. German and Italian aspirations were unlikely to be satisfied by this limited offer, but it would remove their 'legitimate grievances', deprive them of the sympathy of neutral countries—that of the United States in particular—and satisfy the moral sense of the British people.[27]

The League's lack of universality could be eased by the adherence of either the United States or of Germany and, in view of US isolationism, McDougall turned to Germany, 'that powerful and dangerous country'. The existence there of some moderate leaders and an apparent desire for good relations with Britain offered hope for success of 'a great gesture' in the form of a general treaty of cooperation. A treaty could end Germany's feeling of 'economic encirclement and give her immense scope for the beneficial employment of her vigour, in the general improvement of her standards of living'.[28] It could also be a means of restoring Europe as a market for agricultural exporters and Britain.[29]

Policies likely to reactivate the economic life of Europe, though involving some sacrifice of 'present advantages', could benefit Australia.[30] A small loss in colonial trade resulting from an open-door policy would be offset by better access to European markets; abandoning protection should give colonial people cheaper goods and so stimulate trade generally.[31] Nations should explore the possibility of rationalised primary and secondary industries on a world scale, perhaps between industrial countries like Britain, Germany, Czechoslovakia and Belgium, and between the dominions, India and 'foreign agricultural countries'.[32]

---

26 Ibid., pp. 14–15.
27 Ibid., pp. 15–20.
28 Ibid., pp. 21–3, 25–6.
29 NAA/CSIR, A10666, [5], 'Notes on Economic Adjustments and Political Tensions', 4 April 1936, pp. 3–4.
30 Ibid., 'Australia and Europe', 1 April 1936, p. 11.
31 Ibid., 'Economic Appeasement', 21 December 1936, pp. 23–4.
32 NAA/CSIR, A9778/4, M1 N43, 'Note on Australian Economic Policy', 30 April 1936, pp. 3–4.

Rivett, like many thinking men of goodwill in 1936, saw no 'sound ethical ground' for refusing Germany her former colonies, but doubted she would be satisfied merely with better opportunities for trade.[33] McDougall replied that opinion was 'moving rather fast' on the question of transferring mandated territories to former colonial powers, a possibility he would not rule out.[34] Writing from Kenya, shortly after the failure of sanctions against Italy, Elspeth Huxley identified the problem with the treaty idea more bluntly. Other 'have-not powers' might well prefer to emulate Mussolini and 'take what they want if they can, instead of being given less than they want as a favour'. She doubted Germany would be satisfied with anything less than outright ownership, but also whether Germany would in fact press the colonial question: 'She knows very well they have a negligible value.'[35]

These views, and many of McDougall's suggestions, reflected widespread debate. The distinctive feature of his solution was his linking of political settlement with expansion of consumption. Underlying this argument was a completion of the shift in frame of reference that had begun with his engagement with the wheat problem. In the 1920s he had been attempting to resolve trade problems within the British Empire, with careful attention to its various producer interests; now he was tackling the threats of communism and fascism to international security, with his political antennae attuned to the unmet needs of consumers.

Much of McDougall's approach was in accord with significant bodies of opinion in London and Washington that had argued since 1919 that lasting peace required measures allowing German economic recovery.[36] Terms such as 'treaty revision' and 'general settlement' (that is, of problems remaining after Versailles) were used for suggestions to provide German access to colonial trade and raw materials. Supporters of such policies extended beyond those vilified in 1937 as 'appeasers' in the 'Cliveden Set', led by Viscount Waldorf Astor and his wife, Nancy.[37] Advocates in Whitehall of a softer line with Germany included the chief of the newly formed Economic Relations Section of the Western Department in the Foreign Office, F. T. A. Ashton-Gwatkin, and, from late 1935, his deputy, Gladwyn Jebb, just returned from a posting in Rome.[38] Before the 1932 Ottawa Conference, Ashton-Gwatkin had predicted that a high protective

---

33 Ibid., Rivett to McDougall, 19 February 1936.
34 NAA/CSIR, A10666, [5], McDougall to Rivett, 9 April 1936.
35 NLA, MS6890/2/6, Elspeth Huxley to McDougall, 13 May 1936.
36 Most notably by J. M. Keynes in *The Economic Consequences of the Peace*, and by *Observer* Editor, J. L. Garvin, in *The Economic Foundations of Peace*, both published in 1919.
37 For an account of the smear by a former *Times* journalist, Claud Cockburn, and an assessment of the views on Germany of the Astors and their associates, see Derek Wilson, *The Astors 1763–1992, Landscape with Millionaires*, Weidenfeld & Nicolson, London, 1993, Chapter 10.
38 Keith Hamilton, 'Gwatkin, Frank Trelawny Arthur Ashton- (1889–1976)', *ODNB*, 2004, <http://www.oxforddnb.com.virtual.anu.edu.au/view/article/64923> [accessed February 2005]; Alan Campbell, 'Jebb, (Hubert Miles) Gladwyn, first Baron Gladwyn (1900–1996)', *ODNB*, 2004, <http://www.oxforddnb.com.virtual.anu.edu.au/view/article/63251> [accessed 11 February 2005].

tariff and empire preference could diminish British influence over European affairs and create economic antagonism. He favoured economic aid to Germany since 'a weak hysterical individual, heavily armed, is a danger to himself and others'. Jebb described Nazism as a 'cancer' that would yield to 'the radioactive treatment of increased world trade'. But a Foreign Office proposal to offer colonial concessions to Germany in return for limiting armaments, abandoning European expansion and rejoining the League was opposed by the Board of Trade.[39] In 1936, Foreign Secretary Anthony Eden initiated a two-pronged policy of rapid rearmament coupled with the possibility of returning colonies, if 'permanent settlement' could be achieved.[40]

Until mid 1939, politicians and officials continued to explore possibilities for economic cooperation with Germany. Neville Chamberlain, as Chancellor of the Exchequer, supported an approach late in 1936, when he feared the effect of continued rearmament on Britain's adverse balance of payments. Common to all British schemes of 'economic appeasement', in addition to the primary aim of reducing tension, was that of encouraging Germany to import from commodity producers who spent their income on British exports, an aim supported by the City and by the FBI.[41] Cabinet authorised further discussions in April 1937, offering colonial restitution and economic aid in return for Germany's 'good behaviour' in Central Europe; many believed that 'only colonies stood in the way of settlement'. Evidence more recently available suggests Germany's colonial claim was intended to lure Britain into discussions in which Germany would be conceded a right to European expansion. Lacking this knowledge, Chamberlain, early in 1938, proposed facilitating German access to raw materials through management of an area in Central Africa by a consortium of European powers, including Germany, in return for arms limitation and territorial guarantees to Austria and Czechoslovakia. Rejection of that offer changed British policy to one of gradual modification of the German economy through 'the liberalizing influence of increased foreign trade', to pave the way for political settlement.[42] A final attempt at negotiation with a German official, presumed to be an agent of Göring, was sanctioned by Chamberlain in July 1939, in a context of falling British reserves. 'The trade recession dominated his thought on the German problem.'[43]

US policy was influenced by economic considerations, but even more by a desire to avoid entanglement in a European conflict. Many influential

---

39   William Norton Medlicott, *Britain and Germany: The Search for Agreement, 1930–1937*, Creighton Lecture in History 1968, University of London, London, 1969. See, in particular, footnote to p. 5, and pp. 20–1.
40   Andrew J. Crozier, *Appeasement and Germany's Last Bid for Colonies*, Macmillan, London, 1988, pp. 131, 133.
41   C. A. MacDonald, 'Economic Appeasement and the German "Moderates" 1937–1939. An Introductory Essay', *Past and Present*, Vol. 56, 1972 pp. 105–6.
42   Ibid., pp. 107–14.
43   Ibid., pp. 127–30.

Americans, including President Franklin D. Roosevelt, saw sense and justice in reassessing Versailles and, like the British, they hoped to use this as a bargaining tool, though Roosevelt relied on 'universalist emphasis on codes of conduct rather than Britain's bilateral and spheres-of-influence approach to diplomacy'.[44] Early in 1936, Norman H. Davis, 'roving ambassador and friend to Roosevelt and Hull', suggested Germany be given 'a special economic position in South-Eastern Europe, in return for signing an arms limitation agreement and rejoining the League'—a view echoed from Berlin by Ambassador William C. Bullitt in November 1936. Pursuing Hull's freer trade agenda, American diplomats generally believed in the political benefits of reduced trade barriers and increased access to markets and resources, seeing economic appeasement 'both as a diplomatic tactic and a legitimate, publicly acceptable end'.[45]

Roosevelt suggested a non-political conference on economic and social problems in 1936; neither Britain nor Germany was interested. An American proposal late in 1937 for a world conference to establish universal codes governing international relations, including access to raw materials, was abandoned following events early in 1938: Japan's resumption of fighting in China; Hitler's replacement of moderate Neurath with Ribbentrop and recognition of Japan in Manchukuo; Eden's resignation; and the *Anschluss*. In 1939 Roosevelt again appealed for non-aggression pledges in return for parallel economic and political appeasement conferences.[46]

McDougall's eagerness for international conferences and treaties and for use of the League accords more with the American approach than that of the British Foreign Office, where, it has been argued, there were doubts about the effectiveness of the League and resentment of its control from Downing Street rather than Whitehall.[47] The potential threat to imperial trade arrangements in McDougall's approach would also have worried Whitehall.

At a time when Bruce's mind was turning more and more to the possibilities of international action, McDougall's proposals were appealing and matched his own concerns: he became an enthusiastic supporter. Alfred Stirling recalls a long evening in Geneva in May 1936 when McDougall 'poured out to me for many hours the anxieties which the High Commissioner and he shared about the preservation of world peace'. McDougall's 'main theme was the need to re-

---

44 Arnold A. Offner, 'Appeasement Revisited: The United States, Great Britain, and Germany, 1933–1940', *The Journal of American History*, Vol. 64, no. 2, 1977, p. 392.
45 Ibid., p. 378.
46 Ibid., pp. 379–82.
47 Alan Sharp, 'Adapting to a New World? Britain's Foreign Office in the 1920s', *Contemporary British History*, Vol. 18, no. 3, 2004.

activate trade in Europe, and how Australia must produce more foodstuffs and help in the feeding of the hungry of the world. This was vital if war was to be averted.'[48]

Bruce made his views plain in a meeting of British and visiting Australian ministers that same month. Jebb was sent by the junior Foreign Minister, Lord Stanhope, to ask Bruce to amplify his views. With McDougall present, Bruce argued that it was no use attempting to prevent war without trying to eliminate the economic causes 'which inevitably lead to it'. Any declarations made at Geneva or elsewhere about reorganising collective security or reform of the Covenant must be accompanied by a declaration of economic policy. Bruce was less sure about what might actually be achieved, but for 'propaganda value in the world as a whole' there should be a unilateral declaration of intent to abolish imperial preference in the colonial empire and a broadening of the system of awarding contracts. Bruce admitted that without some modification of imperial tariff policy, these measures would do little to expand German exports but insisted that the only hope of preventing Germany's present 'drive to the East' from becoming 'actual military aggression' lay in the British Empire taking more German goods. He favoured Professor Noel Hall's suggestion to permit entry of more semifinished manufactures from Germany, to revive the triangular trade between Europe, the dominions and Britain.[49] Bruce's views must have alarmed officials anxious to preserve Britain's imperial trading arrangements.

Group discussion of an idea over dinner was a favourite McDougall tactic. Almost certainly Bruce approved establishment of such a group early in 1936 to consider 'economic appeasement'. It included Gladwyn Jebb and Australian-born diplomat R. W. A. (Rex) Leeper, known for his views opposing appeasement, as observers from the Foreign Office.[50] Other participants represented the empire, industry and ideas; most had worked with McDougall before. Assistant Editor, R. M. Barrington-Ward, was almost solely responsible for *The Times*' treatment of Anglo–German relations: 'the keystone of his view [on Europe] was the perniciousness of the Treaty of Versailles.'[51] Noel Hall, Professor of Political Economy at University College London and Director of the National Institute of Economic and Social Research, was described by McDougall as 'an old friend of mine'.[52] 'Mike' Lester Pearson was then a Canadian diplomat in London; R. Enfield of the Ministry of Agriculture had been a contact since the mid 1920s; company director and banker Sir George Schuster was an expert on

---

48 Stirling, *Lord Bruce: The London Years*, p. 30.
49 UKNA, FO371/19933, 125552, C4757/59/18, Memo of conversation by Jebb, addressed to Stanhope, 10 June 1936. This appears to be a draft, not all of which was submitted: the part dealing with Hall's suggestion has been crossed through. I am grateful to Sean Turnell for copies of material from this file.
50 Derek Drinkwater, 'Leeper, Sir Reginald Wildig Allen (1888–1968)', *ODNB*, 2004, <http://www.oxforddnb.com.virtual.anu.edu.au/view/article/55366> [accessed 28 January 2005].
51 Franklin Reid Gannon, *The British Press and Germany 1936–1939*, Clarendon Press, Oxford, 1971, pp. 63–4.
52 NAA/CSIR, A9778, M14/37/7, McDougall to Rivett, 13 July 1937.

colonial development. On one occasion, they were joined by Oxford Professor of International Relations and founder of the Institute of International Affairs, Sir Alfred Zimmern, who had identified three causes of war including, significantly, the economic issue of the 'haves and have-nots'.[53] Jebb minuted that both he and Leeper felt the group's approach merited 'the most sympathetic consideration' by government experts. Leeper added: 'we cannot take political initiative in the shape of further commitments. All the more need therefore for economic initiative.'[54]

The group's paper, 'An International Policy for the British Empire', was drafted by Professor Hall and generally referred to as 'the Hall Memorandum'.[55] It called for cooperative leadership of world economic policy from the British Empire, now recovered from depression and in an economic position envied by the 'have-nots', who blamed 'the Ottawa Factor' for contributing to their economic difficulties. An increase in world trade might help 'purchase' a general European settlement, extending even to an agreement on arms limitation. The paper discounted the value of loans, or return of colonies to Germany, prescribing instead a series of coordinated measures based on increased consumption and including a conditional open-door policy to trade in the colonial empire. Its chief focus was the means to increase the empire's triangular trade with industrial Europe, to give 'more elbow room' for dominion producers to trade outside the Empire and to revive British industry, shipping and financial services. Its chief practical suggestion was a revision of the British tariff, but only where it offered prospects of an increase in the triangular trade. Tariff preference by Britain for semi-manufactured goods from Europe, such as car parts and steel sheets, would benefit alike British and European industry, dominion buyers and raw material producers.

Copies of the paper were circulated discreetly. One was sent privately by Barrington-Ward to Tom Jones, Deputy Secretary to the Cabinet and an economist by training. It was 'undoubtedly important', Ward wrote to Jones, 'to start talking the economics of peace as soon as possible'.[56] The group hoped it might lead to 'more vigorous action' by Britain at Geneva at the time of the Tripartite Declaration, an undertaking by Britain, France and the United States to avoid further devaluation of currencies, signed on 25 September 1936. Jebb showed it to S. D. Waley of Treasury and a Mr Brown of the Board of Trade. Neither was encouraging, although Waley admitted to interest in the idea of importing more German goods.[57] Opposition by Walter Runciman, President

---

53 D. J. Markwell, 'Zimmern, Sir Alfred Eckhard (1879–1957)', *ODNB*, 2004, <http://www.oxforddnb.com.virtual.anu.edu.au/view/article/37088> [accessed 26 October 2004].
54 UKNA, FO371/19933, 125552, C4757/99/18, minute by Jebb, 'Origin of the Hall Memorandum', 25 June 1936; handwritten comment by Leeper, 26 June 1936.
55 The National Library of Wales, Dr Thomas Jones C. H. Papers, E1, Items 9 and 10, copy dated July 1936.
56 Ibid., Barrington-Ward's covering letter, 16 July 1936.
57 Jebb, 'Origin of the Hall Memorandum'.

of the Board of Trade, and Chancellor of the Exchequer, Neville Chamberlain, to any modification of British economic policy ruled that out, and the Hall memorandum had no further perceptible effect.[58]

'Economic Appeasement'

The Hall group's discussions must have clarified McDougall's own thinking. By December 1936 he had completed a major revision of his February memorandum, now called 'Economic Appeasement' and 27 pages long. It is possible that McDougall himself coined the term. Jebb writes of 'what McDougall had called the policy of "Economic Appeasement"', noting: 'Incidentally I think [early 1937] was about the first time that the famous word "appeasement" gained any currency and it certainly did not then connote a policy of giving the Nazis everything they wanted; merely a possible means of achieving a "peaceful solution".'[59] Elspeth Huxley recalled McDougall 'discoursing on the need for "appeasement" and that was the first time I heard the word used in the sense that later became so discreditable'. She added: 'it was not then a dirty word and seemed a reasonable approach, since it was the harshness of the Treaty of Versailles that really gave Hitler his chance to come to power.'[60]

The phrase 'economic appeasement' first appears in *The Times* index in mid 1937; McDougall's closeness to editorial figures at *The Times* could have promoted its use. The term is used in subheadings in the Economic Section of the 1937 *Survey of International Affairs*, which states: 'the Australian Delegation to the League of Nations had been especially active in propagating the idea of "economic appeasement", and though its meaning was not always clearly defined, the phrase became almost a commonplace in the economic and political discussions of that year.'[61] Bruce had used it in a speech in Second Committee on 6 October 1936, but in doing so he was commending words of the British delegate who expressed Britain's desire to 'use every method which presents itself to further the economic appeasement of the world'.[62] The noun 'appeasement' was used without the qualifier 'economic' earlier in this context. Turnell points out that in 1936 'one did not speak of the appeasement of Nazi Germany, but of Europe

---

58 UKNA, FO371/21215, 125552, W373/5/50, Minute by Jebb, 9 January 1937. I thank Sean Turnell for copies from these and related files.
59 Gladwyn Jebb, *The Memoirs of Lord Gladwyn*, Weidenfeld & Nicolson, London, 1972, p. 65.
60 Letter to W. Way, 15 April 1986.
61 Alan J. B. Fisher, 'World Economic Affairs', in *Survey of International Affairs 1937*, ed. Arnold J. Toynbee, Royal Institute of International Affairs, London, 1938, p. 65.
62 LN, *Official Journal*, 1936, Second Committee (Technical), Minutes: W. S. Morrison, Financial Secretary to UK Treasury, 5 October, p. 43; Bruce, 6 October, p. 60.

and the world'.⁶³ That is, 'appeasement' did not mean buying off a potential aggressor, but the easing of tension and anxiety. Stressing the importance of economic cooperation as a solution to political problems, Bruce demonstrated its typical use in September 1936, quoting Assembly President, Paul van Zeeland, on 'the feeling of appeasement and relief which would spread through the world if the nations again knew comfort and prosperity'. Bruce went on to make a subtle shift in his own use of the word, expressing a conviction that 'solution of the economic problems that confront us can best be secured [by] the maintenance of world peace, the appeasement of the social unrest and the removal of dangers that threaten all countries'.⁶⁴

'We prefer butter'

McDougall's earlier memoranda had focused on bringing Germany back into the international fold. In the memorandum of December 1936, the German treaty proposal was gone, although the open door to colonial trade remained. The main thrust was now improving standards of living to promote a revival of world trade, soothing the political situation and convincing European governments and peoples that 'the great democracies have both the will and the power to bring about great improvements in the welfare of their own people and of those of other nations which are prepared to co-operate with them'. Britain, France and the United States had all referred, in declarations on signing the Tripartite Declaration, to the need to raise standards of living. McDougall wanted those three countries to go further. People everywhere were realising that 'poverty is not inevitable but due to faults in the productive and distributive system'. McDougall was encouraged in this view by the recent election of reform-oriented governments in the United States and France, which seemed to show increasing pressure for government action to improve the lot of the poor. He therefore called for a 'direct attack upon low standards of living conducted both on the national and [the] international plane'. 'Dynamic economic and social policies' would show democratic countries were better able to achieve comfort and wellbeing for their peoples than the fascist or communist states, and would expand the volume of international trade. Most European states should then rally to the democratic countries and peoples of fascist states would be presented with 'an attractive alternative to preparations for war'.⁶⁵

> The general idea could be colloquially expressed as follows:—Certain dictatorial people have declared their preference for guns rather than

---

63 Sean Turnell, 'Butter for Guns: F. L. McDougall, Nutrition and Economic Appeasement', *Macquarie Economics Research Papers*, no. 14/95, 1995, p. 3.
64 NAA, M104/1, 4, typescript of speech to 17th Assembly, 29 September 1936.
65 NAA/CSIR, A10666, [5], 'Economic Appeasement', 21 December 1936, pp. 1–2.

butter. This unfortunately compels the democracies in their turn to secure a sufficiency of guns. But we prefer butter, and we propose to collaborate with all nations who will join with us in securing more butter.[66]

ILO-based international measures and national policies could achieve increased purchasing power, improved social services and working conditions, and reduction of costs through efficiencies in food production and distribution.[67] McDougall paid particular attention to the contribution of improved nutrition to economic appeasement, using familiar arguments about its potential to affect consumption and trade, and a new addition: the relevance to defence preparations of a policy to 'encourage a maximum diversity of agricultural production'.[68] A section of the memorandum was devoted to reduction of trade barriers, including suggestions that the United States accept more imports and that the United Kingdom accept more goods from Europe. McDougall admitted that this seemed to be a reversal of protection policies advocated in 1931 by many (including, of course, McDougall himself), but European markets were essential for the development on which the import levels of the overseas empire depended. He again urged restructuring of tariff policies to encourage rationalised industrial production. He called for British and dominion governments to take 'parallel action' with Cordell Hull's attempts to liberalise trade by bilateral treaties and a colonial open-door policy, possibly subject to an ILO convention, and perhaps dependent upon reductions in European tariff barriers.[69] In summary, he was proposing that democratic nations resolve to 'revitalize the world's trade', be prepared 'to share their economic advantages' in return for political cooperation and 'hold up to the world the ideal of human progress and culture as the alternative to dreams of national aggrandisement achieved through force or the threat of force'.[70]

---

66 Ibid., p. 3.
67 Ibid., pp. 5–11.
68 Ibid., pp. 12–16.
69 Ibid., pp. 16–26.
70 Ibid., p. 26.

'Any economic proposal emanating from Australia wants watching'

Bruce made a few minor suggestions, but in the next few weeks used all of his considerable status and contacts to place 'private' copies of McDougall's December 1936 memorandum with carefully chosen recipients. He presented it to J. L. A. Avenol, Secretary-General of the League of Nations, as an approach to the political problems of Europe 'from a somewhat different angle from the usual proposals for a frontal attack upon the high tariffs and other restrictions'.[71] Other recipients in London included Anthony Eden, Leith-Ross, Treasury's personal adviser to the Prime Minister, Sir Horace Wilson, former prime minister Ramsay MacDonald and Sir Montagu Norman. A copy later reached Runciman. Copies were sent to the British Ambassador in Berlin, Sir Nevile Henderson, and to the Prime Ministers of Belgium and the Netherlands. At the League of Nations, besides Avenol, copies went to Alexander Loveday, to Paul Stoppani and to H. B. Butler, Director-General of the ILO.[72] A copy sent to US Ambassador Robert Bingham could thence have reached Cordell Hull, causing Eden to write, on a note to that effect, 'Is not this unusual?'[73]

Figure 14 F. L. McDougall with J. L. A. Avenol, Secretary-General of the League of Nations.

Source: E. McDougall.

---

71 NAA, M104, 5/2, Bruce to Avenol, 1 March 1937.
72 Ibid.
73 UKNA, FO371/21215, 125552, W6363, minute by Jebb, 17 March 1937.

Foreign Office officials commented at length upon 'Economic Appeasement'. Gladwyn Jebb, strongly supported by Ashton-Gwatkin, minuted that, in common with Leith-Ross, the Economic Section believed it contained some good suggestions including negotiation on the basis of standards of living; association of the ILO with any possible action; the tariff revision proposals from the Hall memorandum; countering the claims of 'have-not' nations with an open-door trade policy; reconsideration of imperial preference; and the use of League machinery to further 'economic appeasement'. The paragraphs on nutrition, however, were 'anathema to certain Government Departments here', and Leith-Ross was sceptical of their utility. Jebb suggested that the proposals be recommended to the Board of Trade as offering 'a distinct—perhaps the only—hope of avoiding or moderating further political complications in Europe'. They might be discussed unofficially at a meeting between Leith-Ross and Dr Hjalmar Schacht, the moderate German Minister of Economics replaced by Göring in 1936 but still in Cabinet as President of the Reichsbank, as an attempt to divert Germany's colonial demands and as a means of gaining time by 'economic appeasement'. 'It may well be that this is impossible; I do not rate the chances very high; but it is my considered belief that unless we do something very soon we shall wake up one Sunday morning to find ourselves with no time left to gain.'[74]

In the ensuing debate within the Foreign Office, officials tended to judge McDougall's memorandum in two contexts: the debate on policy towards Germany; and the sour legacy of Australian trade diplomacy since Ottawa, in which McDougall had played a significant role. William Strang, head of the Central Department responsible for policy on Germany, was critical of both style and content of the memorandum, which was 'not well arranged...much too long...[and] contains a number of questionable statements'. Insuperable difficulties were likely to prevent a complex three-power economic agreement; if it were achieved, the example of improved living standards would be slow to take effect. The problem should not be oversimplified: 'As between politics and economics it is not easy to say which is hen and which is egg.' The Nazis had not attained power through economic forces alone, nor would economic action alone 'exorcise the Nazi devil'. Higher standards of living would have uncertain appeal in a Germany where 'what they value, in theory at any rate, is the creative struggle for something higher than mere bread'. Many of the suggestions, particularly those from the Hall memorandum, were worthwhile in themselves, but he doubted the time was right for an approach to Schacht. The German propaganda machine had been 'working full blast on the theme of colonies: it may soon begin to proclaim the iniquity of British commercial policy and to arouse indignation among Germany's fellow-sufferers. There is no

---

74 Ibid., W373/5/50, Minute by Jebb, 9 January 1937.

doubt some case to be made against us.' The post-1931 tariff changes had forced trade into 'unnatural channels'. Sooner or later talks like those proposed by Jebb would probably be necessary, but Strang thought Britain should wait until 'more authoritative and more official approaches' were made by Germany.[75]

J. M. Troutbeck of the American Department questioned Bruce's motive in giving a copy directly to the US Ambassador: 'No doubt it will find its way to the State Department and encourage Mr Hull to believe that he has the Australians behind him in his endeavour to wean us from our present commercial policies.' Indeed, the 'rather turgid rhetoric of the memorandum seems designed to appeal to Washington rather than to Whitehall'. Troutbeck doubted whether 'the Australians have an exaggerated regard for the interests of the UK, or indeed for anything outside Australia's particular welfare. Throughout the long drawn out discussions on meat, for example, they have clung to their pound of flesh and more beside.' If Bruce were genuine then his suggestion was timely, since the Americans would demand some dominion sacrifices in return for a commercial agreement with Britain. But the immediate beneficiaries of the scheme would be the dominions and foreign countries, and 'any economic proposal emanating from Australia wants watching'.[76]

Jebb admitted that Troutbeck's suspicions and Strang's misgivings were justified, but thought them not sufficient to destroy the value of McDougall's proposals. He nevertheless modified his earlier recommendation, suggesting that a minute to the Board of Trade should simply mention Bruce as one of several authorities recommending revision of British tariff policy.[77] Again, he was supported by his chief, who gained Strang's agreement to the proposed minute. Ashton-Gwatkin believed there was a movement towards a liberalisation of UK tariff policy, which 'may improve the chances of peace in Europe (& perhaps the Far East, too). It is for us to see if we can push the movement in this direction.' He instructed Jebb to draft the proposed letter to Runciman, adding 'there is much wisdom in [Strang's] minute—and much blather in McDougall's memo. But the latter is on the side of the angels.'[78]

In the event, the Foreign Office made no recommendations on McDougall's December 1936 memorandum, beyond mentioning it in a longer paper dealing with the effect of British tariff policy on foreign relations, and recommending establishment of a ministerial committee.[79]

---

75 Ibid., 13 January.
76 Ibid., 14 January. In a brief search in US National Archives, I found no record of the memorandum reaching the State Department, but cannot rule it out.
77 Ibid., 15 January.
78 Ibid., 17 January.
79 UKNA, FO371/21215, 125552, W6363, minute by Jebb, 17 March 1937.

'The revival of world trade is of first importance'

Bruce was not yet defeated. Empire leaders gathered for an Imperial Conference in May 1937. Briefings for the Australian delegation by officers of the newly formed Department of External Affairs, including two who had been in close contact with Bruce and McDougall, former Liaison Officer in London F. K. Officer and his successor, Alfred Stirling, reported wide discussion of means to ease German economic difficulties as a 'safety valve to ensure peace'; action could begin with a pact with Australia and New Zealand to import more German manufactured goods.[80] The hands of McDougall and Bruce could be discerned in a section of Lyons' opening speech on economic policy, which included the statement that markets beyond the Empire and an increase in world trade were essential.

> Economic policy, however, also has profound effects upon the political relations of the countries of the world. Today we are confronted by the picture of a world in which science has made possible standards of living for all countries far in advance of anything previously experienced and yet in which poverty and unemployment have led to grave political discontents.
>
> There is thus urgent need for wide policies of economic appeasement if our endeavours to bring about peaceful conditions in the world are to be successful.
>
> For this purpose the revival of world trade is of first importance.[81]

Canada's Prime Minister, W. L. Mackenzie King, had been persuaded by Cordell Hull to plead the case for liberalising trade through a series of bilateral agreements, in the hope that Axis nations would perceive the benefits, join in and open the way to discussion of political problems. King duly told the conference that political tensions would not lessen without economic appeasement and 'abatement of the policies of economic nationalism and economic imperialism'.[82] Eden acknowledged that many believed economic appeasement provided 'the key to our difficulties'; most political problems had 'an economic problem inextricably entwined'.[83] The conference's final statement included a declaration of readiness to 'co-operate with other nations in examining current difficulties,

---

80 *Documents on Australian Foreign Policy 1937–49. Volume I: 1937–38* [hereinafter *DAFP* I], eds R. G. Neale, P. G. Edwards and H. Kenway, Australian Government Publishing Service, Canberra, 1975, Document 6, undated memorandum, pp. 29–30; 17, memorandum, 10 March 1937, pp. 47–8.
81 *DAFP* I, 25, Speech by Lyons, 14 May 1937, pp. 69–70.
82 Fisher, 'World Economic Affairs', p. 63.
83 Cumpston, *Lord Bruce of Melbourne*, p. 151.

including trade barriers and other obstacles to the increase of international trade and the improvement of the general standard of living', in the interests of prosperity and international peace.[84]

In rhetoric, the conference was a success for economic appeasement. But these were 'motherhood' statements, swamped by overwhelming attention to foreign relations and defence issues in an atmosphere of urgency and crisis. Bruce understood the reluctance of Australia's government, facing an election later in 1937, to appear to be 'prepared to give up any rights that Australia might at present have in the British market' in the cause of a British–American trade agreement or, presumably, of other measures likely to contribute to economic appeasement, but he was also anxious to avoid any impression that dominion reluctance to discuss concessions was the only obstacle to a trade agreement between Britain and the United States. He suggested a statement asserting that the Government 'would only agree to forgo anything we at present have if as a result we saw an expansion in the markets for our exports'.[85]

Bruce could exercise paramount influence on Australian policy as it was presented in international forums, but the international bias of his thinking became increasingly alien to the mood and realities of Australian domestic politics. Paradoxically, as Turnell points out, he was, with McDougall, 'an initiator of *what was seen to be* the policy of Australia'.[86] 'What was seen' would have encouraged British policy in favour of economic appeasement, if only because it seemed to carry no threat to empire unity. In fact, if 'economic appeasement' were pursued as McDougall and Bruce proposed, it held no imperial relationship sacred: it was an emergent internationalism that transcended the Empire. If it worried some politicians and officials in London, it could have caused much more concern in Canberra.

Jebb later recalled that 'the drive to "Economic Appeasement" began to peter out' from early 1937, though debates on financial and economic assistance and approaches to Germany continued.[87] Neither Bruce nor McDougall was yet ready to abandon the campaign. McDougall continued to develop his ideas in memoranda. The Van Zeeland report, finally completed in January 1938, proposed an offer of financial aid to Germany and Italy in return for undefined political, financial and commercial concessions in a 'Pact of Economic Collaboration'. But Van Zeeland had neglected what McDougall still saw as the key to success: appeals to the idealism and self-interest of ordinary citizens.[88] By then McDougall was prepared to admit that the dictators had succeeded

---

84 *DAFP* I, 45, 14 June 1937, p. 162.
85 NAA, M104, 5/1, Bruce to Lyons, 23 June 1937.
86 Turnell, 'Butter for Guns', pp. 17–18.
87 Jebb, *The Memoirs of Lord Gladwyn*, pp. 64–5.
88 NLA, MS6890/4/5, 'The Road to Economic Appeasement', 31 January 1938, pp. 2, 5.

in 'persuading their peoples that glory, prestige and military power are more desirable than human welfare', but he clung to a hope that the combined effects of autarky and rearmament on standards of living might yet prove to be their 'Achilles heel'.[89] Once again McDougall suggested a declaration of intention to improve living standards by attention to nutrition, housing, efficiency and diversity of agriculture, together with settlement of problems of raw materials, colonial trade and migration, calling for a 'vigorous lead' from Britain and France in association with the League. He added: 'the United States of America appears to be the only country where Government is attempting to interpret its efforts towards international economic co-operation in terms which will appeal to the average citizen.'[90] In spite of his concerns about US isolationism, McDougall had identified the direction in which his own ideas were tending with other strands of thought in the United States.

Bruce felt strongly enough to take the issue to new Prime Minister, Neville Chamberlain, in a letter that followed the lines of 'The Road to Economic Appeasement'.[91] An unidentified Foreign Office commentator described Bruce as 'next to Mr Cordell Hull the principal apostle of economic appeasement, i.e. of a relaxation of protective measures as a necessary prelude to political settlement'. The commentator took issue with a claim in Bruce's letter that attempts to solve economic problems had failed because they did not appeal to the interests of ordinary people. There were other reasons, including governments putting military aggrandisement ahead of economic advantages. Of the democracies, neither France, hampered by financial difficulties and a weak franc, nor the United States, its Congress dominated by isolationists, could play a significant role in international economic action. Great Britain had done all it could: it had no exaggerated protective measures and had modified its trade barriers against manufactures. Its best contribution should be conclusion of a bilateral agreement with the United States. He added that most matters raised by Bruce were already the subject of action by the League of Nations. Bruce had made no specific requests, other than a British–American declaration in favour of peace, limitation of armaments, higher standards of living, fewer obstacles to trade, maintenance of commodity prices and stabilisation of currencies. The commentator doubted the usefulness of a non-specific manifesto to which totalitarian states were not party. It might have some 'psychological value', but the League was dependent on national states for practical action. A chilling conclusion warned that accentuating the difference between living standards in the democracies and those of the totalitarian states might 'make it more difficult for the democratic powers to organise their peoples for the sacrifices that they may be called upon to make'.[92]

---

89 Ibid., pp. 1, 4.
90 Ibid., pp. 6–9.
91 UKNA, FO371/22517, 125552, W3279/41/98, Bruce to Chamberlain, 25 February 1938, pp. 171–7.
92 Ibid., p. 177.

Attention had shifted from avoiding war to preparing for it. Late in 1938, shortly after the Munich Agreement, McDougall joined Orr and Hall in an attempt to persuade the British Government 'to take seriously and put into practice' consumption and nutrition policies, to increase empire preparedness and foreign support. Notes of the discussion reflect an atmosphere of crisis. Having failed to interest the Prime Minister, they proposed to lobby the City and industry. Preparedness and national unity demanded moral leadership and economic strength. Policies to increase consumption would persuade less prosperous Britons that their own social system was worth defending; sound nutrition would contribute to defence and to health. McDougall now used 'economic appeasement' as a broad term for reform of commercial policy to combat subsidised foreign exports; to provide loans and technical advice to assist reorientation of agriculture in Eastern and south-eastern Europe, the Near and Middle East and British colonies; and for any measures to increase prosperity throughout the world. Sympathetic support for Britain would depend on 'really liberal policies in regard to such questions as trade with non-self-governing areas and availability of raw materials, allowing foreign countries the fullest access to markets'.[93] The only answer to a looming US recession, affecting raw materials and world economies, he argued, was increased consumption; nutrition policies would be 'a sure basis for both preparedness for defence and for appeasement…a fresh stimulus to the "peace" trades in this country and a check to the psychology which makes for recession, will react favourably upon our external relations'.[94] Though he urged an agricultural policy adequate for defence, McDougall did not suggest self-sufficiency; losing control of the sea lanes would mean defeat. But the prospect of interrupted sea-borne trade should be faced, and Britain should specialise in products such as fruit, vegetables and dairy products.[95] McDougall was not alone: a study by Viscount Astor and Seebohm Rowntree similarly urged an agricultural policy preparing for war and based on production of protective foods. The authors recommended storage of imported cereals, increased livestock production as 'little granaries of corn' to be slaughtered at need, increased production of protective foods, and 'in the forefront of our proposals a national policy of improved nutrition'.[96]

Economic appeasement, like political appeasement, 'presupposed a rational element in Nazi Germany' seeing that Germany's best interest lay in cooperation; that element existed, but it 'existed powerlessly within an essentially irrational system'. In the months before the Munich Agreement and in his subsequent assurance of 'peace in our time', Chamberlain and those who supported him

---

93 NLA, MS6890/4/5, 'Notes', 22 October 1938.
94 FAO, RG3.1, Series D3, 'Appeasement and Reconstruction', October 1938.
95 Ibid., 'Notes on Agriculture and Defence', 16 November 1938.
96 Viscount Astor and B. Seebohm Rowntree, *British Agriculture: The Principles of Future Policy*, second revised edn, Pelican, London, 1939, pp. 53–8, 265.

failed to understand 'the irrational dynamics of Nazism'.[97] Chamberlain had reached that understanding by 1939. He tried to dissuade Roosevelt from sanctioning Under Secretary of State Sumner Welles' last attempts at negotiation with the Axis powers early in 1940, and spoke to Welles 'with white hot anger' of the German Government's deceit and Hitler's determination to dominate Europe. Welles himself eventually acknowledged Mussolini's 'obsession' to recreate the Roman Empire, and that in Germany 'lies have become truth; evil, good; and aggression, self-defense'. Roosevelt declared on 29 March 1940 that there was 'scant immediate prospect' of establishing peace in Europe, though he privately continued to press the British to state that they wanted only security, disarmament, equal access to raw materials and markets, and not the breakup of Germany. Germany's invasion of Norway and Denmark in April ended any possibility of peace negotiations. The US Administration then depicted the Axis powers as an 'unholy alliance…seeking to dominate and enslave the human race', insisting that experience of the past two years had 'proven beyond doubt that no nation can appease the Nazis' and that there 'can be no appeasement with ruthlessness'. 'Like the British it now knew that the regime in Berlin had to be vanquished.'[98]

Bruce and McDougall would doubtless have endorsed Jebb's account of the mood of the time: 'we all felt in our bones that war was probably coming—it hung all the time like a black cloud at the back of my own consciousness—and that all means of avoiding it must at least be considered, however unorthodox they might be thought to be.'[99] All three had helped provoke discussion on an issue they believed to be vital, and which was already widely discussed. Their efforts became only a tiny strand in the complex web of British policy formulation. In the light of subsequent understanding that their ideas could not have succeeded, 'appeasement', whether economic or political, has lost its sense of peace and comfort. Unfairly, to those who promoted it, the word is now irrevocably tainted with betrayal.

Revitalising the League

McDougall had concluded his memorandum on economic appeasement in late 1936: 'If the nations learn to turn to the League for information, help and advice on economic and social questions, the prestige that has been lost on the political plane may be regained on a firmer basis.'[100] He took up his appointment to the Economic Committee with ideas for 'a wider program [of economic activity]

---

97 MacDonald, 'Economic Appeasement', p. 131.
98 Offner, 'Appeasement Revisited', pp. 589–93.
99 Jebb, *The Memoirs of Lord Gladwyn*, p. 55.
100 NAA/CSIR, A10666, [5], 'Economic Appeasement', 21 December 1936.

than mere [tariff] barrier sweeping'.[101] In July 1937 he predicted that a sense of futility and hostility would inevitably accompany a concentration on political discussions. 'Special prominence' should therefore be given to economic issues in speeches at plenary meetings.[102] There would be no lack of positive material. The 1937 League Assembly would have before it information on progress of a report into means of reducing obstacles to world trade by Belgian economist and Prime Minister, Paul Van Zeeland, commissioned by the British and French Governments, and that of a League committee inquiring into access to raw materials. The Economic Committee's report to the Council would emphasise the importance of measures to improve living standards, and there would be the final report of the Mixed Committee on Nutrition. The Second Committee would have the opportunity to consider his own memorandum on economic appeasement, which was published as an annex to the Economic Committee's report. This body of evidence should present 'a picture of the League of Nations actively concerned to secure a general improvement in human welfare through practical economic measures', resulting perhaps in a 'greater spirit of co-operation than has existed for several years'. The Second Committee should recommend investigation of 'subjects of great significance to economic welfare': avoidance or amelioration of a slump; the standard of living; trade barriers; agricultural credits in peasant Europe; and the extension of the Tripartite Agreement towards progressive decontrol of currencies and other factors affecting monetary policy.[103]

The Economic Committee's report in September, however, was less positive. Removal of obstacles to trade had been slow; self-sufficiency remained a problem; and any trade improvement derived largely from rearmament, which must ultimately 'frustrate any tendency towards the raising of the standard of living'. There remained some hope for improved economic activity. The Mixed Committee had shown 'the unwisdom of maintaining high food prices and limiting exports'. International action might include a joint statement, affirming general objectives that should include peace, prosperity and improved standards of living.[104]

McDougall had hoped in vain for determination in Second Committee to cooperate on reducing trade restrictions. Van Zeeland's report had not yet been completed; it was to prove disappointingly lacking in practical proposals for joint action when it was produced early the following year. The committee's rapporteur gloomily summed up discussion: 'Countries did not seem disposed

---

101 LN, 1933–40, 10A, 27571/357, McDougall to Stoppani, 5 February 1937.
102 NAA, M104, 5/1, 'The Keynote of the Next Assembly', 20 July 1937.
103 Ibid., pp. 2–5.
104 LN, C358, M242, 1937 IIB, Economic Committee: 46th Session, Report to Council, 'Remarks on the Present Phase of International Economic Relations: the Carrying out of the Program of the Tripartite Declaration', 10 September 1937.

to try and find suitable means…to enable them to emerge from their isolation and enter the world flow of trade once again.'[105] McDougall's own paper drew almost no comment. Enthusiasm for economic appeasement was muted. He did succeed in amending a sentence about the effect of improved economic relations on political relations from 'there *might be some* justification for hope' to 'there *is indeed* justification for hope'.[106] Anodyne resolutions reflected the equivocal nature and outcome of the debate. requesting the Economic and Financial Committees to continue their studies into ways of removing trade barriers and exchange controls, appealing to 'all countries concerned to lend every possible support…in order to arrive at practical results'.[107]

Plans to extend the work of the Economic Committee were more successful. Bruce told the Assembly he could not accept a view that political appeasement must precede economic cooperation: 'our political difficulties arise…in a considerable measure from economic causes. With poor and insecure living standards, with low incomes, a poor scale of consumption, with fear of unemployment ever present, individual and family life becomes depressed and hopeless.' In these circumstances 'people are driven to seek distraction or inspiration in exaggerated forms of nationalism and in dreams of national aggrandisement'. Bruce reminded the Assembly that economic and financial problems affected people's lives, 'the way which human beings are fed, clothed and housed, and on the quality of health and general welfare which is possible for them'. The response to his invitation to the Assembly to consider the question of nutrition had exceeded anything he had dared hope for, but had also revealed the extent of malnutrition and its effects; he now called upon the League to 'face realities' and to exercise 'constructive and heartening' leadership on economic and social questions. 'By these means we shall best restore the prestige of the League in the eyes of the world and carry out the great responsibilities that rest upon us.'[108]

As McDougall had suggested in 'Keynote of the Next Assembly', Bruce followed this remarkable speech by successfully proposing a series of League inquiries into means of raising standards of living, of prevention or mitigation of depressions, and into agricultural credits with special reference to Central and Eastern Europe. In the first step towards the Bruce Report, another resolution invited the council to inquire into the structure and functions of the Economic and Financial Organization.

---

105   Ibid., *Official Journal 1937, Special Supplement 171*, Minutes of the Second Committee, M. René Brunet of France, introducing his draft report 30 September 1937, p. 95.
106   Ibid., p. 97.
107   Ibid., Annex 5, p. 126; *Commonwealth Parliamentary Papers: 1937–38–39–40*, Vol. III, no. 42, 'Report of Australian Delegation to the Eighteenth Assembly of the League of Nations, September–October 1937', Appendix D.
108   Ibid., Appendix A, speech by Bruce, 21 September 1937.

A leading article in *The Times* commended Bruce's constructive approach at the 1937 Assembly, linking it with Cordell Hull's bilateral treaties campaign and the idea of cooperation in the cause of peace between nations within and outside the League. The editorial commented that Bruce 'clearly had in mind' McDougall's 'instructive memorandum' of which it had published a full summary a few days earlier.[109] The leader writer doubtless knew exactly whence Bruce drew his inspiration, given McDougall's close ties with Printing House Square. Turnell suggests that from the nutrition campaign until after World War II, *The Times* gave 'saturation coverage' to McDougall's and Bruce's ideas.[110]

Standard of Living Inquiry

The Economic Committee's agenda in December 1937 included preparation of a regime of permanent guarantees for free circulation of raw materials; demographic questions and emigration; and preliminary discussion of the standard of living inquiry proposed by Bruce. On McDougall's recommendation, Professor Noel Hall was invited to draft a substantial paper to guide that inquiry, subject to discussion with a small subcommittee, which included McDougall. Stoppani expected that directions for the major paper would be prepared jointly by Hall and McDougall.[111]

Hall's memorandum, 'National and International Measures to be employed for raising the Standard of Living', was circulated to the Economic Committee as a basis for discussion. Many of McDougall's ideas are represented within it, not least his emphasis on the central importance of raising standards of living. But whereas McDougall's aim was always to simplify and persuade, this paper—the work of a professional economist—stressed the complexity of the problem. There was no easy answer. On the contrary, there was danger in promoting a general international or national policy without proper consideration of all its aspects. A 'practical study of the economics of consumption' was needed, for both national and international levels. Problems of the organisation and control of production, of finance and of international trade should be combined with a study of consumption, the constituents of real income and the capacity of the individual to increase his standard of living. Price movements had variable and complex effects on consumption. He noted that higher prices lowered consumption of 'protective foods' to a greater degree than that of energy foods, and referred to McGonigle's much quoted study, 'Poverty, Nutrition and the Public Health'. The paradox of poverty in the midst of plenty could only be

---

109   *The Times*, 22 September 1937, Summary of 'Economic Appeasement', 18 September 1937.
110   Turnell, 'Butter for Guns', p. 16.
111   LN, 1933–40, 10A, 31123/31123, part 1, McDougall to Stoppani, 13 October; Stoppani to McDougall, 30 October 1937.

resolved by painstaking study of prices, of demand and of effects upon costs and earnings of changes in the relative quantities of different goods produced. All of these touched on government policies, but 'policy can only be rightly framed when all aspects of the problem have been considered'. Hall therefore proposed a study 'distinctive in the fact that it will emphasise the inter-relationships of previous sectional inquiries', but focused on the problem of raising the standard of living. 'It is this problem which is capable of providing a new dynamic to economic policies both national and international.' The inquiry would involve 'pioneer work in a subject which is co-extensive with the whole international economic system'.[112] In at least one sense this was McDougall's document: it proposed exactly what he hoped the Economic Committee would do. Sadly, time ran out before Hall's plans could be achieved.

Following the report of the first committee led by Bruce to inquire into the structure of the League Economic and Financial Organization in 1937, officials in Whitehall discussed a proposal by League Secretary-General, Joseph Avenol, for an 'economic superman' to coordinate the work of the Economic and Financial Committees and to 'inspire and activate the various inquiries' then under consideration by both committees. The records demonstrate a general approval of Bruce the man, if not of his economic views, but little enthusiasm for McDougall. Ashton-Gwatkin noted that the League Secretariat wanted Bruce for the role, and that F. P. Walters, a senior league official, believed an otherwise very busy Bruce would be able to manage, 'with McDougall's help'.[113] Leith-Ross had 'great admiration for Mr. Bruce's personality' but thought he lacked competence to deal with technical questions, adding, 'you know McDougall as well as I do, and I can only say he has just joined the Economic Committee himself'. It would be a mistake, he argued, to put the work of technical experts under a political director.[114] R. M. Makins suggested McDougall hoped for the position of an economic *éminence grise*. The Junior Foreign Office Minister, Lord Cranborne, noted:

> No doubt Mr. McDougall will move heaven and earth to get [Bruce] to accept. Mr M. is, and always has been, an intriguer of the first order… but whatever happens, we should not risk hurting Mr. Bruce, who has very definite views on economic subjects, to which it is probably true that he feels that HMG have not always been over-sympathetic.[115]

The offer was made, but Bruce did not accept it.[116]

---

112   Ibid., League Document E.1008, 30 November 1937.
113   UKNA, FO 371/22516, note by Ashton-Gwatkin, 10 January 1938.
114   Ibid., letter from Leith-Ross to Ashton-Gwatkin, 12 January 1938.
115   Ibid., memorandum by R. M. Makins, 20 January 1938.
116   Lee, *Stanley Melbourne Bruce*, p. 124.

The Bruce Report

The League's 'technical' work strengthened as its political authority waned in the 1930s, and greater collaboration between League bodies and others increased to that end. The Health Organization took initiatives of its own; the Mixed Committee on nutrition broke through boundaries and drew on experts within and beyond the League, encouraging and cooperating with national nutrition committees around the world. A similar pattern was undertaken for work on housing. Yet, at the top, technical agencies remained subject to a Council whose diplomats and politicians lacked the expertise to do more than authorise circulation of technical reports for debate in the Assembly. A system suited to promoting formal agreements between governments was 'totally inadequate' to new forms of action. 'What was needed was a directing body of responsible ministers corresponding in authority to a Council of Foreign Ministers and meeting…regularly and frequently.'[117]

In 1937 a committee chaired by Bruce therefore suggested changes in control of the Economic and Financial Organization of the league. These proposals were elaborated by the Committee for the Co-ordination of Economic and Financial Questions, which met in May 1938—again with Bruce as Chairman. A year later, he chaired the Special Committee on the Development of International Co-operation in Economic and Social Affairs, generally remembered as 'the Bruce Committee'. Its report, 'The Development of International Co-operation in Economic and Social Affairs: Report of the Special Committee', was submitted in August 1939 and known as the 'Bruce Report', recognising 'the decisive contribution made to it at every stage by the Australian statesman'.[118]

The Bruce Report recommended a permanent change in terminology and in structure. It argued the inadequacy of the phrase 'technical problems' and of the distinction made in the League between 'political' and 'technical': 'So called "technical problems" are in every country political questions, frequently the cause of internal controversy and often necessitating international negotiation.' The committee therefore used the term 'Economic and Social Questions'. It pointed to

> the growing extent to which the progress of civilization is dependent upon economic and human values. State policies are determined in increasing measure by such social and economic aims as the prevention

---

117 Walters, 'The League of Nations', pp. 36–7.
118 Ibid., p. 38. The report was officially published as LN, A.23. 1939. A copy included as a special supplement to the *Monthly Summary of the League of Nations*, August 1939, is in NAA, M104, 7/1. The following page numbers are cited from this copy.

of unemployment, the prevention of wide fluctuations in economic activity, the provision of better housing, the suppression and cure of disease…

Modern experience has also shown with increasing clearness that none of these problems can be entirely solved by purely national action. The need for the interchange of experience and the co-ordination of action between national authorities has been proved useful and necessary time after time in every section of the economic and social fields.[119]

The League was not simply a body to prevent war: more than 60 per cent of its budget was devoted to economic and humanitarian work, benefiting both members and non-members. All countries faced similar problems and could benefit from sharing experience with other states. All were concerned with the economic welfare of their citizens and with nutrition, housing and health: 'And all these questions are subject to scientific treatment. What is required therefore, and what is being accomplished, is a joint and intensive study of those common problems on which the security of all nations and all classes of the population depends.'[120]

The report argued for fuller participation of non-member states and more central direction in view of the 'growing intertwining of the different branches of the work'. It sought greater publicity and a platform for discussion to make use of 'the only really potent instrument of progress—an enlightened public opinion'.[121] It proposed that economic and social work be overseen by a new 'Central Committee for Economic and Social Questions', comprising representatives of 24 states and up to eight experts, with member and non-member states participating on an equal footing.[122] The report was approved by the League's last full Assembly on 15 December 1939. Its principles were to be 'effectively embodied' in the Economic and Social Council of the United Nations (ECOSOC).[123]

The Bruce Report stressed the interconnection of social and economic problems: 'social welfare, the care of the child, and the protection of the family, link up directly with the problems of better housing and of better feeding. These in turn are in many ways dependent on economic conditions, on transport facilities and on methods of taxation.'[124] It has been suggested the report 'smacked of a radical humanism rather foreign to the international organisations set up in the 1940s', and that Australian policy on economic questions in the 1940s, such

---

119   'Bruce Report', p. 6.
120   Ibid., p. 8.
121   Ibid., p. 17.
122   Ibid., pp. 19–22.
123   Walters, 'The League of Nations', p. 38.
124   'Bruce Report', p. 15.

as the full-employment approach—sometimes seen as 'essentially novel'—was foreshadowed in the 1930s by men such as McDougall and Bruce, who were aware of the 'potentially radical significance of what they were doing'.[125]

The Last Assembly

Bruce had done what he could to help bring McDougall's vision into being, but there were limits to optimism. By 1938 he gave much of his attention to matters of rearmament in Britain and in Australia. He oversaw preparations to protect occupants of Australia House against gas, splinters and incendiary bombs. An air-raid shelter for 500 people was created in the basement, with air filtration, a decontamination centre, internal emergency lights and power supply, sandbags and protective window coverings. Beyond that, he reported to Canberra, protection of the building against high explosives was not practicable.[126]

Bruce and McDougall did not abandon the League on the outbreak of war. In October 1939, Bruce, Stirling and McDougall agreed with Australian Prime Minister, R. G. Menzies, that the League's political activities 'should…be put into cold storage', but there should be 'a strong expression of the Commonwealth Government's appreciation of the value of the economic and social work of the League' and a decision to implement the Bruce Report, which might well 'tend to strengthen the general position of the League and may even tend to attract back support of countries such as Italy, Hungary, Yugoslavia and Spain'.[127]

In December 1939, during the uneasy calm of the 'phoney war', the Soviet Union invaded Finland and three days later Finland appealed to the League. Although Britain and France had decided nothing could be achieved by a meeting except a further demonstration of League weakness, Council and Assembly were now obliged to meet. Despite public sympathy for the Finns, Britain and France were not willing to risk strengthening Soviet ties with Germany or further discrediting the League. Bruce therefore advised Canberra that 'delegates of secondary importance' should attend; he was leaving representation at the League's Twentieth Assembly to McDougall, who was already in Geneva.[128]

---

125   Hudson, *Australia and the League of Nations*, pp. 178–9.
126   NAA, A1608, AA15/1/1, cable from Bruce to Prime Minister's Department, 16 March; letter from Bruce to Lyons, 5 October 1938; undated memorandum, 'Air Raid Precautions—Australia House'.
127   DFAT, Stirling Diaries, 28 October 1939; *Documents on Australian Foreign Policy 1937–49. Volume II: 1939* [hereinafter *DAFP* II], eds R. G. Neale, P. G. Edwards, H. Kenway and H. J. W. Stokes, Department of Foreign Affairs, Australian Government Publishing Service, Canberra, 1976, Document *312*, cable 594, Bruce to Menzies, 28 October 1939, pp. 358–9.
128   Ibid., *413*, Cable 710, Bruce to Menzies, 8 December 1939, pp. 453–4.

This 'last piece of meaningless ritual' must have been a dismal experience.[129] McDougall was instructed to support an Assembly resolution condemning the Soviet action, which was passed with some abstentions. He was to point out that Commonwealth Government policy was not to impose sanctions unless applied by the whole League membership, but the question of sanctions was not raised. He should consult Canberra and work with the British if the Assembly were asked to give an opinion on the expulsion of the Soviet Union. The Assembly was not consulted; Council members voted in favour with some abstentions.[130]

With hindsight, ironies can be seen in other business undertaken. In debate over a new scale of contributions, McDougall protested against the relative treatment of Australia and Argentina, and the Fourth Committee agreed the scale should apply for one year rather than three. McDougall stressed the need to reduce Secretariat expenses, especially on political activity. The total League budget was cut by one-third, and League and ILO staff agreed to accept salary reductions ranging from 2 to 20 per cent. A policy of staff reduction would continue. A new Council was elected, but would never meet. Delegates returned home to face the real war; staff of the technical organisations relocated to the United States. 'The immense palace of the League was empty and silent.'[131]

Bruce and McDougall continued to push for reformation of the League. Bruce acted promptly to gain British reassurance that a rumoured intrigue by the ILO would not affect British support for implementation of the Bruce Report.[132] In February 1940, McDougall attended a meeting of the organising committee for the recommended new Social and Economic Central Committee. Before that meeting he wrote to J. H. Willits, Director of the Social Science Division of the Rockefeller Insititution, seeking help for a shortfall in the League's budget. A second letter to Willits, written in very different circumstances on 5 June 1940, maintains an obstinate hope:

> ...there may appear to be little justification in believing that it is an urgent matter to consider the problems of post-war reconstruction, yet it may well prove that from next October...those of us who feel vitally concerned with this subject may be convinced that there is little time to lose. I feel these things so strongly that I have asked Mr. Bruce to consult with the British Government...and I hope that, as a result, decisions

---

129   Hudson, *Australia and the League of Nations*, p. 94.
130   *DAFP* II, *414*, Prime Minister's Department to Bruce, 11 December; *415*, McDougall to Department of External Affairs, 12 December; *423*, High Commissioner's Office to Prime Minister's Department, 20 December 1939, pp. 454–6, 468–9.
131   Walters, *A History of the League of Nations*, quoted in Hudson, *Australia and the League of Nations*, p. 94.
132   NAA, M100, 1940, records of interview with Junior Foreign Minister R. A. Butler, 2 and 5 February.

may be taken to preserve a strong economic and social organization of the League and with that part of the ILO which can make effective contributions to the subject of post-war reconstruction.[133]

Conclusion

The efforts of Bruce and McDougall to rejuvenate the League were defeated in the short term by time and circumstance. But the recommendations of the Bruce Report, and its aspirations, were reflected in the establishment of the Economic and Social Council of the United Nations.

McDougall's attempt to deal with the problem of Germany had expanded the 'marriage of health and agriculture' to something much broader: an agenda to draw the poor and the hungry away from the forces of totalitarianism and communism. His thinking had moved beyond the British view of 'economic appeasement' as a renewal of old ideas of 'treaty revision' to contain Germany's territorial ambitions and encourage a reversal of autarky, to something closer to the US version, which added to those aims a questioning of existing structures, including the British Empire and its preferential systems, looking instead to a new internationalist social order. Washington wanted to create a 'New Deal for the World'. McDougall and Bruce were ready to embrace that vision.

---

133   FAO, RG3.1, Series 1, file 1.C, McDougall to Willits, 30 January and 5 June 1940.

Part C: A New Deal for the World

Prologue

McDougall visited Washington with Bruce briefly in 1938, and for longer periods alone in 1941 and 1942. There his ideas were well received; the ground had already been prepared.

In nineteenth-century America, 'Social Darwinism' inspired Andrew Carnegie's *Gospel of Wealth*, which argued in favour of accumulation of wealth, but also that wealth must be used for the benefit of society. Inherent in this view was capitalism without restriction. 'Progressivism' developed as an informal movement aiming to mitigate the harsh effects of unregulated capitalism through measures including democratic reform, regulation of corporations and monopolies, labour rights and social justice. Many progressives campaigned for professionalism, rationality and efficiency in government. Scientific progress and training were essential to that efficiency.

Progressivist thinking underlay much of the 'New Deal' promised by Roosevelt in 1932; the influence of progressivism can be seen in the careers and thinking of many key figures in his administration. In the early 1940s, some of the most senior members would influence and be influenced by McDougall. Foremost amongst these was Vice President Henry Wallace, a former Republican who campaigned for Roosevelt in 1932. As a precocious student, Wallace's interests had ranged from plant genetics and agricultural economics to quantitative analysis. He developed statistical correlation techniques in pioneering corn-hog ratio studies; experiments with corn cross-breeding brought him wealth and fame. From Thorstein Veblen, he took up the idea of 'cultural lag': the inability of economic and social institutions to keep pace with advancing technology. In his first book, *Agricultural Prices*, published in 1920, Wallace 'hoped that contemporary business civilization would soon give way to the rule of "production engineers" and "statistical economists"'. This reflected the ideas of 'evolutionary positivism': a belief that 'progressive social change occurs naturally as more information becomes available, enabling men to use their reason more effectively to control environment. Eventually the data become the mechanism of social change, their mastery and administration the solution to social problems.'[1] McDougall, the self-taught collector of statistics and proponent of thorough investigations, might not have described his views in so theoretical a fashion, but this accorded closely with his approach to effecting change.

---

1 Norman D. Markowitz, *The Rise and Fall of the People's Century: Henry A. Wallace and American Liberalism, 1941–1948*, Free Press, New York, 1973, pp 12–14.

In March 1933 Wallace was appointed Agriculture Secretary, a position earlier occupied by his father, and became one of the Administration's foremost advocates of economic planning. He persuaded Roosevelt to adopt the domestic allotment plan, production control through voluntary crop reduction, which became the basis of the first Agricultural Adjustment Act (AAA). Wallace's book *New Frontiers* portrayed the New Deal as 'an adventure in planning and popular participation—a movement towards economic democracy'; he defined the New Deal as 'an attempt to mediate between the extremes of total security and total freedom'. Like most New Dealers, Wallace stressed the need for mass purchasing power and high levels of consumption to achieve recovery. Administration supporters portrayed him as 'an idealist and philosopher, a dreamer who epitomized the best hopes of the New Deal'.[2] By the late 1930s, his Department of Agriculture led the way in social programs, providing loans to tenants and farmers and relief for the rural poor, production and general agricultural planning, 'ever-normal granary' measures of crop insurance and storage, and food-stamp programs that aimed at both relief for the poor and disposal of surplus produce. He was also held responsible for failures of some programs to achieve their intended goals and for 'destruction of ten million acres of cotton and the slaughter of six million little pigs' under AAA crop-reduction programs. But, like Bruce and McDougall, Wallace thought that 'to have to destroy a growing crop is a shocking commentary on our civilization... made necessary by the almost insane lack of world statesmanship' from 1920 to 1931.[3]

During McDougall's visit to Washington in 1941, Wallace was appointed head of an Economic Defense Board, which was described as a 'sort of ministry of economic warfare'. McDougall sent Bruce a cutting referring to the position giving Wallace 'jurisdiction over post-war planning' and putting him 'in an advantageous position to develop a program which might later serve as a platform for his Presidential candidacy'.[4] Wallace suggested McDougall see economist Winfield Riefler, who was Adviser to the board, following 'a significant career' with the Federal Reserve Board. He had been Economic Adviser to the Executive Council during the New Deal, advising both Treasury and the Department of Agriculture, and had also been Chairman of the Central Statistical Board.

Riefler was a graduate of experimental teaching programs at Amherst College and at the Robert Brookings Graduate School, where students were exposed to the 'institutional approach' to economics. Institutionalism saw economics as a study of the nature and functioning of the 'economic order' and as relevant

---

2   Ibid., pp. 20–1.
3   Ibid., p. 26; Walter J. Samuels, *Henry A. Wallace and American Foreign Policy*, Greenwood Press, Westport, Conn., 1976, p. 37.
4   NAA, M104/1, 9(6), letter, 1 August.

to 'the problem of control' rather than exercises in 'formal value theory'. The graduate school, established to prepare graduates for policy research and senior public service positions, taught 'the art of handling problems rather than... accumulated knowledge'. Its aim was to produce 'craftsmen who can make contributions to the intelligent direction of social change'. It emphasised 'self-motivation, wide-ranging intellectual curiosity, and...the analysis and creative solution of social problems'.[5] McDougall knew Riefler as a member of the Financial Committee of the League of Nations and again when Riefler became Minister in charge of Economic Warfare at the US Embassy in London in 1942. They, too, had attitudes in common. The Amherst–Brookings approach could well describe much of McDougall's own. Graduates of these programs formed an influential network in the 1930s and 1940s. Many occupied key positions in government and in movements for social and institutional reform and international cooperation.[6] These included Stacy May, Assistant Director of the Rockefeller Social Sciences Division until 1942, whom McDougall met in New York in 1938; Mordecai Ezekiel, Economic Adviser to the Secretary of Agriculture from 1933 to 1944 and later Special Assistant to the Director-General of FAO; and Isador Lubin, Commissioner of Labor Statistics and a Special Adviser to Roosevelt during World War II.

Assistant Secretary of State Adolf Berle was an important mentor to McDougall during his 1942 visit. Berle was an original New Dealer, a 'brain-truster' in Roosevelt's 1932 election campaign and known to be personally close to the President. A brilliant corporate lawyer, he was co-author of 'one of the most influential books of the twentieth century', *The Modern Corporation and Private Property*, written with the assistance of a young economist, Gardiner C. Means, and published in 1932. It argued for government responsibility to ensure that corporate power worked for the benefit of individuals—a middle way between socialism and capitalism—and was described by *Time* magazine as 'the economic bible of the Roosevelt administration'. It laid the foundations for much of the industrial, banking and finance legislation of the New Deal.[7] Roosevelt appointed Berle Assistant Secretary in the State Department early in 1938, suggesting the department needed 'an adventurous mind'.[8] In 1938 there were fears of renewed depression; Berle was, writes his biographer, probably the highest-ranking official in Washington conversant with Keynes's ideas. He acted in the State Department as 'a free-lancer with special access to the President', his primary mission 'to focus Roosevelt's and Washington's economic thinking...on

---

5 Malcolm Rutherford, 'Walton Hamilton, Amherst, and the Brookings Graduate School: Institutional Economics and Education', (Draft), September 2001, <http://ideas.repec.org/p/vic/vicddp/0104.html> [accessed 2006], pp. 2, 16, 17.
6 Ibid., pp. 24–6.
7 Jordan A. Schwarz, *Liberal: Adolf A. Berle and the Vision of an American Era*, Free Press, New York, and Collier Macmillan, London, c. 1987, pp. 55–62.
8 Ibid., pp. 110–11.

recovery schemes consistent with business realities and America's place in the world'. He advocated increasing America's national wealth by means of 'pump-priming' through an enhanced Reconstruction Finance Corporation (RFC).[9] Berle had predicted that the United States might find it necessary to use 'the enormous resources of the Federal Reserve system as a means of rebuilding the shattered life' of Europe after the war. He and Roosevelt shared a conviction that 'the time for making peace was during war'. A firm anti-imperialist, Berle believed that

> Britain must realize that her postwar world influence did not rely upon empire but rather 'on her moral and intellectual ability to bring about common action among a great number of nations'…A poorer Britain could not afford a great world role, but the gradual relaxation of its imperialism would encourage American generosity that surpassed even lend-lease.

Foreign aid to Europe and undeveloped countries made sense in Washington, which needed a postwar 'system of open finance so that no country shall find itself short of supplies because it is short of exchange'.[10]

Nutrition had become an important field of study in US academic and philanthropic institutions. The Rockefeller International Health Board was concentrating on methods of assessing nutritional deficiencies and their significance to health. The Wall Street-funded Milbank Memorial Fund, a medical research body directed by Dr F. G. Boudreau, formerly of the League Health Organization and an old ally of McDougall, devoted much of its resources to nutrition studies. Its journal in the late 1930s carried articles on various aspects of the subject, including a review of the League Mixed Committee Report.[11] Boudreau had contributed one on the international campaign for nutrition, which concluded:

> The relation of better nutrition to peace may seem very remote. But there is no single road to peace, and if in the attempt to improve national nutrition, the governments succeed in promoting a fuller measure of social justice and in doing away in part at least with economic nationalism in the interests of health, it may be that the real objective of the League and of the Labour Office will not seem so remote as it appears today.[12]

Boudreau, says Daniel M. Fox, was committed to social medicine and focused on means to promote positive health; his 'strongest personal interest was in policy for food and nutrition'. Fox also notes a longstanding policy endorsed

---

9   Ibid., pp. 118–19.
10 Ibid., pp. 211–12.
11 H. D. Kruse, 'Ever Normal Nutrition', *Milbank Memorial Fund Quarterly*, Vol. XVI, no. 1, 1938, pp. 110–18.
12 F. G. Boudreau, 'The International Campaign for Better Nutrition, *Milbank Memorial Fund Quarterly*, Vol. XV, no. 1, 1937, pp. 104–20.

by Albert G. Milbank, President of the fund board, that 'the fund should only advocate policy that already had some support among leading public officials; to be radical was to risk irrelevance'.[13] Boudreau presumably knew that others agreed with his views.

Postwar planning had been receiving attention in Washington well before Pearl Harbor. Influential people from the President down believed that the mistakes of 1919 could only be avoided by early attention to planning. The Roosevelt Administration sought to avoid the errors made by Woodrow Wilson in 1918–19, including a failure to involve political opponents, particularly isolationists, in discussion, and delaying consideration of the peace until the war was won. A lesson in line with McDougall's arguments was the need 'to seek a more integrated vision of collective security through combining political and military co-operation with economic security…[a] precept based on the perceived successes of some of the League of Nations-affiliated agencies…as well as the harsh experience of interwar economic diplomacy.'[14] Many must have felt, like Berle, that American resources would have to be used.

McDougall's ideas appealed to New Dealers. Their approaches to planning and consumption accorded with his, and the Amherst–Brookings institutionalist approach gave a theoretical basis to his methods. Speeches by Wallace, Welles and others, urging attention to social justice, prepared his way. As far back as 1934 Berle had written: 'A question has been asked why, in a civilization over-full of material things, more than able to supply every human need, the organization of economics leaves millions upon millions of people in squalor and misery?'[15] McDougall asked that question in memorandum after memorandum. Herbert Hoover, who had headed a national food authority during World War I and was chief organiser of relief food supplies for Belgium, was 'among the earliest and most forceful proponents of a novel strategic concept that linked security to social welfare'. Hoover argued that 'famine breeds anarchy. Anarchy is infectious, the infections of such a cess-pool will jeopardize France and Britain, [and] will yet spread to the United States'; nationalism and Bolshevism would not cure hunger or unemployment, but 'desperate populations would take up radical creeds'.[16] McDougall used that argument too.

While isolationism persisted as a significant, and indeed strengthening, force in US politics in the early 1940s, Elizabeth Borgwardt argues that 'a new iteration of the New Deal was becoming nothing less than America's vision for the postwar

---

13 'The Significance of the Milbank Memorial Fund for Policy: An Assessment at its Centennial', *Milbank Quarterly*, Vol. 84, no. 1, 2006, <http://www.milbank.org/quarterly/8401PN.html>
14 Borgwardt, *A New Deal for the World*, pp. 14–15.
15 Schwarz, *Liberal*, p. 105.
16 Nick Cullather, 'The Foreign Policy of the Calorie', *The American Historical Review*, Vol. 112, no. 2, 2007, pars 26–7, <http://www.historycooperative.org/journals/ahr/112.2/cullather.html> [accessed 3 July 2008].

world'. Proponents included journalists, social welfare activists, academics, professionals and church leaders as well as elected political leaders and bureaucrats.[17] She identifies three 'idioms' in the US approach to international problems. The 'legalistic idiom' sought solutions in codes, conventions and arbitration. A 'moralistic idiom' after World War I brought declarations of principle, rather than the 'neutral amorality' of the legalistic idiom that had proved 'tragically inadequate'. Yet the moral approach was partly responsible for the weak international response to Japanese aggression in Manchuria, as 'ineffectual intellectual moralism met studied American indifference'. The 'New Deal idiom' looked for 'sweeping institutional solutions to large-scale social problems'.[18] The domestic New Deal had persuaded Americans that central government was able to tackle 'seemingly intractable problems'. It had also shown 'a connection between individual security and the stability and security of the wider polity' and that government must help individuals achieve that security.[19] 'A groundswell of public opinion…had developed internationalist and interventionist sensibilities on an epic scale, by means of direct personal experiences' so that 'by the end of the 1930s it was a truism that turning one's back on the international economy would have unfavourable domestic repercussions'.[20] McDougall was to encounter this aspect of American politics in Washington.

---

17   Borgwardt, *A New Deal for the World*, p. 50.
18   Ibid., pp. 61–70.
19   Ibid., pp. 77–8.
20   Ibid., pp. 86, 96.

8. The War of Ideas

Summary

Two influences helped reshape the nutrition campaign in the early years of World War II. One was McDougall's new responsibility at Australia House for 'political warfare'. The other was America. McDougall's thinking developed along three intertwining paths: the idea for an international body to promote nutrition; the overwhelming need for British–American cooperation; and ways to harness the nutrition idea to the needs of wartime propaganda.

Bruce and McDougall visited the United States briefly in 1938, and discussed the idea of McDougall, Orr and Hall for an international food institute. The experience gave McDougall some understanding of how the approach might appeal to Americans and the importance of American–British collaboration. In London both men worked to incorporate the nutrition approach into positive and constructive war aims, and into British domestic policy. McDougall's major memorandum of 1940, 'Notes on the Re-statement of Our Aims', written after the fall of France, stresses the need for Anglo–American collaboration in the new world order and draws upon ideas from the United States.

Roosevelt's call for 'freedom from want' provided a new impetus for their arguments. The US Ambassador, John Winant, became a valuable ally in London and in Washington. When McDougall returned to Washington in 1941, he had the support of Vice President Henry Wallace, and senior officials in the Departments of Agriculture, Health and State. Plans for a formal agreement and a conference between US and British Empire experts in 1942 were agreed, but prevented by events following Pearl Harbor. McDougall's ideas were influenced by a briefing in London on political warfare and the need for effective material to persuade peoples of occupied and enemy countries that the allied nations offered something positive. His seminal memorandum 'Progress in the War of Ideas' was written in response to that need.

The Approach of War

A proposal by McDougall in 1938 for an organisation to promote nutrition policies owed much to the models of the Empire Marketing Board and Imperial Agricultural Bureaux in the 1920s. In discussions with Orr and Hall, a new idea was added: the involvement of financial and industrial groups in an

'International Food Institute...to promote the increased consumption of food along the lines indicated by the newer knowledge of nutrition'. The functions of the institute would include interpreting in economic terms and then popularising scientific literature on nutrition and 'seeking ways to reconcile national policies on agriculture and commerce with policies for sound nutrition'. It would work closely with the Economic Organisation of the League of Nations, the Imperial Bureau of Nutrition at Orr's Rowett Institute and the national nutrition committees being formed as recommended by the League resolution on nutrition. It would not engage in publicity, 'indeed its work would have to be objective', but development of 'nutrition consciousness' and stimulation of demand would benefit many interests, including banks, shipping companies, food exporters and suppliers of fertilisers, tin plate and gas. All might contribute to its support, much as they might allocate funds for advertising.[1]

McDougall took this idea with him on a visit to Washington in December 1938, accompanying Bruce on the first leg of a visit to Australia. Apart from a brief time in New York after the Ottawa conference, it was McDougall's first experience of the United States.[2] They found the Administration 'prepared to take a marked interest' in the nutrition approach. Bruce detected 'latent keenness' in talks with Henry Wallace and with Norman Davis, foreign policy and financial adviser to the President and head of the American Red Cross. McDougall met key officials in the Departments of Labor, Agriculture, State and Health. State Department officials were 'prepared to see the significance' of a larger US role in an international campaign to increase nutrition consciousness in Europe and its likely impact on greater consumption. Wallace understood its implications for the reorientation of agriculture and US exports, but his officials were more interested in its relevance to disposal of surplus produce. McDougall learned about experimental schemes to dispose of surpluses to the poor by means of subsidised low prices, cheap milk deliveries to certain areas, school lunches and food stamps. Surgeon-General, Dr Thomas Parran, hoped 'our visit would enable him to take up the question with the President and he hoped to make health the main objective and yet to secure support of Mr Wallace and his Department'. On his way home, McDougall visited philanthropic institutions in New York. Stacy May of the Rockefeller Institution suggested the food organisation idea might be put to the institution's European headquarters in Paris, and stressed the importance of demonstrating clearly its objectivity. He offered to fund a visit by Orr to the United States if it were requested by a US government body.[3]

---

[1] NLA, MS6890/4/5, 'Food Policies in Relation to Economic Activity and Appeasement, part 2', 21 November 1938.
[2] Mr Hume Dow, son of then Australian Trade Commissioner in New York, D. McK. Dow, recalled taking McDougall to see the Empire State Building in 1938. Discussion with W. Way, 1985.
[3] NLA, MS6890/3/2, undated letter to Orr and Hall, written in mid-Atlantic and typed in London.

McDougall spent the return voyage to London considering how progress might be quickened and the idea widened to appeal to a transatlantic audience. The proposed organisation might halve the time needed to achieve adequate food policies: an estimated 25 years for more advanced countries might be reduced to 10, and a much longer period for less developed areas could be reduced to 25. In deference to concerns encountered in New York, he now added that it would be 'essential to secure a governing body of such eminence as to ensure the complete objectivity of the Commission's work'.[4] Fresh from his encounters with the US Department of Agriculture, he emphasised that, although public health was the primary justification, increased economic activity would 'produce a general reorientation of European agriculture…the only permanent solution of the world wheat problem'. And, in a nod to the policy of the Secretary of State to liberalise international trade, he wrote:

> A revival of international trade in foodstuffs will favourably affect transport, international finance, and generally all the industries that supply farmers. The need for increased consumption is the justification for policies intended to increase international trade. The policies here advocated are indeed the true fulfilment of those of Mr Cordell Hull.[5]

'Closest collaboration'

Soon after Germany annexed Czechoslovakia in March 1939, McDougall wrote, in a long letter to reach Bruce during his return journey, 'your days in Washington might be made of the utmost importance to the whole future of the British Empire and of our civilisation'. Roosevelt and Hull seemed to be influencing public opinion with their efforts to preserve peace: 'I feel that you perhaps might initiate large scale policies which, if you found they appealed to Washington, you could urge upon London and this might lead to the closest collaboration between the Empire and USA, whether it is to be war or peace.'

In the event of war, Britain should immediately invite allied and neutral countries to draw up the basis for a just peace. Preparation should be made before hostilities began:

> If the U.S.A. can see that the British Empire is not only engaged in fighting for her own 'vital interests' but also for a peace based on justice, and if there is the clearest evidence of this from the very first moment of the war, then we can rely even more emphatically than is now the case upon American support and participation.

---

4   NLA, MS6890/4/5 and NAA, CP43/1/1, bundle 14A/1943/792, 'A Proposal for Accelerating the Movement towards Better Nutrition and Higher Standards of Living', 18 January 1939, p. 15.
5   Ibid., pp. 4, 6.

In a period of tension short of war there should also be a declaration by 'the great "have" countries', combining firm resistance to aggression with a declaration of intent to remove the legitimate grievances of poorer countries.

McDougall added new thoughts to this familiar ground. Henry Wallace had secured legislation in 1938 for a domestic 'ever-normal-granary': stockpiles to prevent food shortages in times of poor crops, ensuring fair prices in times of surplus and protecting consumers from high prices in times of shortage by releasing stored surpluses. Wallace continued to promote the idea for use on an international scale.[6] McDougall suggested buffer stocks of problem raw materials and foodstuffs be internationally administered and financed, and used to moderate prices. His talks in Washington had broadened his perspective, beyond redress of colonial grievances and increased food consumption to the 'far more important' question of bringing about 'a correlation between the productive capacity, either of separate countries or of the world, and the physical requirements of populations'. Establishment of the food commission could give practical expression to a declaration that the governments of Britain, the United States and other creditor nations were prepared to assist other countries develop resources and access external supplies. McDougall concluded from his discussions in Washington that the proposal would find considerable support in the United States. Declaration of such a policy could 'prove of the utmost importance' to peace. While it would not please the Nazis, it could help 'a Germany that had turned away from aggression towards co-operation' and it might be acceptable to Mussolini.[7]

In 1939 there was no separate Australian representation in Washington. Bruce's talks, on instruction from Canberra, were largely limited to the more urgent business of seeking US assurances for support for the British Empire in Europe and in the Pacific. Cordell Hull did show 'he was aware of the work I was doing on the economic side and expressed his admiration of it', while Norman Davis expressed admiration for Orr, who was then in Washington. Davis promised to put Bruce's suggestion for a committee to report on nutrition to Roosevelt, 'when he got the opportunity for a long and quiet talk with him'. Bruce limited his points to the domestic advantages of a nutrition policy. He did not canvass McDougall's suggestions for British–American cooperation and a food organisation.[8]

---

6 Samuels, *Henry A. Wallace and American Foreign Policy*, pp. 46–7.
7 NLA, MS6890/3/2, McDougall to Bruce, 14 April 1939.
8 For Bruce's report to Canberra, dated 8 May 1939, and his records of interview with Davis and Roosevelt on 3 and 4 May respectively, see *DAFP* II, *82*, and attachments 1 and 2, pp. 108–12. A record of his conversation with Hull is in NAA, M104/1, 7(4) (1939).

Political Warfare

At Australia House, the High Commissioner presided over representatives of many Federal Government departments. The External Affairs Liaison Office, located in premises close to the British Cabinet Offices, had been created in 1924 by Bruce as Prime Minister, to be his 'eyes and ears' in London. The first Liaison Officer, R. G. Casey, reported informally and directly to Bruce, though he was officially responsible to the one-man External Affairs Branch of the Prime Minister's Department.[9] By 1939, the Liaison Officer, A. T. Stirling, was part of a nascent diplomatic network and reported to the External Affairs Department established in Canberra late in 1935. An Australian Liaison Officer had been appointed within the British Embassy in Washington in 1937; Australian Legations in Washington and Tokyo and a High Commission in Ottawa were established during 1940. Stirling and his deputy, John Hood, shared premises with the Committee of Imperial Defence in Richmond Terrace, Whitehall, and undertook routine gathering of information and preparation of cables. Bruce was their Head of Mission, but, unlike other heads of mission, he reported directly to the Prime Minister. Being Bruce, he assumed an authority and access that were readily granted both by British ministers and by Canberra. His was the guiding hand on all Australian foreign policy as it operated in the northern hemisphere, although he paid lip-service to Canberra.

McDougall's position had always been an anomaly in Australia House. Representing no single department, his responsibilities grew haphazardly, usually on his own initiative and largely beyond control of earlier high commissioners. Under Bruce, there was discussion and cooperation, each man working at his own level to a common agenda. When war came, McDougall was given responsibility for 'political warfare' at Australia House.

This was at first the old agenda with a new name and some change of emphasis. It drew from the economic appeasement campaign the need to shore up allied, neutral and particularly American support, and to appeal to moderate elements in Axis countries. Once there was a war, however, it was necessary to prepare for peace. The lessons of Versailles must be applied: the next peace settlement had to be 'an instrument of social and economic betterment', demanded not only by justice but also by expediency, to prevent revolution, communism and the crippling costs of a potential 'Pax Anglo–Gallica': 'Neither this country, Australia nor France will tolerate the indefinite postponement of the long overdue attack upon the problems of poverty.'[10] Plans should be made immediately for a peace settlement based on 'international economic co-operation on a scale never

---
9 Hudson and North, eds, *My Dear P. M.*, pp. xi–xiv.
10 FAO, RG3.1, Series D 3, McDougall memorandum: 'Can we Maintain a Pax Anglo–Gallica?', 23 November 1939. I am grateful to Sean Turnell for copies of his notes from this file.

previously envisaged', including a coordinated attack on problems of poverty to secure a progressive rise in the standard of living, international solutions to commodity problems, reduction of trade barriers and policies to facilitate creditor assistance to other nations.[11]

In the 'phoney war' period, the British Government was obliged to formulate its postwar vision in responses to offers of peace from Hitler and mediation by the Dutch and Belgian monarchs. The view of the French and of some Britons, including Churchill, was that the postwar world should be similar to the prewar one, but with a disarmed Germany. Others, like Bruce and McDougall, visualised 'a new world resulting from a peace settlement which had faced the vital problems of disarmament, territorial adjustment, colonies and the economic needs of all nations, in which Germany would play an appropriate part as a great nation'.[12]

This view was supported by the governments of Australia and New Zealand, and Bruce and McDougall must have been delighted by the words of South Africa's Jan Smuts: 'No peace is worth while which does not result in raising the living standards of the people.'[13]

'Oh the boredom of war!'

McDougall's new responsibility and an increasing diplomatic workload meant that he spent much of his time in the External Affairs Office. Bruce, Stirling and McDougall worked long hours together over important cables to Canberra, on at least one occasion right through the night. Stirling recalls in his published memoirs that in the early war years Bruce himself spent most mornings and afternoons at Richmond Terrace, where McDougall would often come to see him towards evening. He remembers 'McDougall bursting in very soon after the declaration of war, saying as he came, "Oh the *boredom* of war!"'. In his unpublished diary, Stirling records that remark on 15 September, having previously explained on 10 September that McDougall 'is very genuinely seized with what he calls the "boredom" of war when there is so much else to be done—his new League scheme, for instance'—possibly a reference to the Bruce Report.[14] McDougall may have experienced uncharacteristically low spirits for a period in 1940. On 7 February, Stirling recorded in his diary: 'At

---

11   Ibid., 'The Peace Settlement—Which is the More Practical—a Co-operative or an Enforced Peace Settlement?', 27 November 1939.
12   *DAFP* II, *308*, cable 586 from Bruce to Menzies, 26 October 1939, pp.353-5.
13   *DAFP* II, *311*, *318*, *326*, Menzies to Chamberlain, 28 October, Jan Smuts, 31 October, and Michael Savage (New Zealand), 5 November, to Commonwealth Government, pp. 357–8, 365, 369–71.
14   Stirling, *Lord Bruce*, pp. 126–7; DFAT, unpublished Stirling Diaries, 10 and 15 September 1939. I am grateful to Jeremy Hearder for copies of relevant extracts from the diaries.

McDougall's request I took Mr Bruce E. H. Carr's *Years of Crisis*. It has shocked McDougall greatly.'[15] Carr's 'realist' approach to international affairs perhaps gave McDougall some reason to doubt his own belief in reason and cooperation. His output of memoranda slowed: none seems to have been written in the first nine months of that year—an unusually long gap.[16] Pressures of work, uncertainty in the early war months and the shocks of the fall of France and the Blitz probably forced a less vigorous prosecution of broader campaigning. The day after Italy entered the war, McDougall fell in the blackout, breaking his collarbone. Stirling recalled that he looked 'badly shaken up' and that he was 'deeply affected by the situation in France—still quoting endlessly the sombrest passages of Shakespeare'. On 22 June, McDougall was more positive, appearing at the External Affairs Office 'very eager for some action to transfer part of the League of Nations to America'.[17]

Bruce's new responsibilities were far greater than those borne by McDougall. His personal files record incessant representation to the British Government on war-related issues: strategic matters, military appointments, supplies to and from Australia, and Australia's claim to a voice in top-level decision making, the last leading to his being given the right to attend the War Cabinet. He carried responsibility for the safety of a large workforce in Australia House and concern for the welfare of Australian troops in Britain. He pursued personal hobbyhorses, particularly the efficacy of air power against sea power, a subject of debate in the early war years. He saw anyone who might have useful information or ideas and carefully chose targets for his own lobbying. Stirling recalls a long meeting with Bruce and McDougall 'trying to think of outstanding Englishmen for various missions'.[18] Bruce had virtually unlimited access to British ministers, the City, the diplomatic community and Commonwealth representatives. His career and reputation carried weight in all those circles.

Despite his workload, Bruce continued to lobby for the nutrition approach domestically and internationally. Before the length and geographical spread of the war could be predicted, he shared McDougall's fear that the postwar trading situation might repeat that of the interwar years: surplus primary production from overseas producers, particularly the United States, which Britain and Europe could not afford to buy, and a trade war spelling disaster to transatlantic relations. His considerable efforts on behalf of a peace plan with broad appeal embodying social justice had good reason to draw upon the nutrition approach.

---

15   Ibid. The words 'but not the HC' appear to have been added later.
16   FAO files list 10 memoranda from 1940 that I have not seen. Most deal with war aims and the peace settlement and are unlikely to have been written in the lean period.
17   DFAT, Stirling Diaries, 14, 17 and 22 June; Stirling, *Lord Bruce*, pp. 156–7.
18   DFAT, Stirling Diaries, 4 October 1939.

Figure 15 High Commissioner, S. M. Bruce, and Mrs Bruce with Australian troops in London.

Source: *Argus* Newspaper Collection of Photographs, State Library of Victoria.

Those efforts met with little success. He began in 1939 with Lord Hankey, Australian-born Secretary of Cabinet and of the Committee of Imperial Defence until 1938, then Minister without Portfolio in the War Cabinet. Hankey told Bruce his colleagues in the War Cabinet laughed at him when he suggested a nutrition standard be considered in relation to any rationing scheme; he 'was forced to the conclusion they attached no importance to the matter and were completely ignorant of its importance'. Bruce reminded him of the importance of food shortages, largely brought about by the British strategy of blockade, in destroying German civilian morale in 1914–18. Later scholarship has supported Bruce's view, suggesting that changes to traditional dietary patterns played a 'critical role': 'great mental fatigue' and 'real' and 'psychological deprivation' depressed German civilian morale and 'affected military motivation'.[19] Bruce recommended that Orr and Andrew Cairns, Canadian-born Secretary of the International Wheat Advisory Committee, be attached to a broader authority, possibly the Committee of Imperial Defence, to ensure cooperation between food, health and agricultural bureaucracies. Like Orr, Cairns was *persona non grata* in Whitehall, particularly in the Ministry of Agriculture, where he was

---

19 Avner Offer, *The First World War: An Agrarian Interpretation*, Clarendon Press, Oxford, 1989, p. 2.

blamed for a draft international agreement on wheat to which the ministry objected.[20] Hankey feared 'Orr would be regarded as something in the nature of a crank'. Bruce deplored the 'lack of imagination and vision' of the Ministry of Health, but modified his suggestion to a small advisory committee to the War Cabinet, to include Orr.[21] Hankey doubted he could convince his colleagues. Bruce also recommended Orr to W. S. Morrison, Minister of Food, and to Home Secretary, Sir Samuel Hoare, who, as 'head man of the Home Front', agreed to talk to Orr.[22]

There were many nutrition scientists in Britain. Orr's distinction was that he had become associated with outspoken advocacy of change; official wariness of him, and of McDougall, perhaps helps to explain why Bruce's recommendation could not have succeeded at this stage. Policies were gradually undertaken in accordance with nutritional thinking, including giving priority to milk production. Much has been written of the beneficial effects of wartime dietary measures in Britain such as the greater extraction rate of flour.[23] The official war historian gives credit to Orr for spearheading the nutrition movement of the 1930s, which provided both pressure for and acceptance of these measures, but notes that Orr's *Food, Health and Income* was referred to, in the 1937 *Report of the Advisory Committee on Nutrition*, in a manner 'so covert as to be unrecognisable'.[24]

'The Re-statement of Our Aims'

In August 1940, Clement Attlee, Leader of the Labour Party and Lord Privy Seal from May 1940, effectively (and officially from February 1942) Deputy Prime Minister in the National Government, was appointed chair of a Cabinet subcommittee on war aims. Bruce considered development of a clear statement of aims a high priority and immediately offered help, suggesting 'possibly McDougall could do some valuable work'. Attlee noted the name, but his assistant Harold Laski subsequently confirmed Bruce's fears that 'the committee was in fact making very little progress and that Attlee himself has hardly sufficient drive for the job'. An outspoken left-wing critic of Attlee's apparent timidity within the National Government, Laski published an 'Open Letter to the Labour Movement' in October 1940, urging Labour to demand a statement of war aims from the Government. He was later forced to apologise in the National Executive

---

20 R. J. Hammond, *Food. Volume I: The Growth of Policy*, History of the Second World War: United Kingdom Civil Series, ed. W. K. Hancock, HMSO, London, 1951, p. 352, fn.
21 NAA, M100,1939, record of conversation with Hankey, 19 October 1939.
22 Ibid., records of conversations with Hankey, 27 October, Morrison, 13 November, and Hoare, 1 December.
23 See Lizzie Collingham, *The Taste of War: World War Two and the Battle for Food*, Allen Lane, London, 2011, pp. 384–405.
24 Hammond, *Food: The Growth of Policy*, p. 219.

Committee.²⁵ Laski did commend 'the short memorandum we had put in [as] much the best of all the contributions he had read to Attlee, and [said] that both he and Attlee would be prepared to subscribe to it entirely'.²⁶ Laski, Bruce and McDougall discussed ways of interesting Churchill in the question. The war aims committee recognised the demand for an early statement of the general principles and objectives for which the British were fighting, but agreement on its terms proved difficult. Various drafts were prepared, but after Churchill vetoed the most promising one 'the committee had decided to postpone its meetings and faded away'. A biographer argues that Attlee's acquiescence was not 'timidity', but concern for the fragile cohesion of a government still including many Chamberlain supporters, compelling Attlee to avoid a potentially divisive statement.²⁷

The Bruce–McDougall 'short statement' to Attlee's committee has not been identified. A 29-page memorandum called 'Notes on the Re-statement of Our Aims' is dated 22 October. It refers to an earlier version called 'War Aims' and reiterates some of its content. It seems reasonable to conclude that the October memorandum reflects much of what had been submitted earlier to the Attlee committee, and possibly content of a document referred to by Alfred Stirling, on 28 July, as a 'Bruce–McDougall memorandum handed to [the Foreign Secretary, Lord] Halifax last week', which related to rejection of any peace offer by Hitler. Bruce had cabled Canberra on 21 July that it was necessary to counter Nazi economic propaganda with 'a positive and constructive policy' and Prime Minister Robert Menzies had agreed.²⁸ Stirling also records McDougall preparing a 33-page memorandum on war aims in October 1940, its progress closely watched by Bruce.²⁹

'Notes on the Re-statement of Our Aims' is McDougall's most important memorandum of 1940. The title refers to the fall of France. Until then, most people had expected that 'the keystone of post-war reconstruction would be Anglo–French collaboration'. After the fall of France, British–American collaboration was the only alternative to Nazidom, hence the need to rethink war aims. From this time on, both McDougall and Bruce believed in the absolute necessity of British–American cooperation for the new world order: 'In the post-war political reconstruction it will be essential to secure the full and permanent participation of America.'³⁰

---

25 Trevor Burridge, *Clement Attlee: A Political Biography*, Jonathan Cape, London, 1985, pp. 140, 144.
26 NAA, M100, records of conversation with Attlee, 27 August 1940, with Laski and McDougall, 21 October 1940.
27 Burridge, *Clement Attlee*, pp. 144–5. A note suggests the postponement dated from January 1941.
28 W. J. Hudson and H. J. W. Stokes, eds, *Documents on Australian Foreign Policy 1937–1949*. Volume IV: *July 1940 – June 1941* [hereinafter *DAFP* IV], Department of Foreign Affairs, Australian Government Publishing Service, Canberra, 1980, Document *30*, Menzies to Bruce, 22 July, pp. 40–1.
29 Stirling, *Lord Bruce*, p. 171.
30 NLA, MS6890/4/6, 'Notes on the Re-statement of our Aims', p. 27.

'Notes on the Re-statement of Our Aims' repeats an assertion from the missing earlier paper that to save Europe from tyranny, 'we must fearlessly decide upon profound adjustments in the political, economic and social spheres' and must lead 'a beneficent revolution by associating with our defence of liberty and International Law effective proposals for securing economic freedom and social justice'. In the later paper, McDougall began to consider the nuts and bolts of framing and organising such a peace settlement, although he made 'no attempt… to put forward ideal solutions'. He aimed simply to indicate 'the minimum of change needed' for security, equity between nations and greater social justice between classes.[31]

Under the heading of security, McDougall acknowledged the need to reconsider 'doctrines of national sovereignty' in view of the 'inter-dependence for security' and possible 'substitution of international for national forces', as well as control of civil aviation and international industrial cartels. Methods should be devised to prevent 'violent breaches of the international system' more effectively, and methods for modification and change of that system: political machinery rather than arbitration was needed.[32] This section perhaps owes something to McDougall's reading of E. H. Carr, who had condemned the 'pathetic fallacy that international grievances will be recognised as just and voluntarily remedied on the strength of "advice" unanimously tendered by a body representative of world public opinion'.[33] Carr argued that interwar internationalism had failed to take account of the importance of power, yet power—military, political and economic—had consistently determined outcomes.[34]

A second section of 'Notes on the Re-statement of Our Aims' dealt with another new idea: the duties, as well as the rights, of man. Again, McDougall seems to have been influenced by Carr, who wrote of the complexity and impersonality of modern industrial society: 'The real international crisis of the modern world is the final and irrevocable breakdown of the conditions which made the nineteenth century order possible. The old order cannot be restored, and a drastic change of outlook is unavoidable.'[35] McDougall wrote: 'In the middle decades of the twentieth century we cannot return to the individualism of the nineteenth century but must rather seek new definitions of the relationship of the State and the individual. This will involve the consideration both of human rights and [of] human duties.' The weakness in totalitarianism is that

> once the State is regarded as an end in itself, its glory, and its power superior to all other considerations, the virtues of pity, forbearance,

---

31 Ibid., pp. 1–3.
32 Ibid., pp. 3–6.
33 Carr, *Twenty Years' Crisis*, p. 210.
34 Ibid., pp. 102–8.
35 Ibid., pp. 236–7.

charity, come to be regarded as weaknesses. Ruthlessness in the service of the State becomes a virtue and cruelty, that intolerable insult to the dignity of man, is the inevitable consequence.

McDougall wanted a 'statement designed to emphasise individual responsibility towards collective well-being' based, he suggested, on sources including the medieval theologians, modern constitutions such as those of the Weimar Republic and some Latin American states, and the literature of the 'Corporative state'.[36] The last two suggest the influence of his American experience.

Sections on economic relations included much of McDougall's previous thinking on avoiding uneconomic forms of industry and agriculture, removal of trade barriers and social policies to improve living standards. On colonial policy, he suggested that an international commission, administering an international fund for active policies to improve living standards, would give 'all nations some degree of responsibility for non-self-governing areas'. The experience of the Empire Marketing Board had shown the effectiveness of grants 'made with imagination' and building upon expertise already existing in the receiving country. An 'International Commission' controlling an international fund 'mainly subscribed by the creditor nations' could provide financial assistance and technical advisers, thus allowing for increasing international responsibility and control 'without any formal transfer of administrative authority'.[37] The idea of a supervisory commission for colonial territories had evolved into provision of practical assistance on the ground, becoming closer to what were to be the functions of FAO.

A section headed 'Social Justice' reiterated a call in the earlier paper for 'positive action on the boldest scale to associate our defence of liberty with the re-dress of economic inequalities'.[38] The wartime economy saw 'all resources…harnessed to the war effort'; in the postwar world, 'our effort should not be slackened by peace but at once deflected to the tremendous task of solving the problem of poverty'. Controversies between capitalism and socialism should be avoided. McDougall hoped 'British political instinct will find the way to increase the powers of the State to secure a far wider conception of economic security and yet to retain in large measure the flexibilities and initiative of private enterprise'. In other words, he aspired to what was sometimes called in the United States 'a middle way'. McDougall concluded with tentative suggestions about an international organisation: it should be based on the experience and machinery of the technical organisations of the League of Nations and the Bruce Committee recommendations.[39]

---

36 'Notes on the Re-statement of our Aims', pp. 7–8.
37 Ibid., p. 11.
38 Ibid., p. 21.
39 Ibid., pp. 22–3, 29.

With this memorandum, the elements of McDougall's postwar vision were largely in place. Predominant were American–British collaboration in the context of internationalism, machinery to enable individual and collective responsibility for assisting less fortunate peoples and states, and association of the defence of liberty with social justice. What remained to be established was a strategy for achieving them.

'Freedom from want'

On 6 January 1941, a key part of that strategy was provided. In his third Inaugural Address, President Roosevelt pledged increased production of armaments, foreshadowed the Lend–Lease program to assist Britain and called for cooperation from all sectors of the American community. He also declared 'the mighty action that we are calling for cannot be based on a disregard of all things worth fighting for'. He listed the foundations of a healthy democracy, and he undertook to remedy deficiencies at home. Then he continued:

> …we look forward to a world founded upon four essential human freedoms.
>
> The first is freedom of speech and expression—everywhere in the world.
>
> The second is freedom of every person to worship God in his own way—everywhere in the world.
>
> The third is freedom from want, which, translated into world terms, means economic understandings which will secure to every nation a healthy peacetime life for its inhabitants—everywhere in the world.
>
> The fourth is freedom from fear, which, translated into world terms, means a world-wide reduction of armaments to such a point and in such a thorough fashion that no nation will be in a position to commit an act of physical aggression against any neighbour—anywhere in the world.
>
> That is no vision of a distant millennium. It is a definite basis for a kind of world attainable in our own time and generation.[40]

Bruce and McDougall saw immediately that the worrying possibility of a postwar clash of economic interests between the United States and the British Empire could be avoided by the realisation of 'Freedom from Want'. Civil servants on both sides of the Atlantic would subject the 'Four Freedoms' to close analysis, searching for precise meanings. McDougall was happy to provide his own simple

---

40 <http://www.fdrlibrary.marist.edu/4free.html> [accessed 8 October 2008].

interpretation of the third: 'Freedom from want means sufficient food, adequate housing and clothing, reasonable leisure and the means for its enjoyment.' Problems of oversupply would then dissolve and increased standards of living might be made affordable for two reasons. First, Roosevelt had also promised freedom from fear; if that could be achieved in the peace settlement, resources formerly used for armaments could be devoted to social welfare. And second, industrial countries could not afford to neglect measures to raise standards of living; adequate markets for their products would depend on them.[41]

A set of typed notes, found in Bruce's personal files and apparently written in the early months of 1941, spells out a new nutrition campaign based on giving 'practical expression to the third of President Roosevelt's four freedoms…and to expand slightly what he means by it'. Roosevelt should 'strengthen the Allies' hands in the war of ideas by adding to the concept of freedom that of economic welfare and social justice'. To persuade him, it would be necessary to emphasise that 'the achievement of these aims is physically practicable but requires some revolutionary thinking' and that unless the British Empire and the United States 'do some of this thinking jointly, we may find ourselves poles apart'. Roosevelt need only commit to joint British Empire–US examination of agricultural, trade and health factors involved and the national and international action required to translate the general objective into terms of practical policies.[42] The presence of these unsigned notes in Bruce's own file suggests that he, like McDougall, had seized upon the Four Freedoms as the rhetorical basis for their campaign: although British cooperation remained essential, both men now believed that the best hope of action lay with Roosevelt.

McDougall wrote several short papers on this theme. One proposed a formal US–British agreement to adopt policies of 'diets adequate for health' at home and in relief and reconstruction policies.[43] Others stressed the importance of the United States as a potential competitor in world export markets and the consequent importance of its placing 'the standard of living in the forefront of economic policy'. Another worried that commitment to a prosperous British agriculture, as sought by agricultural interests, could 'prejudice the whole position' of basing reconstruction on increased consumption and reorientation of agriculture.[44]

---

41 Speech, 'Empire Primary Products in Relation to Post-War Reconstruction', given on 1 April to a meeting of the Royal Society of the Arts, chaired by Bruce, printed in *Journal of the Royal Society of Arts*, 25 July 1941. Copy supplied by E. McDougall.
42 NAA, M100, May 1941, untitled, unsigned and undated paper.
43 NLA, MS6890/4/6, 'Freedom from Want, a beginning', undated, one-page version.
44 Ibid., 'United States Agriculture and World Markets', 26 March 1941; 'Reconstruction Problems: the Urgency of British–American Understanding', 29 March 1941; 'United Kingdom Agriculture and the Future', 10 March 1941.

The American Ambassador

Bruce had a valuable ally in the person of the new American Ambassador, John Gilbert Winant, who replaced the outspoken, defeatist and unpopular Joseph Kennedy early in 1941. Winant was a Republican three-time Governor of New Hampshire, sometimes considered a potential presidential candidate, and an idealist and independent thinker. He supported the New Deal and Roosevelt's re-election for a third term in 1940 and was personally close to the President and his adviser Harry Hopkins.[45] Winant saw his mission as facilitating understanding between Britain and the United States; he cultivated a wide range of contacts and reported in detail not merely through normal State Department channels, but directly to the President and Hopkins. His allies in Washington included Eleanor Roosevelt, Interior Secretary Harold Ickes, Secretary of War Henry Stimson, and Justice Felix Frankfurter. A notable exception was Cordell Hull.[46] Winant's Economic Advisor in London, E. F. Penrose, has written:

> To him the outbreak of war was an occasion not for suspending public concern with economic and social questions but rather for re-examining the existing order to determine how far past shortcomings had contributed to present strife. He did not believe that political settlements in themselves could provide an enduring basis for peaceful international relationships, and he knew that economic settlements would require long and arduous preparation if the errors made after the First World War were not to be repeated…he took every opportunity, from an early stage of the war, to advise the State Department and sometimes the White House to take constructive action on a variety of postwar international economic matters.[47]

Winant also had experience in international organisation: he was Assistant Director of the International Labour Office in Geneva for a few months in 1935, including the time when the ILO Conference carried the Australian resolution on nutrition. He returned to Geneva in 1937, becoming ILO Director in 1939. In 1940 he oversaw its move to Montreal, whence he was summoned by Roosevelt to his new appointment. In Geneva Bruce had 'close associations with him'.[48]

---

45 Nina Davis Howland, 'Ambasssador John Gilbert Winant: Friend of Embattled Britain, 1941–1946', PhD thesis, University of Maryland, 1983, pp. 3–12; Sylvia B. Larson, 'Winant, John Gilbert', *American National Biography Online*, February 2000, <http://www.anb.org.virtual.anu.edu.au/articles/06/06-00727.html> [accessed 8 October 2008].
46 Howland, 'Friend of Embattled Britain', pp. 14–15, 19–32, 47–8.
47 E. F. Penrose, *Economic Planning for the Peace*, Princeton University Press, Princeton, NJ, 1953, p. 12.
48 NAA, M104, 9(1), Bruce to Fadden, 25 September 1941.

Figure 16 John Winant with F. L. McDougall in Geneva. Winant (centre) seated next to McDougall (with pipe).

Source: E. McDougall.

Bruce discussed the nutrition campaign with Winant in May 1941. He found him 'generally receptive to everything I had been saying…we can I think take it that our ideas are well sold to Winant'. In particular, Winant agreed with Bruce's view that 'so far from detracting from the President's determination to aid the Democracies it would strengthen it, if he had the picture of a practical policy in his mind that was going to help to realise all that he stood for once peace was achieved'. They agreed that Bruce should continue his nutrition campaign in London.[49]

To press the importance of transatlantic cooperation to the Australian and British Governments, Bruce wrote an untitled memorandum, which he labelled 'A', and, with McDougall's 'enthusiastic and useful assistance', another, labelled 'B'. 'A' dealt with the need to reinforce physical force against Germany with the 'psychological factor' of assured higher standards of living. 'B' was given a title: 'British–American Understanding.' It stressed the importance of economic cooperation between Britain and the United States to avoid repetition of inter-

---

49 Ibid., M100, May 1941, record of conversation with Winant, 22 May 1941.

war problems and outlined the nutrition campaign.[50] Bruce distributed them widely in London: recipients included Winant, Anthony Eden, press baron Lord Camrose, Home Secretary Herbert Morrison, shipping magnate Sir Alan Anderson, Labour Cabinet member Arthur Greenwood, Lord President of the Council Sir John Anderson, Orr and Sir George Schuster. They were also submitted to the War Cabinet's Committee on Post-War Economic Problems and Anglo–American Co-operation. They were sent to Menzies on 18 July, but did not reach him before he lost office; Bruce then wrote to the new Australian Prime Minister, Arthur Fadden, recommending he read them. They were also sent to economist Dr Roland Wilson and Rivett in Australia.

Winant believed Britain's desperate situation was not understood in Washington. At the end of May 1941 he returned at his own request to argue for greater material support. He was invited to stay at the White House so that he could have easy access to Roosevelt and Hopkins, and he saw many other officials. Most of the discussion concerned immediate military assistance to Britain.[51] Winant also took copies of Bruce's memoranda 'A' and 'B', and told Bruce he had discussed them with Roosevelt and Vice-President Wallace, who were both 'very receptive and anxious that some definite steps should be taken'.[52] Back in London at the end of June, he told Eden and the Canadian and South African High Commissioners of his Washington talks on the subject. Bruce was delighted by Winant's account and with his enthusiasm and grasp of the subject. Winant suggested Bruce send a cable to Wallace expressing his interest in Winant's report.[53]

Efforts by the International Wheat Advisory Committee to replace the failed 1933 Wheat Agreement had stalled in 1939, but negotiations between Argentina, Canada, Australia, Britain and the United States were resumed in Washington in July 1941.[54] McDougall was to attend. Bruce sent a letter to Wallace with McDougall, stating that 'McDougall has my complete confidence' and hoping Wallace would discuss questions of nutrition and agriculture with McDougall personally. He also cabled Wallace on 26 June: 'extremely interested in Winant's report of his conversations with you.'[55] Bruce told Winant of his hope 'that something to tidy it up might be accomplished while McDougall was in America, and a possible line of action determined upon'.[56]

---

50 The papers are in NAA, M103; information about their distribution is in NAA, M104, 9(1).
51 Howland, 'Friend of Embattled Britain', pp. 107–17.
52 NAA, M104, 9(1), Bruce to Menzies, 18 July 1941.
53 Ibid., M100, record of conversation with Winant, 24 June 1941.
54 Dunsdorfs, *The Australian Wheat Growing Industry*, p. 310.
55 University of Iowa, Henry Wallace Diaries, microfilm copy in Library of Congress, Washington, DC [hereinafter HWD]. A copy of Bruce's letter, dated 3 July, is in NAA, M104, 9(1).
56 NAA, M100, record of conversation with Winant, 24 June 1941.

A New Idea Each Morning

'I like these Americans'

McDougall travelled to Washington in the summer of 1941. Bruce explained the journey to Fadden, first recounting his concerns about postwar policy and the importance of British–American cooperation, the history of the nutrition question and its advantages for Australia, his discussions with Winant, and Winant's recent discussions in Washington. He continued:

> Soon after the Ambassador returned to London it was decided to send McDougall to Washington to represent Australia at the Wheat Conference. As you know McDougall has worked in very close co-operation with me for many years on these questions and a great part of the results which have been achieved are due to his initiative, industry and perseverance. I decided to take advantage of his visit to America to put him in touch with the Vice President who is playing an increasing part in regard to questions of post war reconstruction, and after consultations with [Australian Minister in Washington R. G.] Casey to sound out American official opinion on these subjects.[57]

Wartime travel across the Atlantic was dangerous, subject to long delays and to permission from the governments involved. Without the official reason of the wheat negotiations, McDougall might not have made the journey. He left London by air in the first week of July and was forced to spend three 'trying hot days' in Lisbon before crossing the Atlantic.[58]

Washington before the New Deal, according to one observer, had been 'a small town, mildly important as the seat of the national legislature, a place where President Coolidge habitually had taken a nap after lunch'. The New Deal increased the tempo and the range of government activity: 'Bright young men and women flocked to Washington…They put in uncounted hours with a gusto.'[59] The looming war accelerated industrial recovery; planning to effect Roosevelt's undertakings of aid to Britain increased the attraction and power of the capital. Journalist Marquis Childs wrote that the atmosphere of wartime Washington was 'in many ways a reprise of the early New Deal era'.[60] McDougall found its energy exhilarating. The Americans he met were enthusiastic about his ideas and keen for Empire–US cooperation. He was busy and confident: 'I've not

---

57 Ibid., M104, 9(1), Bruce to Fadden, 25 September 1941.
58 Ibid., M104, 9(6), McDougall to Bruce, 19 July 1941.
59 Extracts of Helen Hill Miller's unpublished 'Washington Observed: A Reporter's Notebook', <http://www.cosmos-club.org/web/journals/1995/miller.html> [accessed 19 January 2006].
60 Borgwardt, *A New Deal for the World*, p. 70.

touched a golf club, indeed apart from one swim I've not had a moment before 10 p.m.…The humidity is very great and I feel I could do with a few days quiet in a cool place but it's all extraordinarily interesting. I like these Americans.'[61]

The most important American to McDougall in 1941 was the Vice President, Henry Agard Wallace, who, as Bruce had hoped, acted as his mentor. McDougall saw Wallace several times in his office and at social occasions.[62] Roosevelt had explained his choice of Wallace as running mate in 1940 in terms both of farm states' support and of ideology: Wallace would reaffirm the Administration's domestic and foreign policies. 'He is no isolationist, he knows what we are up against in this war that is rapidly engulfing the world.' Norman D. Markowitz writes that Wallace was 'possessed of a keener intellect than Roosevelt, [but] lacked the charm and cunning with which to build a career of his own in a politics increasingly dominated by personality. His vision, however, remained large and his imagination open.'[63] Walter J. Samuels describes a figure out of place in the increasing sophistication of Washington: 'Never a glad-hander or back slapper, he was basically a shy person who felt uncomfortable with strangers and often appeared detached and aloof.' Although Wallace had 'many friends and a group of capable advisers in the Department of Agriculture, he was essentially a loner who arrived at positions more on the basis of his own thinking than on the advice or suasion of others'.[64]

Wallace advocated a middle course between isolationism and internationalism. He believed that US policies of the 1920s, limiting the ability of debtor countries to earn dollars, had played a major role in creating world depression and war. In 1940 he set out in a letter to Roosevelt his ideas for postwar economic programs to create stability and improve 'conditions of life among the common peoples of the world'. They included continuation of Cordell Hull's reciprocal trade agreement program to ensure freer flow of goods and services, international commodity agreements, a world ever-normal granary to stabilise prices and encourage consumption, and US credit to rebuild war-torn countries, given in such a way as to make debt repayment possible: 'I would hate to see us again commit the various errors that we committed during the twenties and early thirties.' 'The overthrow of Hitler is only half the battle', he wrote, shortly before Pearl Harbor; Americans should 'think hard and often about the future peace'. Determining national boundaries and creating a new international organisation would not guarantee a durable peace. Economic planning and a concerted effort to promote industrialisation and improved living standards throughout the world were also necessary.[65]

---

61   NAA, M104, 9(6), McDougall to Bruce 19 July 1941.
62   HWD. For the relevant period in 1941, the diaries record only dates of meetings, without detail or comment.
63   Markowitz, *People's Century*, pp. 28–31.
64   Samuels, *Wallace and American Foreign Policy*, p. 35.
65   Ibid., pp. 83–4.

Bruce had instructed McDougall to be guided by Wallace. At their first meeting on 17 July, McDougall found him 'in good form and spirits and clearly quite keenly interested in our approach'. Wallace suggested McDougall see Assistant Secretary of State Dean Acheson, Dr M. L. Wilson of the Department of Agriculture, Surgeon-General Thomas Parran, and economist Winfield Riefler.[66] McDougall also carried a letter of recommendation from Bruce to Under Secretary of State, Sumner Welles. Welles was interested enough in his ideas to comment, but went no further on their first meeting. McDougall met other senior State Department officers: Assistant Secretaries Dean Acheson and Adolf Berle, and Harry Hawkins, head of the Trade Treaties Section. Most expressed interest and even enthusiasm, but McDougall remained unsure of Acheson's views. It was not until his visit the following year, and after Orr had also talked to Acheson, that he discovered how far Acheson was prepared to support the nutrition campaign. Following his meeting with Orr later in 1941, Acheson dictated a note supporting 'the nutrition program recently discussed in Washington by Sir John Orr and Mr McDougall'. He thought the approach would 'ultimately involve nearly all the most thorny problems of domestic and international finance and economics', but that efforts 'should not…be diluted into ultimate problems at the expense of developing the core of the nutrition program', the success of which might facilitate solution of more difficult problems. He suggested steps for approaching the problem. This paper was intended for McDougall, Orr, Parran and Boudreau. It did reach Boudreau's office and was filed by a secretary, but was not received by the other three. McDougall commented ruefully in 1942 that as Acheson was the Assistant Secretary directly involved in commercial policy, the paper 'would have been of great help in dealing with Whitehall'.[67]

Washington's political and official life extended well beyond its office buildings to evening functions where informal discussion of substantial issues could take place. Wallace took McDougall to a 'party' to discuss Empire–US cooperation and presided over a 'discussion dinner' with 'the people engaged in the National Nutrition drive and the two most active scientists. I had to open the show and then for two hours had a barrage of questions and points fired at me. The interest was very keen and the evening most useful.'[68] Wallace asked McDougall, Wilson and Parran to draft a possible agreement between the United States and British Empire countries 'on the lines of our suggestion', and also wanted notes on a possible speech on the subject. He also suggested the three see Eleanor Roosevelt 'thus to strengthen his own approach to the President'. Casey gave a dinner for Wallace, his guests including British Ambassador Lord Halifax, Minister at the

---

66 NAA, M104/, 9(6), McDougall to Bruce, 19 July 1941.
67 Ibid., M104, 10, McDougall to Bruce, 4 October 1942. Acheson's paper, 'A suggested Program of Work to Develop the Plan of Advanced Standards of Nutrition', 7 November 1941, is attached.
68 NAA, M104, 9(6), McDougall to Bruce, 19 and 22 July.

British Embassy Dr H. B. Butler, Acheson and Parran: 'We started serious talk at 9.30 and went straight on to 12.30.'[69] McDougall dined with the nutrition scientists and Parran and Boudreau for more discussion.

A US–British Empire conference

Throughout July the Wheat Conference occupied most of McDougall's time. By early August a draft Memorandum of Agreement, which McDougall had helped prepare, was ready for submission to governments, allowing the conference to adjourn for 10 days. 'The wider international business' was also progressing well. The Americans McDougall met were

> extremely keen on the early getting together of Empire–US interests both because they feel the need to start now if we are to be successful and also because they seem clearly to realise that there is urgent need of an Anglo–American declaration of post-war purposes if Europe is to be rallied against Hitler.[70]

The McDougall–Parran–Wilson draft was completed. It aimed 'to test the practicability' of an agreement between the United States and British Empire countries to adopt policies to improve the diets of their own peoples and to offer practical assistance to other nations for improved nutrition. Acheson's staff were keen, but McDougall was dubious about Berle's suggestion that Latin America be included. He hoped to be able to 'leave the issue in the Vice President's hands, put as clearly as possible and in the best form for his use with the President'. McDougall had agreed with the Chairman of the Advisory Committee on Nutrition, Dr Russell Wilder, Parran and others that in March or April 1942 the United States should call a conference of scientists, economists and agriculturalists from the United States, Britain, Canada, Australia and New Zealand to consider the possibility of such an agreement and its application during the reconstruction of Europe. He intended to discuss this idea with Wallace.[71]

The conference idea was warmly supported by Agriculture Secretary Claude Wickard, Sumner Welles and Harry Hawkins. Welles said he was 'intensely interested' in the food proposals as a means of preventing postwar conflicts of interest; when McDougall asked for permission to repeat those words, Welles 'remarked that I must go further and state that he was "determined to use such influence as he possessed to try and bring the [conference] proposal to a

---

69   Ibid., McDougall to Bruce, 26 July and 1 August.
70   Ibid., McDougall to Bruce, 19 July.
71   Ibid., Letters of 1, 8 and 11 August.

successful fruition'".[72] Hawkins told McDougall that the State Department was 'inclined to attach considerable importance to the idea of putting food policies in the forefront of post-war reconstruction', and had 'put one of their men on to examine the whole picture'. He suggested McDougall return early in 1942 to help prepare for the conference. Hawkins, Riefler and Agriculture officials all asked McDougall for 'suggestions for a programme of work'.[73] The State Department was liaising with Health and Agriculture, where Wickard instructed H. R. Tolley, Director of the Bureau of Agricultural Economics, to keep in touch with Wilson on the matter. Wallace 'strongly supported' the idea, but added that the British attitude would determine its future.[74] Thus, to cooperation and a published agreement between the United States and the British Empire, McDougall could now add the idea of a conference at government level.

The five-government agreement drafted by McDougall, Parran and Wilson declared intentions to 'enable all sections of their own populations to secure a full sufficiency of those kinds of food needed to meet all human physiological requirements', to pool experience and to establish a 'joint consultative commission' to report on necessary 'adjustments of agriculture and commercial policies' and steps necessary to help other countries achieve the same ends.[75] McDougall elaborated in memoranda on the national and international benefits of such an agreement; the importance of surveys to establish national resources and demand; desirable adjustments in agriculture; economies possible in distribution; international obstacles to achieving adequacy of food in various countries and to trade in foodstuffs; and a suggestion for a 'Council on Food Problems' representing each contracting party.[76]

One of these Washington memoranda had an important effect on his subsequent work. The British Ambassador, Lord Halifax, 'challenged me to put the whole idea on one sheet of paper. This I have done and enclose the highly concentrated results.'[77] One wonders how much more effective McDougall and his campaigns might have been had Bruce thought to give him this common bureaucratic instruction much earlier in his career. The 'highly concentrated' memorandum makes use of every inch of the single sheet, but loses nothing in scope and gains considerably in comprehensibility.[78] The discipline seems to have been lasting. Subsequent memoranda tended to be shorter: the average length of memoranda written in 1941 is some four pages, compared with nearly 14 pages in 1936.

---

72 Ibid., M104, 9(1), 'Note on Food Policy Discussions in Washington', 11 September 1941.
73 Ibid., M104, 9(6), handwritten letter to Bruce, 19 August.
74 Ibid., M104, 9(1), 'Note on Food Policy Discussions in Washington'.
75 Ibid., M104, 9(6), undated draft.
76 Ibid., M104, 9(1), memoranda including 'Food Policies in Reconstruction: Program of Work', 25 August; 'Freedom from Want: the Food Problem', n.d.; both written in Washington.
77 Ibid., M104, 9(6), handwritten letter to Bruce, 19 August.
78 Ibid., M104, 9(1) and MS6890/4/6, 'Freedom from Want a Beginning', undated.

Wartime shortages of paper and staff may have contributed to the change but it seems likely that Halifax's challenge, and the experience of working closely with American bureaucrats, finally taught McDougall the value of brevity.

American–British Cooperation

In July, Winant, prodded by Bruce's memoranda 'A' and 'B', had suggested US–British–dominion talks on economic cooperation, beginning with agriculture and nutrition. Both Winant and Bruce lobbied ministers.[79] Bruce was pleased with their progress, telling McDougall in mid July: 'position developed considerably since you left.' He was even more optimistic in mid August: 'there may be interesting developments following on joint declaration by President and Prime Minister.'[80] He therefore instructed McDougall to return to London, subject to Wallace's agreement. McDougall obtained a reservation on a bomber leaving Montreal on 28 August and was back in London on 30 August.[81]

Bruce reported to Canberra that he was convinced by McDougall's reports and by his own discussions with Winant that 'the United States Administration is intensely interested in the idea of placing policies with regard to food in the forefront of post-war economic reconstruction'. It was likely that the United States would shortly approach the United Kingdom, which in turn would consult the dominions about holding a conference in the early months of 1942.[82] But Bruce's habitual pessimism about British ministers, frequently expressed at the end of a record of what otherwise had seemed a useful conversation, was justified; nothing came of his and Winant's midyear campaign for American–British discussions.

As the year neared its end and the United States was finally drawn into war, a philosophical McDougall set out in memoranda his understanding of the differences between the views of the two allies on postwar planning. There were 'encouraging' signs in Washington that Americans realised that 'security and welfare cannot be attained in isolation'; that 'America's best interest will be served by the generous use of American resources to promote reconstruction in other countries'.[83] The United States was thinking in 'large scale terms' of schemes such as a Tennessee Valley Authority for the Danube Valley, of proposals to stabilise world agricultural and raw material prices and to raise the standard

---

79 Ibid., M104, 9(1), Bruce to Menzies, 18 July.
80 Ibid., M100, July 1941, cable from Bruce to McDougall, 12 July; M104, 9(6), 15 August.
81 Ibid., M104, 9(6), handwritten letter to Bruce, 19 August.
82 Ibid., M104, 9(1), Bruce to Fadden, 25 September.
83 Ibid., M104, 10, 'British–American Co-operation', 21 January 1942, p. 4.

of living in raw material producing countries. Cooperation with the United States would achieve 'an era of security in which all countries will be able to engage in a campaign against poverty and in the organisation of well-being'.[84]

Whitehall, on the other hand, was obsessed with the difficulties Britain would face after the war. Treasury worried about postwar balances of payment; the Ministry of Agriculture and farmers wanted to protect high-priced domestic agriculture; imperial enthusiasts feared the threat to imperial preferential tariffs in US multilateralism. McDougall believed nevertheless that a much wider section of official opinion—including some Treasury figures, economists in the War Cabinet Secretariat, principal officials of the Board of Trade, Labour members of the War Cabinet, and *The Times*, *Manchester Guardian* and the *Economist*—was convinced the difficulties must be overcome. The alternative for Britain was balance-of-power politics in a Europe dominated by the USSR, heavy military commitments undermining material welfare and strain on the cohesion of the Commonwealth:

> There is really no choice. The advantages of co-operation with America are overwhelming, the disadvantages of its failure are so patent that every Empire Government will determine that nothing must stand in the way of joint action to secure for ourselves and for the world...Freedom from Fear.[85]

McDougall's Washington experience had led him by then to abandon ideas of reworking Geneva organisations, which would not be acceptable to the US Congress or to the USSR. League of Nations and ILO staff and expertise, however, should still be used.[86]

Progress in the War of Ideas

By 1942 McDougall's definition of political warfare was precise:

> Its object is to achieve the political purposes of the State, either without bloodshed or, after hostilities have commenced, to ease the tasks of the Armed Forces. Its purpose is to undermine enemy morale, to secure Allies, and to strengthen the will to victory of our own people and those associated with us. It is the war of Ideas.[87]

---

84   Ibid., pp. 8–10.
85   Ibid., p. 10.
86   Ibid., M104, 10, 'The I. L. O. and Reconstruction', 10 February; 'International Mechanism for Reconstruction', 13 February; 'International Mechanism for Reconstruction Plans', 3 March 1942.
87   Ibid., A 2937, 'Political Warfare', secret memorandum, 4 May 1942.

In the immediate aftermath of Pearl Harbor, American interest in postwar matters was temporarily diverted. Riefler warned a small dinner group including McDougall that 'America was today an angry nation'; effective public discussion of postwar reconstruction projects would have to wait until there had been some military successes. American–British agreement was urgent, nevertheless, so that plans could be announced as soon as those victories had been achieved, and Americans would welcome early action on nutrition.[88]

McDougall's own sense of urgency was soon jolted. In April 1942 Bruce arranged for him to be briefed on the British political warfare organisation. He visited its headquarters, met regional directors for German, French and Scandinavian areas, and attended a meeting of the Political Warfare (Japan) Committee at the Foreign Office. He made it clear he did not want to know the techniques of 'black' warfare, but 'desired to understand the major problems…the lines of policies adopted, and the particular gaps which existed in the political warfare armory'. He found that the British authorities 'considered that they now knew as much about how to conduct war propaganda as Goebbels. What they really lack is the most effective material.' Political warfare used weapons based on fear or hope. 'So far as the United Nations were concerned, fear was not a factor which could operate'; the authorities therefore needed 'hopes which are both substantial and definite'. But each official McDougall spoke to

> said that they are, in effect, carrying on with their right hand tied behind their backs so long as there is no predominantly British plan for the future of Europe. Europe, in effect, wants to know what we are prepared to do with our power once we have broken Nazi domination.

On economic and social issues, the Atlantic Charter was 'too vague to be of much value'; the need was for definite plans for the welfare of particular areas; undertakings 'for co-operation of a practical kind to maintain full employment, to improve standards of living, to reduce social inequality'. He noted the statement of a former Bulgarian official: 'we know what Germany will give us and we don't like it, we know what Russia will give us and we are doubtful whether we like it. We have no idea what Britain and America will give us and so we feel compelled to choose between Germany and Russia.'[89]

While none of this was really new to McDougall, the sense of urgency and of the desperate need for 'ammunition' in the 'war of ideas' and the unanimity of the demand for action all galvanised and reshaped his campaign yet again. Much of what he learned, and in particular the sense of urgency, reappeared in a new memorandum called 'Progress in the War of Ideas', which exists in several locations and forms. The first version was written hurriedly after discussion

---

88 Ibid., M104, 10, 'Dinner to Sir William Jowitt', 24 March 1942.
89 Ibid., A 2937, 'Political Warfare', secret memorandum. 4 May 1942.

with Bruce in July 1942 when McDougall was about to set off on another visit to the United States. He jotted down two and a half pages summarising their conclusions, which began, in part:

> The vigorous prosecution of the War of Ideas is essential because: —
>
> Our *political warfare* directed towards enemy-occupied countries lacks ammunition…
>
> Our *own peoples* need the stimulus of positive ideas if they are to give full support for an all out offensive effort…
>
> *To win the peace*, we must secure agreement now regarding many of the methods of international co-operation which will be essential. If we delay until the end of the war, we are almost certain to fail.

The tone was urgent. Concrete proposals were needed now to give body to the vague terms of the Atlantic Charter and President Roosevelt's 'Four Freedoms'. Severe winter food shortages were predicted for Europe in its third year of war: 'Men's minds everywhere will be concerned with food. This will be the time for the United Nations to present to the world the picture of how they propose, on the food front at least, to secure "Freedom from Want, everywhere in the World".' Some economic issues would take time to solve, but for others immediate steps were practicable. Food was essential and the need was measurable: 'there are many advantages in starting the United Nations' campaign against poverty with a limited and realisable objective.' Planning should begin for assessing all countries' food requirements and resources, special problems of backward areas and methods of safeguarding the interests of both producers and consumers of key commodities.[90]

The nutrition campaign had reached solid ground. The combined impacts of the American visit and the briefing on the needs of political warfare had given it purpose and definition. The urgent need for 'Progress in the War of Ideas' would be the basis of McDougall's discussions during his next visit to Washington.

---

90   Ibid., M104, 10, 'Progress in the War of Ideas', undated three-page version. Sending a later version to Stafford Cripps, Bruce wrote that it had 'originated from a paper which McDougall drafted after discussions with me here'. W. J. Hudson and H. J. W. Stokes, eds, *Documents on Australian Foreign Policy 1937–1939*. Volume VI: *July 1942 – December 1943* [hereinafter *DAFP* VI], Australian Government Publishing Service, Canberra, 1983, Document 42, 9 September 1942, pp. 85–90.

9. 'A Keen Outsider'

Summary

McDougall returned to Washington in 1942, bringing with him rudimentary notes on 'Progress in the War of Ideas'. He came at a time when prominent members of the Administration were calling for a peace based on higher standards of living. Vice President Wallace arranged introductions for him, including a meeting with Eleanor Roosevelt, which led to his dining with the President. He was also encouraged by Under Secretary of State, Sumner Welles, and Assistant Secretary of State, Adolf Berle.

Although deeply divided, the State Department was responsible for postwar planning; McDougall was briefed on progress by key officials and asked to facilitate discussions with the Department of Agriculture. During the process of rewriting 'Progress in the War of Ideas' in consultation with these officials, the memorandum was read by many influential people, including the Roosevelts. McDougall was also asked to prepare a draft program for action on food and agriculture. Under his direction, the Department of Agriculture prepared 12 papers that were discussed by an 'assessment group' of senior officials in the Agriculture and State Departments. A final draft of 'Progress in the War of Ideas', renamed 'Draft Memorandum on a United Nations Program for Freedom from Want of Food', became the fourth of the seminal documents subsequently published by FAO.

This chapter considers possible reasons for the privileged treatment McDougall was given in Washington: that he was seen as Australian rather than British; that his ideas on social justice and nutrition were mirrored in the United States; and that he came at a propitious time. Finally an attempt is made to assess the extent to which McDougall's work influenced Roosevelt's decision to call the conference at Hot Springs early in 1943.

McDougall's Third Visit to Washington: August to October 1942

Wheat negotiations had continued in 1941 after McDougall left Washington, and the resulting Memorandum of Agreement to attend an International Wheat Conference and to establish an International Wheat Council was initialled by government representatives, including Australia's Edwin McCarthy. Thus McDougall had an official reason to return in 1942. As Bruce later reported to Labor Prime Minister, John Curtin, who had taken office in October 1941:

> ...you cabled to me suggesting that Mr F. L. McDougall should go to Washington to attend the first meeting of the International Wheat Council. I greatly welcomed this suggestion as...it afforded an opportunity of ascertaining, relatively at first hand, the trend of thought in the United States with regard to post-war reconstruction through the contacts made when I took McDougall with me to Washington in 193[8] which were renewed, in his case, when he went there last year.[1]

McDougall left Bristol on 31 July for a 16-hour flight across the Atlantic to New Brunswick via Eire and Newfoundland, reaching New York late on Saturday, 1 August, and Washington the next day. The Wheat Council was established soon afterwards with Paul Appleby as Chairman and Andrew Cairns as Secretary, but McDougall did not return to London until 29 October.

McDougall took with him notes headed 'Progress in the War of Ideas', urging the need to establish concrete allied plans for the postwar world to be used as 'ammunition in the war of ideas' in the winter of 1942–43, when Axis morale was expected to be vulnerable. As in 1941, he also took with him letters of commendation to Vice President Henry Wallace, and Under Secretary of State Sumner Welles.

The previous May, in an encouraging sign that the US Administration had recovered from the shock of Pearl Harbor, which temporarily paralysed thinking about postwar planning, Wallace had made his most famous broadcast, saying:

> The peace must mean a better standard of living for the common man, not merely in the United States and England, but also in India, Russia, China and Latin America—not merely in the United Nations, but also in Germany and Italy and Japan.

> Some have spoken of the 'American Century'. I say that the century on which we are entering—the century which will come out of this war—can be and must be the century of the common man...No nation will have the God-given right to exploit other nations. Older nations will have the privilege to help younger nations get started on the path to industrialization, but there must be neither military nor economic materialism...

---

1 *DAFP* VI, 78, Bruce to Curtin, 22 November 1942, pp. 157–8.

> Those who write the peace must think of the whole world. There can be no privileged peoples…we cannot perpetuate economic warfare without planting the seeds of military warfare. We must use our power at the peace table to build an economic peace that is just, charitable and enduring.[2]

Soon afterwards, Sumner Welles, whose words were generally taken as an indication of Roosevelt's thinking, spoke of the 'frontier of human welfare'. He declared there must be no repetition of US failure to ensure reconstruction after World War I. He also acknowledged need for an international police power in the postwar world, and that the United Nations would become the nucleus of a future world organisation. For constitutional reasons, 'United Nations' was the preferred US term for all countries fighting against the Axis. Welles denounced discrimination on grounds of race, creed and colour, and said 'the age of imperialism was ended'.[3] From the British Embassy, Isaiah Berlin reported to London that Welles chaired a secret State Department advisory committee on postwar reconstruction; the speech was likely to have been inspired by committee members. Berlin understood that views already established in the Administration and particularly likely to appeal to Americans included abolition of colonial empires and equal access for all democracies to raw materials. He understood that Roosevelt talked privately in similar fashion to Welles's speech.[4]

More encouraging speeches followed, including one by Dean Acheson, who also declared determination to avoid the mistakes of the past and to reject 'special privileges and vindictive exclusions'. United action was necessary to achieve increased production, employment, trade and consumption, and to meet

> a need such as we have never known to move goods between nations—to feed and clothe and house millions whose consumption has for years been below minimum requirements, to restore devastation, to build and rebuild all the means of production, and, in the years beyond, to move that far greater volume of goods required by the standards we are determined to achieve.[5]

While to some British ears such statements veiled an attack on Britain's interests—in particular, on the Ottawa tariff preference system—they also signalled a welcome rejection of isolationism and a vision very much in accord with that of Bruce and McDougall. On 18 July, the *Economist* published what McDougall

---

2   NAA, M104, 10, 'The Price of Free World Victory', text of broadcast address, 8 May 1942.
3   H. G. Nicholas, ed., *Washington Despatches 1941–1945, Weekly Political Reports from the British Embassy*, Weidenfeld & Nicolson, London, 1981, 4 June 1942, p. 42.
4   Ibid., 11 June 1942, pp. 43–4.
5   NAA, M104, 10, 'Building in War for Peace', text of speech at the Institute of Public Affairs, University of Virginia, 6 July, sent to Bruce 24 July 1942

described as 'a very remarkable article' summarising recent speeches by leading Americans on postwar reconstruction. The tenor of the paper's comments is reflected in the following passages scattered throughout:

> War is seen as part of a continuous process whose roots lie deep in poverty, insecurity, starvation and unemployment.
>
> Here is a new thesis, a far deeper view of social and international responsibility than was ever enunciated at the close of the last war…
>
> Above all, it seeks to put an end to isolationism, not only on the grounds of comradeship and world solidarity, but also in plain self interest.
>
> The key is expanding markets…a future planned 'in terms of increase and not curtailment'.
>
> Let there be no mistake about it. The policy put forward by the American Administration is revolutionary. It is a genuinely new conception of world order. It is an inspiring attempt to restate democracy in terms of the twentieth century situation, and to extend its meaning in the economic and social sphere.

The sting in the article was criticism of British failure to make 'a concrete and creative reply' to these speeches from 'the most progressive elements in America'.[6] The auspices nevertheless seemed good for McDougall's visit.

Bruce's letter to Wallace began with congratulations on his May speech:

> I need hardly tell you how greatly I am in accord with what you then said. The series of speeches by responsible men in the United States which have followed your broadcast are most encouraging…McDougall knows my views on economic and social questions and he will be able to convey to you much that I could not write.

He wrote in similar terms to Welles.[7]

Bruce's hopes were amply fulfilled. Henry Wallace and Sumner Welles acted as mentors to McDougall for the next three months. They introduced him to useful people, they suggested strategies, they even sought his help and advice. Through Welles, he gained access to high levels of State Department thinking on postwar planning. Wallace gave him access to other parts of the Administration and to the White House. And they used him as a facilitator to achieve cooperation between parts of the Administration generally at loggerheads.

---

6   Ibid., A989, 43/735/658, typed copy of article in the *Economist*, 18 July 1942, sent by McDougall to Canberra.
7   Ibid., M104, 10, letters to Wallace and Welles, 29 July 1942.

Henry Wallace

McDougall first had a brief meeting with Wallace, who said his coming was 'timely and very welcome…he asked me to see a number of people and then see him again. He told me to seek an interview with Mrs Roosevelt.'[8] Wallace probably made that arrangement himself as he and others used Eleanor Roosevelt as a conduit for raising issues with the President. McDougall subsequently reported to Bruce that his luncheon with Mrs Roosevelt occurred 'owing to a suggestion by the Vice President'.[9]

Wallace noted in his diary that McDougall

> thinks the time has come for the President to speak out clearly with regard to how the United Nations are going to make the world ability to produce abundantly, work in terms of a higher standard of living for all the people. He thinks if something of this sort is said very clearly and very strongly, it will have a decisive effect on the people of Germany.[10]

At a much longer meeting a few days later, McDougall found Wallace

> for the first time…very easy to talk with and he showed the keenest interest in the Food approach and saw at once its significance for political warfare. He wants to see if it is possible to go right ahead but considers that this mainly depends upon whether we can get the State Department to take the right line.

> He gave me his views of leading personalities here and said that the key people at the State Department for these purposes were Sumner Welles, Berle and Acheson…

> Wallace asked me to see the people in the Office of War Information about the political warfare aspects of the Food approach…

> Wallace was also keenly interested in the Security aspects of buffer stocks or as he prefers to say 'ever-normal granaries'.[11]

Wallace wrote in his diary: 'McDougall was visioning a committee to work on this problem of international food planning.' They discussed this to the extent of listing suitable members.[12] Tolley recorded this discussion in the first sentences of his memoirs regarding the Hot Springs conference: 'On August 12,

---

8 Ibid., letter to Bruce, 11 August 1942.
9 Ibid., 19 August 1942.
10 HWD, 6 August 1942.
11 DAFP VI, 22, pp. 43–5, McDougall to Bruce, 13 August 1942.
12 HWD, 12 August 1942.

1942, Henry Wallace and Frank H. McDougall [sic] decided that a good United Nations food committee would consist of Alvin Hansen, Dr Thomas Parran… Winfield Riefler, Dean Acheson and myself.'[13] Alvin Hansen was a Harvard economist at the centre of a group known as 'the American Keynesians' who had been opposing 'pump priming' by private capital, advocating instead programs of public spending, redistributive taxation and full employment.[14]

The State Department

> The State Department in 1940 was overwhelmed by intrigue and fatigue…Secretary of State Cordell Hull
>
> complained openly and often about prima donnas of the department. [Assistant Secretary Adolf] Berle ranked high on that list because he showed 'extreme egotism and ambition', worked almost exclusively to the President and covered too much ground in the Department. Hull also distrusted Welles…[who] increasingly acted independently of Hull in 1940 because he anticipated that in 1941 either Hull would be President and sack him or Roosevelt would be President and sack Hull. It seemed inconceivable that Hull, Welles and Berle would be around for a third Roosevelt term.[15]

All three were still there in 1942 and, if anything, relations were worse. Ranged on the side of Hull were Assistant Secretaries Breckenridge Long and Dean Acheson. Acheson later wrote that Hull, 'suspicious by nature…brooded over what he thought were slights and grievances…His hatreds were implacable… long cold ones'. Hull's relations with Roosevelt were strained from the first by economic problems, beginning with the London Economic Conference of 1933, 'torpedoed by the President with the Secretary on the bridge': Roosevelt's decision to float the dollar was announced just before the conference began. Hull's determination to liberalise international trade was thwarted by policies like Wallace's AAA, while Roosevelt turned more and more 'to other, more energetic, more imaginative, more sympathetic collaborators', to the extent that Hull, 'the senior Cabinet officer, became one of the least influential members at the White House'. Even in his own department, 'the Secretary's influence and authority were diluted', particularly after Welles became Under Secretary. Unlike the 'slow, circuitous, cautious' Hull, Welles, with his 'incisive mind and decisive nature…grasped ideas quickly and got things done. More and more he

---

13 Hollis Library, Harvard [hereinafter HLH], 'The reminiscences of Howard R. Tolley', microform, p. 578. I am grateful to Tim Rowse for locating this and similar material in the Hollis Library.
14 Borgwardt, *A New Deal for the World*, p. 137.
15 Schwarz, *Liberal*, pp. 134–5.

took over liaison with the White House on international political matters. Mr Hull rankled under what he believed to be Welles's disloyalty and the President's neglect.' The department became divided into 'Welles men' and 'Hull men'.[16]

Acheson and Adolf Berle were both responsible for aspects of postwar planning. Acheson records that his own relations with Welles were good, but that he and Berle disliked one another: 'for four years…we maintained a wary coexistence on the second floor of old State, separated by the offices of the Secretary and Under Secretary, who side by side managed to do the same thing.' The dysfunction spread downwards. Bureaucratic power rested essentially with division chiefs and advisers who were also 'constantly at odds, if not at war…obscurity in lines of command of the assistant secretaries permitted the division chiefs to circumvent them at will and go directly to the Secretary or Under Secretary'.[17]

By the time of his second meeting with Wallace, McDougall had also seen Welles. Sumner Welles had a formidable reputation. He was described by columnists as 'one of the most-discussed human enigmas in Washington' and as 'this tall, powerfully built, beautifully-tailored man with the glacial manner, and an expression which suggests that a morsel of bad fish has somehow or other lodged itself in his mustache'.[18] Welles was close to Roosevelt, ran the State Department and wielded major influence in US foreign policy. Remarkably, Welles himself sought a meeting with McDougall, who reported to Bruce:

> He could not have been more cordial, it was quite remarkable. He spoke of you with great admiration. He said that my coming over delighted him and that he hoped I should be here for some time. He then gave me a brief outline of the work being done on post-war problems in the State Department and said that while he did not propose to discuss this work with the U.K. at this stage he would discuss it with me after I had seen a number of people. I said that if he regarded me as your personal representative and not as an accredited representative of the Commonwealth Government I thought I could discuss any post-war question. He asked me to see Berle and Acheson and get other names from them and then he would devote much time to a general discussion of American ideas of post-war. He was impressed with the need of concrete plans for political warfare and said that he thought you and he were very close in ideas.[19]

---

16 Dean Acheson, *Present at the Creation: My Years in the State Department*, W. W. Norton & Co., New York, 1969, pp. 9–12.
17 Ibid., pp. 14–16.
18 Joseph Alsop and Robert Kintner, quoted in Benjamin Welles, *Sumner Welles: FDR's Global Strategist: A Biography*, St Martin's Press, New York, 1997, p. 209.
19 NAA, M104, 10, handwritten letter, 11 August 1942.

McDougall added as a postscript: 'Whatever the level of your stock elsewhere it's very high here.'

McDougall understood that external economic relations had been the subject of a bureaucratic battle between the State Department and the Board of Economic Warfare (BEW), headed by Wallace, with Milo Perkins as his deputy. Early in 1942, BEW had sought control over Lend–Lease (a move soon dropped) and the Reconstruction Finance Corporation (RFC) purchasing of strategic supplies. Opposition came particularly from Secretary of Commerce, Jesse Jones, head of RFC, and a determined opponent of apparent BEW aims to combine its operations with social justice policies, thus ignoring the needs of US business in sourcing strategic materials. Conservative opinion was further alienated by accusations of a 'communist front' in BEW and an earlier obscenity case against a BEW employee. A Presidential Order in April required automatic approval of BEW funding requests, but left RFC subsidiaries under the control of Jones. The order, however, also gave BEW authority to advise the State Department on terms of Lend–Lease agreements, and for BEW representatives to negotiate with economic warfare agents of foreign governments, leading Welles to protest against the creation of 'a second State Department and a second foreign service'. Roosevelt subsequently authorised a series of compromises making BEW personnel abroad responsible to local chiefs of mission, thus handing control to the State Department. The war between Jones and Wallace was to continue until the very public 'Battle of Washington' in mid 1943, arising from charges by Wallace that the RFC had procrastinated in procurement of vital war materials, including quinine. The result was an executive order abolishing both BEW and the procurement functions of RFC, establishing a new Foreign Economic Administration under Leo Crowley, and publicly humiliating Wallace. It has been suggested that Hull feared BEW powers might be expanded to include negotiation of a postwar economic settlement, which would be a reasonable move 'if the administration had seriously considered creating any administrative vehicle for the world New Deal'. But, adds Norman D. Markowitz:

> Roosevelt the broker had always dealt on the basis of existing power. From the outset the BEW's plans to combine the war emergency with model programs for economic development and improved labor and welfare standards abroad became subordinate to the superior power of the State Department and the President's need to retain the appearance of unity within his administration.[20]

The State Department was effectively declared in control of postwar planning. As a result, wrote McDougall, the department was 'spurred into quite vigorous activity': committees had been established on all aspects of political and economic

---

20   Markowitz, *The Rise and Fall of the People's Century*, pp. 66–74.

reconstruction, chaired by Assistant Secretaries of State and including senior officials from other departments and some non-officials. Responsible to these were subcommittees and a research section directed by Leo Pasvolsky, 'a rather key person since he has Mr Hull's special confidence'. The general approach seemed 'very international and liberal', but progress varied: a committee on 'Monetary and Banking' under Assistant Secretary Berle was drafting its report and work had been progressing on world transport questions. On food and agriculture, Paul Appleby, Under Secretary of the Department of Agriculture, was 'regarded as a rather key man', but little progress had been made because key men from the Bureau of Agricultural Economics had been moved to war agencies. Now, after several meetings at Wallace's request between Bureau head, Howard Tolley, and McDougall, Tolley was 'working out a tentative scheme'. State Department preparations were being kept pretty secret. It is only because Mr Welles definitely authorized talks with me that I am being given information. The [British] Embassy has not, so far as I know been given more than hints at the scale of the State Department's operations.' In fact, Isaiah Berlin had reported on 11 June that a State Department advisory committee on postwar reconstruction, presided over by Welles, met weekly, but 'discussions, and indeed the existence of this committee, are strictly secret'. There was, wrote McDougall, a strong view that Winant had erred in giving the impression in London that the United States was ready for talks: 'the State Department is determined to clear its mind before proceeding to Anglo–American discussions.'[21]

Adolf Berle

After Sumner Welles, Assistant Secretary Adolf Berle was McDougall's most important contact in the State Department. Berle's reputation was equally formidable; indeed, Welles has been described as 'one of the few people in Washington who could match Berle for intellectual arrogance'.[22] Berle was 'a ganglion of complexes. Known to be short-tempered, difficult, abusive, snobbish and elitist, [he] was unexpectedly genial and generous with his time with strangers who dared to seek him out in spite of his forbidding reputation.'[23] This, and doubtless his genuine interest in McDougall's message, meant it was the genial Berle whom McDougall encountered. During 'a long evening' together, Berle 'said that Mr Welles had asked him to give me a clear idea of the way in which the State Department was approaching the whole reconstruction problem'.[24] On 7 September, McDougall had a two-hour meeting with Berle,

---

21   NAA, M104, 10, McDougall to Bruce, 19 August 1942; Nicholas, *Washington Despatches*, p. 43.
22   Schwarz, *Liberal*, p. 189.
23   Ibid., p. 203.
24   NAA, M104, 10, letter to Bruce, 19 August.

who showed him, in confidence, State Department papers on postwar planning. Berle also gave McDougall some tactical advice, suggesting that he make a point of seeing Cordell Hull since his close association with Wallace made his approach to the Administration 'rather lop-sided'.[25] McDougall did not manage to see Hull, but did meet Hull supporter Dean Acheson, and worked closely with Harry Hawkins, head of the Trade Treaties Section and a close associate in Hull's multilateral trade campaign.

Because of the link McDougall was making between political warfare and food policy, he also saw Archibald MacLeish, second-in-command of the Office of War Information (OWI), and officers of the Office of Strategic Services (OSS). On Wallace's instruction, he read a draft relief and rehabilitation policy, later emphasising to Milo Perkins and Riefler the importance of coordinating short-term relief measures with long-term reconstruction policies.[26]

The White House

The meeting Wallace had suggested with Eleanor Roosevelt took place over lunch at the White House on Saturday, 15 August. 'She was very nice and friendly, and suggested the possibility of getting me to dinner to have a chance of a few words with the President. I should not be surprised if this benevolent intention failed to materialize.'[27] In spite of his doubts, McDougall dined at the White House, seated next to the President, at a 'family dinner' on 24 August. Other guests were Roosevelt's close adviser Harry Hopkins and his wife, who were living at the White House, and 'several young people associated with Mrs Roosevelt's activities'.

> The President was in good form and talked with great animation. I did not attempt to press our ideas too hard but talked to him about the war of ideas and the need for ammunition for that war in the shape of concrete schemes to give real meaning to the Atlantic Charter and to the phrase Freedom from Want. He seemed to like that line of country and could not have been nicer. After dinner we saw a film and then the President retired to work. He is certainly a most remarkable man and his immense vivacity is extraordinarily attractive. I am very grateful to Mrs Roosevelt for arranging this opportunity.[28]

---

25   *DAFP* VI, *40*, p. 82–3, letter to Bruce, 6 and 7 September.
26   NAA, M104, 10, private note for Board of Economic Warfare, 'Relief and Reconstruction, Joint or Separate Agencies, some Pros and Cons', n.d., enclosed with letter to Bruce, 20 August.
27   Ibid., letter to Bruce, 19 August.
28   *DAFP* VI, *29*, p. 62, letter to Bruce, 26 August 1942.

At the request of Mrs Roosevelt, McDougall later chaired a round-table discussion of the International Students Assembly.[29]

Figure 17 F. L. McDougall presenting a copy of the *Story of FAO* by Gove Hambidge to Eleanor Roosevelt in 1955.

Source: E. McDougall.

Rewriting 'Progress in the War of Ideas'

McDougall's memorandum 'Progress in the War of Ideas', brought to Washington in rough note form, underwent a series of revisions. On the weekend preceding his White House dinner, he worked over it with Harry Hawkins and other State Department officials. He sent Bruce a copy of the new version and another to Mrs Roosevelt with his letter of thanks for the dinner meeting, commenting that he thought the case for urgent action was 'overwhelmingly strong'. She replied that she was 'most interested' in the memorandum.[30]

---

29 *DAFP* VI, *40*, pp. 79–80, letter to Bruce 6 and 7 September.
30 Ibid.; Franklin D. Roosevelt Library, handwritten letter to Mrs Roosevelt, 25 August, with copy of seven-page version of 'Progress in the War of Ideas', and Mrs Roosevelt's reply, 26 August.

On 25 August he dined with Wallace, Milo Perkins, Harold Butler, Riefler and a 'remarkable young Englishwoman, [economist] Barbara Ward'.[31] His exposition on that occasion of the ideas in the memorandum is recorded in Wallace's diary:

> McDougall held forth as usual on the psychological value of the weapons which we have in the form of food. His battlecry is 'Freedom from want everywhere in the world'. He would have established a number of technical expert commissions to study and report on the freedom from want problems in the different parts of the world. The commissions would report to an economic council of the United Nations which would make decisions based on the reports, arrange for publicity and in effect become a munitions factory for the war of ideas. There should be the closest possible co-operation with the Office of War Information in the United States and the British Ministry of Information.[32]

The following evening he was taken by Wallace to dine with Francis Miller and his wife. Miller was Chairman of the Donovan Committee on Psychological Warfare. His wife, Helen Hill Miller, was American correspondent for the *Economist*. She had been an executive of the National Policy Committee, founded by her husband, which advocated support for Britain and an end to isolationism. She had served as liaison between the British Ambassador and the Administration and had worked as a writer for Henry Wallace when he was Agriculture Secretary.[33] Also present were staff of OWI and OSS. McDougall talked of timing a psychological offensive to coincide with German morale reaching a new low in the coming winter as difficulties with food and transport increased. After dinner they discussed the latest version of 'Progress in the War of Ideas', Wallace reading aloud and pausing for comment as he went. McDougall made further revisions as a result of these 'constructive comments'.[34] During a meeting lasting one and a half hours, he gave the final version to Wallace, who described it in his diary as

> a most interesting memorandum. The essence of the McDougall approach is that Germany is much weaker internally than we realize, that we must hit her with everything we have now, that the situation is perhaps a little bit like it was in October of 1917. We must remember that Wilson came out with his 14 points in January of 1918 [and although the German Army inflicted defeats on the British and French between March and July] nevertheless the 14 points took hold; the German psychology crumbled.

---

31 *DAFP* VI, *29*, letter to Bruce, 26 August 1942.
32 HWD, 25 August 1942.
33 <http://www.cosmos-club.org/web/journals/1995/miller.html> [accessed 19 January 2006].
34 NAA, M104, letter to Bruce, 1 September.

Wallace lunched with the President that day and gave him a copy of the memorandum, together with a clipping from The *New York Times* urging 'now was the time to strike with single-mindedness on all fronts including the psychological…The President read the four pages of the memorandum very carefully'.[35] There is thus firm evidence that Roosevelt read 'Progress in the War of Ideas'.

Berle read it and agreed that a psychological offensive should be prepared and ready for launching as soon as Roosevelt and Churchill thought the time right. But he also thought that priority should be given to enunciation of policy for an international bank, to reassure Americans that the financial burden of reconstruction would not fall upon them. Standards of living should be placed in the forefront to explain that 'use of US gold reserves to secure economic activity will be reflected in increased demands for U.S.A. goods and hence more employment in factory and farm'. McDougall told Bruce that he found Berle's point about the politics of US public opinion persuasive.[36]

With Welles absent from Washington, Berle thought he might himself discuss the food and agriculture campaign with Roosevelt, though he admitted he did not 'fully understand' it. He gave helpful advice. Under Secretary of Agriculture Paul Appleby had been asked to produce a document, which had 'hung fire'. McDougall should work with Appleby 'in getting the position clear'. Berle gave Hull a copy of 'Progress in the War of Ideas', and Hull told Hawkins 'it seemed to contain much sound sense'. Hawkins told McDougall that Wallace had insisted the President read it.[37]

Thus, by early September, after some five weeks in Washington, McDougall had talked to the Roosevelts and Harry Hopkins, to Wallace, and to much of the senior echelon of the Departments of State and Agriculture. He had met, socially or officially, the second-in-command of OWI, staff of OSS, the Director of BEW and the Chairman of the Committee on Psychological Warfare. He had talked with senior economists Winfield Riefler and Alvin Hansen, and had held 'long talks' with officials at the British Embassy responsible for relevant areas of economic policy, Harold Butler and Redvers Opie. A two-day meeting of the League Economic Committee early in his visit brought him in touch once more with international economic experts, including Britain's Leith Ross, Sir Frederick Phillips, Chairman of the League Financial Committee, Canada's Deputy Minister of Finance, W. C. Clark, and American Henry Grady.[38]

---

35 HWD, 26 August 1942.
36 *DAFP* VI, *40*, pp. 80–3, letter to Bruce, 6 and 7 September.
37 Ibid.
38 NAA, M104, 10, letters to Bruce, 19 August and 11 August [handwritten].

McDougall always enjoyed drafting 'in committee' and made good use of the information and advice he was given. The outcome of all this discussion was a revision of the rough notes on 'Progress in the War of Ideas' he had brought with him. The two and a half pages had expanded to a polished memorandum of more than four, carefully crafted to appeal to Americans.[39]

A seven-page revision follows the outline of the first version, but is clearer and much more telling. Paragraphs are shorter and the text is punchier: 'the military offensive now being demanded by the peoples of U.S.A. and the British Empire must be paralleled by a psychological offensive.' Subject headings have been added: 'The Psychological Factor', 'The Time Factor', 'Ammunition for the War of Ideas', 'An Immediate Programme', and 'Methods of Progress'. New ideas have been added. Historical examples show the value of ideas in winning wars: the French revolutionary army even before Napoleon; the Union Army in 1863 after the Emancipation Proclamation; and Wilson's Fourteen Points in 1918. Two new points address international and US domestic sensitivities. First, ideas should be put forward not as American, or even as Anglo–American, but as of the United Nations, to avoid 'suggestions about American imperialism or of an Anglo–American hegemony'. Second, tackling hunger 'can show our own farmers how their war-time efforts can be correlated with the health requirements of our own peoples and of the rest of the world'.

Most significantly, a vague reference in the first version to assessment of food requirements by 'the best experts' has now become investigation by 'technical commissions' reporting to 'an Economic Council of the United Nations', which would then 'take decisions…arrange for publicity and thus become the munitions factory in the war of ideas'. Ideas could be disseminated by the US OWI and Britain's Ministry of Information, aiming at 'vigorous discussion of our aims [being] reflected to enemy and enemy-occupied countries'. The draft concludes with the words: 'It is essential to draw the peoples of the world into the process of discussion…because our hope of actually achieving the Four Freedoms will depend upon the people wanting them enough.'[40]

A third and final version is the one Wallace showed President Roosevelt on 2 September.[41] Changes from the second version are mostly cosmetic. The

---

39 There are various copies, mostly undated. The first rough notes occupy two and a half single-spaced pages. The seven-page revision, that sent to Eleanor Roosevelt, is double-spaced. The final version, that shown by Wallace to the President, is little changed and exists in at least two forms: a single-spaced copy of four and a half pages and a double-spaced copy of eight.
40 'Progress in the War of Ideas', seven-page version, n.d., in NAA, M104, 10, and in FDR Library, file 'McDougall'.
41 It exists in two forms, one of which is dated Washington, 1 September 1942. A single-spaced dated copy of four and a half pages is in NLA, MS6890/4/4, and another in NAA/CSIR, A9778/4, M1 N43. An undated version, annotated 'further copy' in McDougall's own hand, is in NLA, MS6890/4/4, and another in NAA, M104, 10.

five sections are retained, though the first title, 'The Psychological Factor', is omitted. Much of the text is unaltered, but it has been polished and tightened, with sentences and paragraphs further shortened. It begins:

> The purpose of the United Nations is first to win the war and then to win the peace.
>
> To win the war we must pass from the defensive to the offensive and utterly destroy the German and Japanese military despotisms.
>
> To win the peace we must, during the war, reach agreements which will determine the pattern of the post war world.
>
> The simultaneous prosecution of these two purposes is necessary since the factor of political warfare can be of tremendous importance.

The final version includes one new idea:

> ...current fears of some business men, trade unionists and farmers that the war production programme will be followed by surplus capacity and unemployment is proving a psychological obstacle to an all out war effort. We must convince them that an expanding world economy will bring them freedom from these fears.

In subsequent weeks, many influential people in the Administration read the paper. McDougall told Bruce, for example, that Supreme Court Justices Jackson, Douglas and Frankfurter were all 'greatly interested'.[42] It is not clear how many of the readers, including Roosevelt, knew that the ideas contained in it originated with McDougall. None of the copies cited here bears his name, but the information may well have been given informally. His discussions were so widespread that his ideas must have been generally known even before copies of the memorandum were circulated. Bruce gave him some credit: in sending a copy of the final draft to Stafford Cripps, then British Ambassador to the Soviet Union, he described it as originating 'from a paper which McDougall drafted after discussions with me here', revised in discussion with 'either the State Department or Wallace's people' and 'indicating something of what is in the minds of the Americans'. He explained that McDougall had been 'in touch with Wallace and the State Department on reconstruction questions'. Although these organisations had 'not at all times been on the best of terms', McDougall, 'who has contacts in both has apparently been used to a considerable extent as an outsider to help in composing these differences'.[43]

---

42 NAA, M104, 10, letter, 4 October 1942.
43 *DAFP* VI, *42*, pp. 85–6, Bruce to Cripps, 9 September 1942. Here Bruce claimed to have seen three revised versions from Washington. He may have been counting what I believe to be the first version as a revision, if earlier notes were made of their conversation. There may be other versions I have not seen. Or Bruce may simply have lost count.

'Active preparation' in the Department of Agriculture

By early September progress had stalled, and McDougall was frustrated. Key contacts were away, including Welles and Parran. Paul Appleby recalled his own support for McDougall's ideas, when Agriculture Secretary Wickard was inclined to be cool and suspicious of McDougall. Appleby, 'irritated...bore down on him. I more or less made demands on him, and I dictated to him that this was important business—this was the kind of thing that was wholly desirable that the United States as a great food producer ought to be interested in.' But Appleby identified a weak point in McDougall's campaign:

> In smaller groups we talked over McDougall's plan of action, which was based on the notion that we had to have a whole lot of popular discussion about this kind of thing. McDougall had been having a lot of discussion with individuals, and his method was one, that would have continued discursive, I think, for a pretty long period of time...I happened to think that it was not necessary and actually would be delaying to try to build up popular sentiment, but that some initial action...was in the area of discretion of the executive leaders of the government. I discouraged and was more or less in a position to block the effort to get out and start beating the drums over the country.[44]

Appleby directed McDougall to what he thought a more productive approach. MacLeish of OWI wanted active preparation for psychological warfare in the coming winter, and believed food and agriculture provided 'much the best material'. Appleby suggested he and McDougall collaborate in preparing a proposal for action to submit to the State Department. By 14 September McDougall had a preliminary draft, amended after comment by Tolley. The draft proposed the Department of Agriculture prepare six papers to be discussed by an interdepartmental panel and then submitted to Wallace, Hull and Welles, who would consider submission to the President. Orr was expected in Washington late in September and McDougall hoped to involve him.

Bruce had suggested McDougall return to London by the end of September. McDougall replied that with such interest in the idea of combining reconstruction with political warfare, with food and agriculture in the forefront of the program, his early departure would be damaging: 'I think Wallace and possibly Welles would feel that we had raised very important issues but had not been as helpful as possible in following them up.'

---

44 HLH, 'Reminiscences of Paul Appleby', microform, pp. 263–5.

A handwritten postscript to this letter indicates what 'positive action' the US Government might take:

> 1. The approach to other governments to obtain agreement to establish a technical expert commission. This would be much easier if an Economics Council of the United Nations already existed.
>
> 2. A suggestion of Hawkins of the State Department is that if the President accepts the view that a major political warfare drive should be commenced this winter, then he should decide to call a conference on food and agriculture. The invitations should be accompanied by the papers we are proposing to produce and the conference should take the form of plenary meetings with speeches. The conference to conclude its sittings with a request for the establishment of a Technical Expert Commission to prepare plans for action by the United Nations.

Bruce agreed that McDougall should remain in Washington until about 10 October.[45]

A week later McDougall reported rapid progress by the Agriculture Department: BAE staff were working on the capacities of national agriculture to supply food requirements, methods to bring food within the purchasing capacities of lower-income groups and the role of international trade in food. Staff of the Farm Security Division were applying their domestic experience to international problems. The Division of Foreign Agricultural Relations had produced and discussed with McDougall a paper on Eastern Europe and was now working on Latin America and China. Staff of the Advisory National Committee on Nutrition were 'going into the vital statistics approach'. McDougall must have been very busy: 'I am seeing all these people and discussing the way in which the papers shall be presented.' The assessment committee, which was to begin a series of meetings on 1 October, would comprise, as representatives of the Department of Agriculture, Dr M. L. Wilson, Tolley, Director of Foreign Agricultural Relations Leslie Wheeler, and one other. Berle or Hawkins would represent the State Department together with Parran for Public Health, Boudreau for nutrition, Milo Perkins or his nominee for BEW, Eisenhower, 'brother of the General', for OWI, with Orr and McDougall. 'The main objective will be to get an agreed statement which would be the sort of document the President could send to other United Nations Governments if he decided to go ahead on these lines…I shall try to have a preliminary draft ready before the group meets.'[46]

The Department of Agriculture prepared 12 papers, rather than the six earlier proposed, and McDougall spent the first weekend of October summarising

---

45   NAA, M104, 10, letters to Bruce, 14 and 20 September.
46   Ibid., letter to Bruce, 20 September.

their 120 pages for a dinner discussion of 30 people, headed by Wallace, Berle, Hawkins, Parran and Boudreau. Berle was 'very pleased with the progress made over food and agriculture'. Orr had not arrived: 'It is a great nuisance', wrote McDougall on 12 October. Group discussions had taken place, a small drafting committee was working 'pretty solidly', but he wanted Orr to consider the draft before the final meeting of the whole group. He was confident that Wallace, the State Department and OWI believed 'the Food and Agriculture approach is right and should have a high priority', but there had been no indication of Roosevelt's thinking.

> The Vice President tells me that the President is keeping his own counsel about post-war questions. Wallace feels sure that the President is having some members of his White House entourage work in strict secrecy on the subject and that neither he, nor the State Department can get any inkling of how the President's mind is shaping.[47]

In 'a most satisfactory hour', Sumner Welles told McDougall he

> had been unhappily conscious of a gap in the State Department's preparation on reconstruction because the Department of Agriculture had not found time to do a real job on Food and Agriculture. He had now been told that this gap was being well filled by our joint group's work and that pleased him very much...he anticipated that by the middle of December the committees on all aspects of reconstruction would have reported and that it would then be possible to go to the President and propose methods of consultation with other nations.

Welles agreed with McDougall's view that they might have only 18 months to formulate joint schemes, obtain the approval of governments, 'sell the ideas' to American, British and dominion peoples and 'put them across the air to Europe'. He had read 'and much liked' McDougall's memorandum 'Progress in the War of Ideas', and agreed that political warfare 'might play an important part [if] supplied with proper munitions'. To counter Welles' view that the United States must know its own mind before discussions with other nations, McDougall suggested advantages in early informal discussion: 'it was harder to modify completed proposals than those which were in a less advanced stage.' Welles seemed to agree, and thought 'we should get to a United Nations basis early in 1943 and hoped I should be back in Washington then'. He wrote to Bruce that McDougall's visit had been 'most helpful. It has not only been agreeable but most useful to all of us to have the opportunity of talking with him and of exchanging views.' He repeated his hope that McDougall would return in 1943.[48]

---

47   Ibid., letters to Bruce, 4 and 12 October.
48   *DAFP* VI, *61*, pp. 132–4, handwritten letter, 16 October; NAA, M104, 10, Welles to Bruce, 14 October 1942.

The draft prepared by the group is the memorandum published in 1956 by FAO as the fourth document in *The McDougall Memoranda*, the booklet referred to in the introduction to this study. Orr arrived just in time to give his opinion, and it was approved by the full group on the day McDougall departed for home. Orr urged American business leaders to persuade government of the economic advantages of what he called 'a World Food Plan'. He also discussed it with senior officials, but his memoirs give no indication that he influenced any changes to their approach. He declined a suggestion that he be invited by Mrs Roosevelt to meet the President, but left America 'hopeful that the President would call a conference of the nations to consider a world food policy' and that he would be supported by 'the more intelligent industrialists and financiers'.[49]

After reading the group draft, Wallace 'expressed great interest and warm appreciation'. Hawkins assured McDougall that 'close attention would be given to the paper in the Department'. In sending a copy to Curtin, Bruce warned it

> should be regarded as unofficial and as expressing the personal views of the members of the group. It will, however, be closely considered in the State Department in the hope that it may be found to provide a suitable basis for the formulation of the Administration's policy on this subject.

He went on:

> McDougall tells me that he is satisfied that the Vice President, Mr. Sumner Welles, Mr. Acheson and Mr. Berle are now convinced that the Food and Agriculture approach should be given a high priority in the United Nations programme for reconstruction. He cannot indicate what Mr. Cordell Hull's attitude is likely to be but Mr. Hawkins, the head of the Commercial Treaties Division of the State Department, was a keen member of the group and Mr. Hawkins is a trusted adviser to Mr. Cordell Hull. McDougall found the officials of the United States Department of Agriculture keenly interested and desirous of playing a considerable part in the development of policy along these lines. He has, however, emphasised that he cannot form any opinion as to whether the method suggested in the Memorandum of the setting up of technical expert commissions appointed by a projected United Nations Economic Council will commend itself to the Administration.[50]

---

49 Orr, *As I Recall*, pp. 157–9.
50 *DAFP* VI, 78, pp. 157–60, Bruce to Curtin, 22 November 1942. The full memorandum is published as an attachment, pp. 160–8.

'Draft Memorandum on a United Nations Program for Freedom from Want of Food'

Like many of McDougall's memoranda, this jointly authored paper bears an unwieldy title. Its substance is little more than a polished version of 'Progress in the War of Ideas'. The first words, indeed, repeat those of the earlier memorandum: 'The purpose of the United Nations is first to win the war and then to win the peace.' It moves on quickly to a more vigorous statement of campaign: 'We have promised to our own citizens, and to the peoples of the world, freedom from want through an expansive economy with full employment, better labour conditions and social security. We have, in effect, undertaken to engage in a world wide campaign against poverty.'

Preparation to fulfil those promises for the peace should be made before the war's end, and would require 'many forms of action in the economic and financial fields', but food—'the most essential of human needs', the production of which 'is the principal economic activity of man'—must be given priority and adequate nutrition must be 'the first concern of agricultural policy'. Realisation of 'Freedom from Fear' and of 'Freedom from Want' would require a world authority with both political and economic functions. The memorandum suggests steps for its formation and for 'technical commissions', including one to 'formulate action programs designed to assist the nations to achieve freedom from want of food'. It repeats the call for a pledge by the prosperous countries of the United Nations to institute national policies for adequate nutrition and to assist others towards that objective. Its concluding paragraph stresses the urgent need for action 'if we are to avoid losing the peace', and the universality of its objectives:

> The end of the war will find all peoples impatient for a return to peace conditions. The United Nations must be ready with measures and organizations to carry out their pledges, otherwise national legislatures may adopt ill-considered, short-sighted and nationalistic policies and vested interests will re-entrench themselves…We look forward to a co-operative World Commonwealth to which every nation will make its individual contribution, in which variety of culture will be matched by unity of purpose to secure for all the four essential freedoms, and the right to participate in, and contribute freely to, international counsels for the future welfare of mankind.

'A keen outsider'

Early in his visit McDougall had explained how he might be useful amidst the 'warring principalities' of Washington: 'there is much confusion and a good deal of jockeying for position between Departments. A keen outsider comes into the picture and may be able to get action by bringing people together on an extra-departmental basis.'[51]

That is hardly sufficient to explain why an official on the staff of a dominion high commissioner in London was paid such attention by some of the most awe-inspiring figures in the Roosevelt Administration. What was it that made this 'keen outsider' so acceptable and interesting to his hosts? There are, perhaps, several answers, including the fact that he represented Australia, the coincidence of his ideas with much US thinking, and the timing of his visit.

Representing Australia

There was an identified 'Anglophile' faction in Washington, but almost all the senior figures McDougall saw, including Roosevelt, Wallace, Welles and Berle, were not part of it. They acknowledged that alliance with Britain was necessary to defeat the Axis but remained deeply suspicious of British motives in postwar planning, even expecting to find it easier to work with their Soviet allies than with British statesmen who were probably doing all they could to shore up Britain's empire for the future. New Dealers saw Britain as the centre of bankers' capitalism against which the New Deal was directed. Roosevelt distrusted the British aristocracy, he suspected the Foreign Office of pro-fascist tendencies and thought British representatives generally were determined to trap Americans into defending purely British interests. The imperial preference system established at Ottawa in 1932, which was seen as an obstacle to expansion of US trade, was the object of particular anger.[52] It has been suggested that this distrust was the reason for State Department reluctance to permit negotiations in London or, indeed, negotiations with the British at all, until Washington was adequately prepared.[53]

Welles was prepared to show much of the confidential planning material to McDougall and to seek his comments, because McDougall represented not Britain, but Australia. There were reasons for Americans, both private and official, to think sympathetically of Australia in 1942. America was at war, but

---
51 *DAFP* VI, 22, p. 42, letter to Bruce, 13 August 1942.
52 Robert Skidelsky, *John Maynard Keynes: Fighting for Britain 1937-1946*, Macmillan, London, 2000, p. 92.
53 The frustration by Hull of Winant's efforts to arrange such negotiations are detailed in Howland, 'Friend of Embattled Britain', pp. 202–10.

the only significant fighting in which its forces had a major part had been in the Pacific. General MacArthur had his headquarters there; American soldiers were based there; Australia was in the news and high in American consciousness. Some would have been aware of issues apparently dividing Australian interests from British: Curtin's 'look to America' speech and his differences with Churchill over withdrawal of Australian troops from the Middle East. Official Washington had been intrigued by another matter. Early in 1940 Australia opened its first foreign diplomatic mission, a legation in Washington, led by former Federal Treasurer R. G. Casey, whose energy, charm, skill at public relations and enthusiasm for all things technological (he flew his own plane) quickly won him influential friends at the highest level.[54] Then in March 1942 the British Government appointed Casey its Resident Minister in Cairo. Casey and Roosevelt were happy with the appointment, and Curtin had apparently given his consent, but the manner of its announcement infuriated him and cables between Curtin and Churchill were subsequently published.[55] Isaiah Berlin reported from Washington that 'even in friendlier quarters the treatment of what appeared to be an intra-Imperial dispute showed traces of *Schadenfreude* and gave further evidence that the wave of anti-British feeling is running strongly'.[56]

As a representative of Australia in 1942, McDougall thus represented both America's brave fighting ally in the Pacific and a victim of British imperial high-handedness. It seems clear that Bruce was also well regarded. His recommendation of McDougall appears to have carried weight and considerably eased McDougall's path.

A Coincidence of Ideas

In London Bruce and McDougall had hammered away on doors that for the most part remained closed. Established agricultural and commercial policies made it unlikely that their ideas could be accepted. Public health and food authorities were at best lukewarm. Longstanding antipathies in Whitehall to McDougall and a suspicion that the idea was merely a ploy to sell more Australian produce added to the barriers against them. In Washington there were no barriers. Indeed, McDougall was largely preaching to the converted, his contribution being to bring together problems and ideas already current in a way that was new.

In the United States there was widespread scientific interest in nutrition, particularly in the large foundations, as noted in the prologue to this section.

---

54 Hudson, *Casey*, pp. 117–19.
55 Ibid., pp. 132–3.
56 Nicholas, *Washington Despatches*, 26 March 1942, p. 27.

Work was also being done in the Health Department, although Thomas Parran told McDougall in 1938 that his department should do more.[57] In Wallace's Department of Agriculture much of the focus during the 1930s had been on disposal of surplus in schemes of distribution to the needy. The preoccupation with surplus throughout that decade made agricultural authorities amenable to suggestions of increasing consumption in a way unimaginable in Britain. Howard Tolley recalled that when McDougall came to Washington

> he found a rather surprising agreement among a lot of people…on the importance of more and better food in making and preserving peace after the war. Some of us who had been so greatly concerned about agricultural surpluses in the United States in the pre-war period saw that here might be an opportunity to escape from burdensome surpluses and reduction of production…and turn our efforts and programs toward expanding food production and distribution of food to hungry people and low-income people throughout the world. It had appeal for Henry Wallace. Then the question of getting full U.S. support of course arose, and attention was turned toward the White House and FDR.[58]

The fact that postwar planning was receiving much attention in Washington, and the appeal of McDougall's approach to New Dealers influenced by progressivism and other ideas emphasising science and efficiency, also prepared the way for his contribution.

Timing

McDougall's 1942 visit could not have been better timed. Some military successes had occurred, and the paralysing anger of the months immediately after Pearl Harbor had been transformed into energy not only for waging war but also for planning a just and lasting peace. Officials eager for such action were frustrated by bottlenecks in the planning process and by the difficulty of bridging the gaps between the fortress departments. Cooperation must surely have occurred eventually in some way, but McDougall's arrival at a moment of need suggested a solution.

To have arrived earlier in 1942 would have been too soon, but a few months later could have been too late. Congressional elections in November entrenched the conservative hold on Congress and made the President wary. Henry Wallace's importance diminished accordingly, declining further in the bureaucratic battle throughout 1943, and Roosevelt chose the conservative Democrat Senator Harry

---
57 NLA, MS6890/3/2, letter to Hall and Orr n.d.
58 HLH, 'The reminiscences of Howard R. Tolley', p. 579.

S. Truman as his fourth-term running mate over a marginalised Wallace. Late in 1943 Sumner Welles was forced by personal scandal to resign. He continued to write of his visions of the peace from outside the Administration, but was powerless to implement them. Roosevelt, in failing health, was increasingly constrained by domestic and international pressures.

A Conference on Postwar Food

On 24 February 1943, McDougall opened his *Times* to find a headline, 'Conference on Post-war Food'. President Roosevelt had told his press conference that he was considering the possibility of 'a conference in the spring to discuss postwar food production', not relief, but 'the permanent food supply of the world'. He recalled the 'near success' of wheat conferences that had sought to prevent wide fluctuations in the wheat price and said that this conference would seek 'to prevent both famines and large surpluses, and at the same time to give the producer the assurance of a decent price'. He saw the conference as an 'exploratory investigation'.[59] The Australian Legation in Washington could only tell McDougall that there had been 'a number of public suggestions that a conference of United Nations should be held', including one from New Zealand's representative on the advisory Pacific War Council, Walter Nash, some weeks earlier. Roosevelt had said then that he thought a conference might do more harm than good. Public pressure had continued, and some days before the press statement Roosevelt had made a similar announcement to the Pacific War Council, saying that the conference 'should be confined to what he thought was the safe topic of post-war food, mentioning specifically wheat'.[60]

The 1956 booklet discussed in the introduction to this study, *The McDougall Memoranda*, followed the reproduction of 'Draft Memorandum on a United Nations Program for Freedom from Want of Food' with these words:

> The rest of the story is well known. The President of the United States, through the intervention of Mrs Eleanor Roosevelt, became interested in ideas which in many respects were close to his own. He talked to McDougall and convened the Hot Springs Conference in May 1943, which was followed by preparatory work and the final establishment of the Food and Agriculture Organization of the United Nations at the Quebec Conference, in October 1945.

---

59 NLA, Times Digital Archive, accessed 2 August 2006.
60 NAA, A2937, 'Post-War—Food', cable 9 from McDougall to Washington Legation, 24 February; cable 38 from Washington Legation, 28 February 1943.

Similar versions are given in early histories of FAO.[61] Bruce also made that link, urging Curtin to cable Australia's support, to 'reinforce constructive elements both here and in Washington', and adding that he had 'good reason to believe' that Roosevelt's announcements were 'inspired' by the memorandum McDougall had written with the aid of the committee in Washington late in 1942.

But what evidence exists to support this claim, beyond the words of Bruce and McDougall themselves? What other influences might have been responsible? Contemporary documents support claims made in FAO literature that McDougall met Roosevelt and Harry Hopkins and urged the idea of starting international postwar planning with food and agriculture. They show that Mrs Roosevelt received a copy of the memorandum 'Progress in the War of Ideas', and that the President read it in the presence of Wallace. It is not clear from material I have seen whether Roosevelt read the later 'Draft Memorandum…' resulting from the work of the committees in October. Apart from questioning the primacy given to the 'Draft Memorandum…' in FAO, lack of this evidence signifies little. The difference between the two memoranda is not great: both, in their final form, were the result of extensive consultation in Washington and both argue for urgent action on food and agriculture as a starting point for postwar planning.

Writers about the Roosevelt Administration generally agree that it was never possible to know what was in the President's mind or to do more than guess at reasons for his decisions. Roosevelt's method of government was calculated inscrutability. Contemporary and later accounts suggest that trying to guess at his thinking was a major preoccupation of Washington insiders. Some argue that he found it difficult to make up his mind, particularly when an unpopular or painful decision was to be made. One commentator suggested that 'Mr Roosevelt relies upon opposites to coax him first to one side and then the other, believing apparently that this will ensure a middle course approximately in line with the temper of the average American'.[62] Whatever the reason, his method ensured that he kept decisions firmly in his own hands. Harry Hopkins was his closest adviser, living in the White House, with unlimited access and an unlimited bailiwick. By 1942 the State Department had been given authority in postwar planning but, as noted earlier, Wallace for one believed that postwar planning was also being considered at the White House. This was typical. Roosevelt would authorise planning, sometimes by more than one person or group, without their knowing that others were also at work on the issue. He liked to be given a range of options. He would listen, as he did to McDougall, and read, as he did McDougall's memorandum, but give no indication of his

---

[61] A similar version of these events, probably based on McDougall's own account, is given in Gove Hambidge, *The Story of FAO*, D. van Nostrand, Princeton, NJ, 1955, pp. 45–9. Briefer references to the account are in Paul Lamartine Yates, *So Bold an Aim: Ten Years of International Co-operation toward Freedom from Want*, Food and Agriculture Organization of the United Nations, Rome, 1955, p. 49; and in Orr, *As I Recall*, p. 158.
[62] Marquis Childs, quoted in Borgwardt, *A New Deal for the World*, p. 74.

views. He would allow Welles and others to make speeches that might or might not have his endorsement, and he would judge reactions. When persuaded to a particular course of action, he still waited until he judged the time was right. Amongst many examples are the following comments: Roosevelt 'enjoyed dividing authority among competitors'; he 'tacitly and cynically' encouraged conflicts'.[63] Roosevelt himself said: 'I am a juggler, I never let my right hand know what my left hand does'; 'spreading power among contending forces allowed [him] to retain undisputed authority in his own hands'.[64]

By 1942 Roosevelt was on shifting political ground: he was dealing with momentous issues of war and peace in a complex domestic political system with strengthening forces of opposition. Alonzo L. Hamby argues that having led 'a nation unprepared for the grim realities of international power down a twisting and often devious path to war', Roosevelt represented 'an inconsistent amalgam' of the two dominant strains of twentieth-century American liberalism: Theodore Roosevelt's emphasis on a need for national strength and self-interest, his 'imperative of a forceful American role in the larger world'; and Woodrow Wilson's vision of a 'pacific world community, united in adherence to a supranational body striving to meet the needs of all mankind'. At home, with isolationism still a potent factor, Roosevelt was forced to exercise a leadership of 'constant interaction with an inconsistent self-deceptive public mood'; the majority followed him because 'in his inconsistencies he reflected their conflicting concerns'. In the wider world, he faced the difficult task of dealing with Stalin across an 'unbridgeable ideological gulf'. In that situation, Roosevelt's 'splendid inspirational banner' of the Four Freedoms 'could only interfere with the realistic diplomacy' needed to deal with the world as it was; instead he tried to secure peace through 'atmospherics…compromise and… postponement of differences'.[65] This perhaps explains his decision to begin postwar planning with a 'safe topic'.

Elizabeth Borgwardt gives Roosevelt more credit for commitment to the ideals professed in the 'Four Freedoms', noting that he had written that section of his speech himself. In April 1939 he had suggested discussions with the Axis powers on opening up international trade, stating that every nation was entitled to 'assurance of obtaining the materials and products of peaceful economic life'. Borgwardt suggests this marked 'the transition toward blending economic concerns with the politics of collective security'. The President had been interested in the British debate about social welfare that was to culminate in

---

63 Schwarz, *Liberal*, pp. 88, 159.
64 Doris Kearns Goodwin, *No Ordinary Time: Franklin & Eleanor Roosevelt: The Home Front in World War II*, Simon & Schuster, New York, 1994, pp. 137, 156.
65 Alonzo L. Hamby, *Liberalism and Its Challengers: F. D. R. to Reagan*, Oxford University Press, New York, 1985, pp. 37–46.

the Beveridge Report; he kept a file of press clippings labelled 'economic bill of rights'. The phrase 'freedom from want' had been suggested to him by a journalist at a press conference in 1940.[66]

A confident assessment of McDougall's place in Roosevelt's decision making seems out of the question. There were many voices raising ideas similar to those McDougall was putting forward. Isaiah Berlin noted, at the time of McDougall's visit, 'many writers over a period of time have argued the United Nations need a more definitive statement of war aims'.[67] We know that Winant raised nutrition with Roosevelt in some form before McDougall's 1941 visit, and that an international conference on the subject had been proposed during that visit, only to be lost in the tumult surrounding Pearl Harbor. Given that Welles was a close adviser, and in 1942 was masterminding the committee bridging Agriculture and State, it seems likely that Roosevelt would have known of its work, and might well have seen its final 'Draft Memorandum…'.[68]

The President's interest in food questions may have been heightened in the last months of 1942 when committees in both Washington and London assessed the food needs of liberated areas in the 'emergency period' immediately following liberation. Tolley chaired the Washington committee; in London McDougall was one of 'a distinguished group of British and European agricultural and nutrition experts'. Their difficult and complex task undertaken over some two months produced a report 'widely circulated' in London and Washington bureaucracies. The magnitude of the requirements put forward impressed the supply authorities and helped to correct overly optimistic ideas still lingering from the days of surpluses. E. F. Penrose, chairman of the London committee, notes that the two reports had 'hardly been completed when President Roosevelt called for a United Nations Food Conference'.[69]

In his press conference announcement, Roosevelt specifically excluded emergency relief problems from the ambit of Hot Springs. These, he said, were already being dealt with. The emergency matter might nevertheless have reminded Roosevelt of the longer term problem, or perhaps he judged that it had paved the way for a longer term discussion he had earlier considered. Throughout the war farm

---

66 Borgwardt, *A New Deal for the World*, pp. 48–9, 98.
67 Nicholas, *Washington Despatches*, 5 October 1942, p. 90.
68 Borgwardt writes that Roosevelt had 'actively reviewed' 'Draft Memorandum on a United Nations Program for Freedom from Want of Food', which had 'clearly captured his capricious imagination'. But her identification of the memorandum Roosevelt actually read seems based only on statements in the FAO booklet in which it is reproduced. Borgwardt, *A New Deal for the World*, p. 115.
69 Penrose, *Economic Planning for the Peace*, pp. 133–9.

prices were the subject of constant pressure in Congress; it was not an issue Roosevelt was likely to forget.[70] It is perhaps significant that Roosevelt recalled, in his press conference, attempts in the 1930s to solve the wheat price problem.

The contemporary records provide a rather different impression from the vague accounts circulated through FAO. It is clear that a carefully constructed campaign was waged to attract the inscrutable president to the idea of urgent preparation for international action on food and agriculture. McDougall was undoubtedly a key player, both in bringing together like-minded people from the warring principalities of Washington as the 'keen outsider' and in providing telling arguments to support it, thus giving shape to ideas less focused, though widely accepted within the Administration. But his success depended on the readiness of the key figures to accept those ideas and to provide high-level facilitation, to use him to ends they thought valuable. There remains a question about the role of Bruce, certainly in prompting Winant's discussions, and in using his weight in support of McDougall. How much real influence did he have in Washington? In the end, it is not possible to apportion credit to any one individual with certainty.

McDougall's credit remains undiminished, nevertheless. His relentless campaigning provided a catalyst for action within the Administration. Once the international conference was called, the 12 papers that his committees had produced shaped its agenda and its decisions, and thus the ultimate nature of FAO. In that sense 'Draft Memorandum on a United Nations Program for Freedom from Want of Food', which is also a result of that work, deserves its place as a fundamental document of the organisation.

---

70   Walter W. Wilcox, *The Farmer in the Second World War*, Iowa State College, Ames, Ia, 1947, Chapters 9 and 15. See also Nicholas, *Washington Despatches*, passim.

10. 'A New Era in the World's History'

Summary

Bruce, McDougall and Orr played significant roles in the formative years of the Food and Agriculture Organization of the United Nations, sometimes recalling the ideas of 1935. That their world was changing rapidly and posing new problems does much to explain the difficulties they and their fellow idealists were to encounter.

As the 1943 conference called by Roosevelt approached, Britain and the United States agreed that the new organisation should be an advisory body; they feared a body with executive powers, as envisaged by Orr, would interfere with trade and commodity policies. There was confusion about its purpose: some, like J. M. Keynes, believed more fundamental economic problems should be solved first. Participants hailed the conference itself as a success, and a spirit of optimism prevailed. McDougall joined the Australian delegation at Hot Springs, and was appointed Australian representative on an interim commission established by the conference to prepare for the permanent organisation. His contribution to the work of the commission was significant: he was one of the most influential and active members of Committee C, responsible for preparing for the work of the organisation, and he chaired a 'Reviewing Panel' established to guide specialist committees of experts assisting Committee C.

The Food and Agriculture Organization (FAO) was established by its first conference in 1945. Despite British Government opposition, failure to find another suitable candidate led to Orr becoming FAO's inaugural Director-General; he appointed McDougall one of his two 'special advisers'. The body was constrained by a small budget and by failure of the USSR to join. Its constitution was ambiguous, generally favouring an advisory role, but allowing room for an executive interpretation. In its first year it faced a world food crisis. Orr sidetracked the constitution by establishing a separate body, the International Emergency Food Council, to recommend production increases and direct supplies, and then sought approval to create a World Food Board to provide credit and hold surplus food stocks for distribution in time of need. US and British opposition forced the establishment of a preparatory commission to consider other proposals. Bruce headed the commission, which recommended a World Food Council lacking the executive functions of Orr's proposed board. The elected Council became the new executive of FAO, and Bruce its first chairman.

Orr left FAO in 1948 and was replaced by Norris E. Dodd, who focused on technical assistance. McDougall remained as 'Counselor', providing advice on policy and overseeing relations with other international organisations. His last important memorandum, 'The Challenge to Western Civilization', brought a Cold War interpretation to all that he had previously argued. He saw problems of development as paramount, for political as well as for economic reasons, and continued to believe that FAO could achieve its goals.

The Hot Springs Conference

As hasty preparations were made in Washington for the conference announced by the President, the London *Observer* published an article by John Boyd Orr, who wrote that 'the carrying out of a world food policy based on nutritional needs will bring about revolutionary changes in our social and economic system', changes benefiting 'every class of the community'. The conference would adopt 'a food policy based on needs' and establish a 'Technical Commission' to report on food conditions in all countries. Economists and financiers would be asked to plan 'a financial corporation' to provide long-term credits, and 'an International Agricultural and Food Commission' would probably be set up to 'arrange and control international trade in food and advise governments so that national schemes will fit in to the world scheme'. 'This conference', he concluded, 'must mark the beginning of a new era in the world's history'.[1]

Policy advisers in Britain and in the United States had been considering postwar economic arrangements for more than a year before Roosevelt's announcement of a food conference. Lend–Lease arrangements were made early in 1941; Article VII of the subsequent Mutual Aid Agreement and the Atlantic Charter of August 1941 laid out expansionary aims but no detailed means for achieving them. While they agreed that mistakes of the interwar period must not be repeated, the United States and the United Kingdom differed on the means to attain their agreed objectives. In both countries agriculture departments jealously guarded the right to protect farm incomes—in the United States by subsidising exports, cutting across the State Department's aim of liberalising trade. In Britain agricultural policy focused on restricting imports. An empire lobby believing that Britain should make the most of its imperial advantages persisted. Britain would be a debtor nation by the end of the war, with few remaining assets and little

---

1 *Observer*, 25 April 1943.

trade; the United States would be a creditor; they approached postwar problems from different perspectives. Each blamed the other for the currency problems of the 1930s; many Americans believed problems had been compounded by development of the sterling area, which, together with the Ottawa preference agreements, was a focus of US resentment. In British eyes, the sterling area and the Ottawa agreements formed an interlocking defensive system allowing empire countries to build up 'sterling balances' in Britain, thereby enabling London to remain a financial centre and maintain a positive balance of trade with the United States. British policy emphasised the importance of postwar economic expansion, but there remained 'extreme reluctance to abandon a system which combined imperial sentiment with seemingly solid economic advantages'.[2]

As Maynard Keynes in Britain and Harry Dexter White in the United States worked over their respective proposals for postwar arrangements for currency and trade, discussion between them and their advisers occurred from time to time, but formal intergovernmental negotiations were postponed. The US Government wanted to begin with trade talks, which had broken off in mid 1941; it was not yet prepared for monetary negotiations, while the British were not ready to negotiate trade and wanted monetary discussions first. As a result, there were no official talks on these fundamental questions until 1943. In the meantime, there was to be an official conference on food and agriculture.

Responses to the President's announcement ranged from scepticism to hostility. After the first mention of the conference in the Pacific War Council, the British Ambassador in Washington commented that 'it was an unwise proposal but that nothing was likely to come of it'. In Australia one official commented: 'You cannot develop a useful conference out of this proposal'; and another: 'you may be able to develop a conference that is less useful than it threatens to be.'[3] A common reaction was puzzlement: why hold a conference on food when there were so many bigger problems? Keynes, who had been working for some 18 months on his plan for a clearing union to solve monetary problems bedevilling the previous decade, was 'more dejected than I had ever seen him', recalled Ernest Penrose, then Economic Adviser at the US Embassy in London. Having heard Penrose's explanation of Roosevelt's possible motives, Keynes replied: 'What you are saying is that your President with his great political insight has decided that the best strategy for postwar reconstruction is to start with vitamins and then by a circuitous route work round to the international balance of payments!' Penrose himself thought Roosevelt believed that 'if the subject

---

2 Skidelsky, *Keynes: Fighting for Britain*, pp. 132–8.
3 NAA, A989, 43/735/740, cable S.24 from Dixon, 18 February; minutes of interdepartmental committee, 23 March 1943.

chosen was of common interest throughout the world and might be expected to bring wide agreement at least in aims', it could ease the way towards 'the more difficult political issues later'.[4]

The Washington bureaucracy was caught unawares. Dean Acheson claimed the conference had been 'popped on the State Department', and its officers did their best in the circumstances.[5] Department of Agriculture officials complained that their expertise was ignored and they were not informed of plans.[6] The War Department worried about transporting delegates over air routes across the Atlantic and the Pacific closed to all but military traffic and, with 'an intense battle' being waged in North Africa, could not undertake to give them priority.[7] Although the conference was postponed from the April date first announced, preparations were rushed. Arranging the conference was perhaps the least taxing task. A more difficult one was to define its purpose and scope. Lester Pearson reported to Ottawa that there was 'bewilderment' in Washington about the purpose of the conference, noting Acheson's view that 'the best chance of useful work being done by it would be to take as a basis of deliberation the "McDougall Report" on Nutrition of October 1942'.[8]

The invitation received in Australia stated US belief that it was time to begin 'joint consideration of the basic economic problems' likely to confront the world after 'attainment of complete military victory'; allied and associated governments would exchange 'views and information' on topics including plans for postwar food production, import requirements and exportable surpluses, possible agreements and institutions to achieve efficient production and adequate supplies; trade and financial arrangements; and coordination of national policies for improving nutrition and consumption.[9] Acheson responded to criticism that the agenda attempted to cover too much ground: by taking the broader picture and 'directing attention to an expanding agricultural economy, the danger of concentrating upon restrictive commodity schemes would be avoided'. The purpose was to 'work towards broader objectives and emphasize the new responsibility of governments to see that their people are well fed'.[10] No doubt he also hoped that breadth of subject matter would avoid the difficulty of reaching an agreed US policy; there remained a significant gap between Agriculture's lingering fear of surpluses and instinct for production restrictions, and State's policy of freer trade.

---

4 E. F. Penrose, *Economic Planning for the Peace*, Princeton University Press, Princeton, NJ, 1953, pp. 117, 120.
5 NAA, A2937, Post War Food, Halifax to FO, 10 April 1943.
6 NAC, RG 25, Vol. 2261, 5050-C-40, Scott to Pearson, 1 April 1943.
7 USNA, 550.ADI/189, Secretary of War to Secretary of State, 22 April 1943.
8 NAC, RG 25, Vol. 2261, 5050-C-40, Pearson to Ottawa, 25 March 1943.
9 NAA, A989, 43/735/740, Note, 26 March 1943.
10 USNA, Harley Notter files, Box 80, records of State Department Committees, E Minute 44, 7 May 1943.

The Australian Legation in Washington reported that Roosevelt 'has no acute economic sense, and he did not appreciate the fact that a conference on plans for postwar food could achieve very little unless some degree of prior agreement had been reached on trade and the exchanges'. Some members of the British Government feared the conference could endorse production or export restriction, the method favoured in the 1930s to increase farm incomes, thus raising food prices; others were anxious about a possible threat to imperial preference arrangements. Formally, the British suggested an explicit statement that the conference would be held within the context of the Atlantic Charter and Article VII of the Mutual Aid Agreement, thus favouring expansionary trade policies.[11] No such statement was made, perhaps because of division among policy makers in Washington. Britons and Americans were nevertheless united in wanting to avoid concrete proposals: the conference should be 'technical' rather than political; hence the invitation specified 'a small number of appropriate technical and expert representatives'. It was to be a low-key 'technical wartime meeting'; dress would be 'business suits only', and wives were not invited.[12] An announcement that the press would be excluded raised controversy in Washington.

'A feeling of great trust and faith in the future'

A special train transported delegates to the spa resort of Hot Springs, Virginia, in the Allegheny Mountains, some 250 miles (400 km) south-west of Washington. The conference might have been 'low key' compared with prewar international conferences, but it was a luxurious experience for delegates from war-affected countries, and there were at least some vestiges of pomp and circumstance: 'plenary sessions and flags flying, orchestras playing and state dinners, and cocktail parties and so on.' Most remarkable was the atmosphere. Howard Tolley recalled:

> I had hopes that this would be a great forward step toward lasting peace...We really expected big things out of the conference before we ever went to Hot Springs. As preparations for it went along most of us felt more and more hopeful that great things would come out of it.[13]

Accounts record almost universal goodwill and optimism. Norway's Dr K. Evang commended the 'absence of academic hair-splitting discussion' and found 'after

---

11 NAA, A989, 43/735/740/1, Department of External Affairs comment on Note from Washington Legation, 14 April 1943.
12 NAC, RG 25, Vol. 2261, 5050-C-40, US Legation, Ottawa, to Secretary EA, 10 April 1943; Secretariat of Food Conference, undated note.
13 HLH, 'Tolley', pp. 581–6.

the first 2 or 3 days there was a feeling of strong optimism and after a week a spirit of enthusiasm'. An outstanding feature had been the unreserved cooperation between the United States and the Soviet Union. There was 'a feeling of great trust and faith in the future'. McDougall thought that 'nations are bolder and prepared for more far-reaching and constructive suggestions in 1943 than they were in 1937'. He welcomed the appointment as delegates of 'technical and expert people who really did know something about the subject they were discussing', and their authority for full discussion without commitment.[14] But British economist Lionel Robbins wrote in his diary: 'It is very easy to be friendly in a luxury hotel with all expenses paid and no binding commitments on the agenda. When we did touch on bread and butter questions…opinion was by no means so united.'[15]

Much of the credit for the prevailing optimism must go, nevertheless, to the British delegation. Paul Appleby believed that he had influenced British thinking about 'the McDougall plan' during a visit to London late in 1941. At the request of Welles, backed by Winant, Appleby set out to lobby in favour of liberal trade policies after Keynes had alarmed Washington by predicting that Britain's dire postwar situation would demand exchange controls, trade quotas and limits on consumption in order to maximise exports. Appleby saw a number of ministers and found Lord Woolton, the Food Minister, to have 'a very favourable reaction to the McDougall plan'. He believed Woolton's influence helped ensure the British delegation to Hot Springs was a high-level one and that it was briefed according to Woolton's expansive views.[16] 'Extremely gratified' to find that in fact the United States shared their view that the conference should be limited to exploration of problems and possible means of resolution, the British made every effort to facilitate its success.[17] The State Department for its part was reassured by circulation of a British paper by J. P. R. Maud of the Ministry of Food. Maud argued for expansion of trade and employment, opposed restriction of production and advocated education, pooling of information and coordination of research.[18]

A notable omission from the British delegation was John Boyd Orr, still considered to be an outspoken and politically dangerous 'crank' in Whitehall. Bruce had lobbied British ministers for his inclusion in some capacity, but was forced to report failure despite American wishes that he should attend: 'my impression is [ministers] are afraid of him even as an adviser.'[19] From a diplomatic

---

14   *Proceedings of the Nutrition Society*, 1943, pp. 164, 167, 'The Hot Springs Conference', Edinburgh, 2 October 1943, <http:/journals.cambridge.org> [published online 28 February 2007; accessed 15 May 2011].
15   Skidelsky, *Keynes, Fighting for Britain*, p. 300.
16   HLH, 'Paul Appleby', pp. 267–77.
17   NAA, A2937, Post War Food, FO to Washington Embassy, 20 April 1943.
18   UKNA, FO371/35373, 'Confession of Faith', draft by J. P. R. Maud, 10 May 1943; NAC, RG 25, Vol. 2261, 5050-C-40, 'Confession of Faith', sent to Ottawa by Pearson, 12 May.
19   NAA, M100, April 1943, McDougall to Bruce 22 April; Bruce to McDougall, 29 April 1943.

point of view, the delegation was nevertheless an impressive one. Led by Junior Foreign Minister Richard Law, its 10 delegates, accompanied by a team of advisers, represented no less than eight Whitehall departments. Several—among them Law and Lionel Robbins, an academic economist then Director of the Economic Section at the War Cabinet Offices—were to attend subsequent conferences, including Bretton Woods. P. H. Gore-Booth, of the Washington Embassy, wrote of a determination to 'retain political leadership' despite being surpassed by the United States in 'material achievement and power', by means of 'on the spot merit'. The British paid attention to *minutiae* of procedure and protocol, undertaking to entertain every one of the other delegations to dinners or cocktail parties, and holding regular delegation meetings in order to identify their own internal difficulties and present a united front.[20] After the first full day, they reported to London:

> …the absence of general discussion seems to indicate that no delegation except the Australians and ourselves have as yet a well thought-out consistent and positive policy. The Americans certainly have ideas but they are not yet decided and in any case are concerned not to throw too much weight about…It is noteworthy that neither the Russians nor the Chinese spoke at all today.

J. E. Coulson commented: 'general apathy and indecision not altogether surprising in view of the shortness of time and the general obscurity about objects…Although we probably dreaded this conference more than anyone, we seem to be the only Delegation which has been able to put forward any positive ideas.'[21] The effort put into these is evident in cables and records of daily delegation meetings.[22] Australia's J. B. Brigden found the British delegation 'outstanding' and 'uniformly excellent'.[23]

The Australian delegation played a significant role at Hot Springs. It was led by H. C. Coombs, recently appointed Director of Post-War Reconstruction and advocate of what was described as 'the positive approach' to Article VII of the Mutual Aid Agreement. The positive approach shared with the nutrition initiative an emphasis on improved standards of living as a goal of public policy, but it was more broadly conceived, interpreting the Atlantic Charter and Article VII as a commitment to encourage expansion of trade and higher living standards, including responsibility to assist underdeveloped countries to raise theirs, and national and international commitment to policies of full employment. The approach reversed important strands of traditionally protectionist Labor policy

---

20 UKNA, FO371/35378, P. H. Gore-Booth, draft memorandum, 'Food Conference—United Kingdom Delegation and its Relation with Other Delegations', n.d.
21 UKNA, FO371/35373, British Delegation to FO, 5, 19 May; minute by Coulson, 21 May.
22 UKNA, FO371/3572-8, passim.
23 NAA, A989, 43/735/606, Brigden, 'Some Impressions: UN Conference on Food and Agriculture', 31 May.

in its willingness to question imperial preference and to oblige countries to spend their credit balances, while continuing to uphold Australia's right to extend its own manufacturing industries. The positive approach was to be formally espoused by the Labor Government in 1944; it offered 'an internationalism of [Labor's] aspiration, to substitute for the predatory internationalism that they feared'. It would see Australian delegations at international conferences 'in another kind of Keynesian crusade' attempting to 'use Keynesian analysis with its emphasis on sustaining demand as the basis of reform of the international economic order'. By the time he reached Hot Springs, Coombs had discovered an emphasis on employment policy in London. US officials, however, believed it would be unacceptable to Congress.[24]

McDougall was also a member of the Australian delegation. Coombs found him 'a very interesting bloke, a bit of a charlatan, but a very appealing one with an element of good sense in the line that he was taking…a kind of virtuous con-man'.[25] Coombs noted that McDougall was working 'quite closely in collaboration with the Americans'. 'They appear to be seeking his advice and criticism on the preparation of material and the working out of the plans for the Conference. This is proving quite valuable to us.'[26]

Dean Acheson led the US delegation at Hot Springs, but attended only spasmodically. Delegates remained in constant touch with the State Department by telephone tie line. Howard Tolley thought the United States made less effort to influence policy at this conference than at later ones.[27] They, too, held meetings to decide policy, but on occasion there was open division.[28]

The Hot Springs Resolutions

The conference divided into four sections: I) Consumption Levels and Requirements; II) Expansion of Production and Adaptation to Consumption needs; III) Facilitation and Improvement of Distribution; IV) Recommendations for Continuing and Carrying forward the Work of the Conference. All sections except IV comprised several committees, and all committees prepared resolutions. The Final Act of the Hot Springs Conference included 33 resolutions on the need for international action to increase and promote agricultural production, technical skills and consumption, and on individual governments' responsibility to ensure a sufficient food supply. The text of the resolutions fills nearly 17 long

---

24 Tim Rowse, *Nugget Coombs*, Cambridge University Press, Cambridge, 2002, pp. 119–25; H. C. Coombs, *Trial Balance*, Macmillan, Australia, 1981, pp. 32–6, 43; Coleman, Cornish and Hagger, *Giblin's Platoon*, pp. 222–9.
25 Rowse, *Coombs*, p. 123, citing Coombs in interview with Sean Turnell; Coombs, *Trial Balance*, p. 41.
26 NAA, M448, 11, [3], Coombs to J. B. Chifley, 23 April 1943.
27 HLH, 'Tolley', pp. 591–3.
28 UKNA, FO371/35378; Gore-Booth, draft memorandum, 'Food Conference'.

pages of fine print.[29] Section IV prepared the resolution to establish an Interim Commission to formulate 'a specific plan for a permanent organization in the field of food and agriculture'. The Commission's chief task would be to draft a constitution. The permanent organisation would come into being once that constitution was signed by 20 nations.[30]

As a result of British determination to oil the wheels, the goodwill of all participants and a general aspiration to make use of this unprecedented opportunity to build a better world, the Hot Springs Conference had succeeded against all expectations. Yet the British encountered one difficulty at its conclusion. They had hoped that Section IV would recommend the permanent organisation be limited to collection of information and research, but the Americans would not agree: 'The best we could do was to get agreement that it should not be mandatory for the Interim Commission to include…functions of an executive nature.'[31]

The conference in fact had issued a clarion call, its formal resolutions beginning with a declaration that 'freedom from want of food, suitable and adequate for the health and strength of all peoples, can be achieved'. First, it continued, the war must be won, and in the immediately following critical period, urgent efforts must be made to economise consumption, increase supplies and distribute them to best advantage. Then equal efforts must be made to win and maintain freedom from fear and from want: 'The one cannot be achieved without the other.' Until then there had 'never been enough food for the health of all people. This is justified neither by ignorance nor by the harshness of nature.' Production must be greatly expanded, requiring 'imagination and firm will on the part of each government and people'. 'The first cause of hunger and malnutrition is poverty.' There must therefore be 'an expansion of the whole world economy to provide the purchasing power sufficient to maintain an adequate diet for all'. Primary responsibility lay with each nation, and the committee recommended all governments adopt the conference findings and recommendations, urging 'early concerted discussion of the related problems falling outside the scope' of Hot Springs. The first steps, nevertheless, should be taken without waiting for solution of other problems.[32]

---

29 NAA, A989, 43/735/740, copy of Confidential UK Government Print, PCP (43) 4.
30 Sergio Marchisio and Antoinietta Di Blase *The Food and Agriculture Organization (FAO)*, Martinus Nijhoff, The Netherlands, 1991, p. 10; *American Journal of International Law*, Vol. 38, no. 4, October 1944, 'The Interim Commission on Food and Agriculture and the Food and Agriculture Organization', pp. 708–11.
31 UKNA, FO371/35374, Food Delegation to FO 25 and 31 May.
32 NAA, A989, 43/735/740, copy of UK Confidential Print, 'United Nations Conference on Food and Agriculture', Annex B, Resolutions.

The Interim Commission

McDougall had been rapporteur of Section IV, and the final document probably demonstrates his influence. He was dubbed 'Father of the Conference', but otherwise took no very prominent role at Hot Springs. The Interim Commission, on which he again represented Australia, was a body better suited to his methods; there he operated as he had on the EMB and IEC: not occupying its most prominent position, yet steering its work to a considerable extent by virtue of his diligence, knowledge and ideas. He did not lack opposition. The nascent organisation was already subject to political manipulation from the major powers. British civil servants, and Minister of Agriculture, R. S. Hudson, in particular, had a list of *persona non grata*. They had been able to prevent Orr from attending Hot Springs and made it clear he should have nothing to do with the Interim Commission. At Hot Springs they had also successfully objected to the appointment of the Canadian Secretary of the International Wheat Conference, Andrew Cairns, as secretary to Section IV, a move causing unease in Washington and Ottawa. After much discussion, British civil servants agreed it would be impolitic to make a second objection to Cairns, against whom they really had nothing except his 'enthusiasm and strongly held views'.[33] They were, however, determined that neither McDougall nor Paul Appleby should be appointed Chair of the Interim Commission.[34] By suggesting that neither a Briton nor an American should occupy the chair, they could avoid the problem of Appleby, the precise objection to whom they could not then adequately explain, but which related to US agricultural policy. Despite Appleby's advocacy of expansionary policies in London in 1941, he had been known to favour international planning of agriculture and restriction of production, along the lines of the 1930s US AAA.[35]

Britain would prefer that the Chairman of the Interim Commission represent one of the dominions. Australian participation at Hot Springs had been the most impressive, but suggesting an Australian occupy the chair might lead to McDougall's appointment. After Coombs, McDougall had been the member of the Australian delegation 'most in evidence' and 'the fact that he had played some part in the promotion of the Conference and has certain ambitions connected with the outcome…rather complicated matters at times'. Alarm increased as the British Embassy in Washington reported that McDougall had circulated a memorandum proposing the commission meet quickly to 'begin work on all aspects of the food problem', and suggesting formation of a series of expert committees including, possibly, one on commodity arrangements, which was

---

33 UKNA, FO371/35380, minutes by Coulson, 4 and 11 July; Ronald, 5 August; FO to Washington for Twentyman, 12 August; notes by Twentyman, 'IC4', 26 July.
34 Ibid., FO371/35377, minute by Coulson, 5 July; Coulson to Clutterbuck, 7 July; FO to Washington, 6 July.
35 Ibid., FO371/35373, Sixth meeting of UK Delegation, 12 May.

'obviously dangerous'.[36] The British were overly fearful: in the event the task of chairing the commission went to Canada's Lester Pearson, for whom McDougall, who had never wanted it, had lobbied energetically.

The international organisations formed at the end of World War II are now taken for granted as part of the world order. It is difficult to appreciate the magnitude and complexity of the task facing their founders. The Interim Commission had three main objectives: first to draft a 'formal declaration or agreement' recognising governments' obligations to raise standards of living and nutrition, to improve agricultural efficiency, to cooperate with other governments on these matters and to report on progress; second, to prepare a specific plan for the new organisation; and third, to initiate preliminary statistical investigations and research into problems likely to face the organisation, so that once established it could begin work immediately.[37] The commission, comprising representatives of the 44 nations at Hot Springs, and occupied Denmark, was formally inaugurated on 15 July 1943, only six weeks after the conference concluded, showing, in Acheson's words, a 'clear indication of the earnestness with which the United Nations have undertaken their commitments and the vigor with which they are prepared to carry them out'.[38]

The Interim Commission existed for more than two years, located from October 1943 in its permanent home: 2841 McGill Terrace, a mansion in a leafy area of central Washington. Most rooms became offices, while 'the commission and its committees and advisory panels met in the dining room around a huge oak table'.[39] The full commission met infrequently, but members were entitled to attend any of its committee meetings [40] This meant that an enthusiastic member like McDougall could be involved in nearly all of its activities. Most members were Washington-based diplomats, with heavy workloads, particularly as the pace of planning for postwar organisations increased. McDougall remained on Bruce's staff in London. He spent long periods in Washington, returning to London whenever Bruce needed him. When in Washington, he took an interest in other areas of postwar planning, and in some leftover business of the League of Nations, but he was able to devote most of his time to the commission.

McDougall told Bruce:

---

36 Ibid., FO371/35378, draft memorandum, 'Food Conference'; FO to Washington, 4209, 25 June; telegrams 2771 and 2772 from Washington, 16 June, and minutes thereon.
37 NAA, A2937, Post War Food, 'Suggestions on Methods of Work of the Interim Commission', n.d. This document appears to be part of McDougall's memorandum, discussed in the preceding paragraph and summarised in telegram 2772 from Washington to FO, 17 June.
38 NAC, RG 25, Vol. 2261, 5050-B-40C, part i, Address of Welcome by Acheson, 15 July.
39 Hambidge, *The Story of FAO*, p. 52.
40 NAA, A2937, Post War Food, McDougall to Bruce, 22 July 1943.

>...the Interim Commission does not go well when I am away. There is no driving force and not too much knowledge. The staff is hard working but not competent and of the Commission only Pearson, Twentyman and Appleby are really competent and none of them devote much time to the work of the Commission.

Its committees included 'a lot of good experts', but 'they only see part of the picture'. 'All the hard work [on the commission report] has been done by a drafting sub-committee aided by Cairns and a very good American, [Gove] Hambidge, both of whom were lent for this purpose.'[41] The records show that Pearson and McDougall, with the Agent-General for India in Washington, Sir Girja Bajpai, the US representative, Appleby for the first year and then Howard Tolley, and Edward Twentyman of the United Kingdom, constituted an informal inner circle, exchanging and assessing ideas. When they agreed, as they usually did, a suggestion was likely to go ahead. McDougall had known and respected Bajpai since their days together on the IEC; Tolley found him 'an especially good parliamentarian', impressive in his chairmanship of committees at Hot Springs.[42] Twentyman, a senior Treasury official, was based at the British Food Mission in Washington. He was described as a 'fearless' and 'dynamic' administrator, of somewhat 'unconventional' dress and manner; 'some of his best work was done fishing in troubled waters'.[43] His death in an air accident while returning to Washington in March 1945 is a reminder of the discomforts and dangers faced by all who travelled constantly in the cause of the new world order.

Committee C

Decisions of the Interim Commission were taken by three committees; their scope was bounded by the Hot Springs resolutions. Committee A drafted the formal agreement by all participating nations to raise levels of nutrition, improve agricultural efficiency, cooperate with other nations and submit periodical reports on progress. Committee A's work was largely complete by August 1943. Committee B devised the structure and constitution of FAO, dealing with some contentious issues, particularly concerning financial contributions. Committee C provided some assistance to Committee B, but its chief task was to prepare for the work of the new organisation, including initiation of preliminary statistical investigations and research. It did not commence serious work until late 1943 when Committee B's work was nearing completion. On the bones created by

---

41　Ibid., M104, 12(4), McDougall to Bruce, 28 June 1943, 16 January 1944.
42　HLH, 'Tolley', p. 588.
43　*The Times*, obituary, 17 April 1945.

B, C was to develop flesh. McDougall was well placed to influence the shape of the new organisation. Formally, he was a member of Committee C, but he also participated in the drafting subcommittee of Committee A.[44]

It seems to have been McDougall's suggestion that the commission form two advisory panels of experts, on science and on economics, to assist with their planning. Bajpai subsequently agreed that each of these should be able to appoint subcommittees.[45] This decision to form panels of technical advisers would set the pattern for the work of the commission throughout 1944. There was a sense of urgency in discussions by then; Interim Commission members expected the 20 acceptances necessary to form the new organisation well before the end of the year, after which it must begin work without delay. Late in 1943 McDougall had listed possible preparatory tasks to be undertaken by Committee C, from which the Commission Secretariat selected the three most urgent in January 1944. The first task was to represent the interests of food and agriculture at any UN conference: FAO must not be caught unprepared by a sudden summons to an international conference. The second was to enable the Director-General and the Council of FAO 'to envisage the tasks confronting them and to determine on questions of relative urgency'; this preparation should begin with a general survey of the food and agriculture position in each country. The third task was to facilitate the work of FAO by measures such as establishing links with research and academic institutions, particularly with a view to drawing on their personnel. Committee C agreed on 2 February 1944 that the most urgent of the three was preparation of conference papers. A joint committee of Committee C and the Commission Secretariat subsequently recommended preparation of memoranda for conferences on agricultural credit, commodity control, commercial policy, monetary stablisation, employment problems and transport. Committee C also recommended that commencement of country-by-country surveys begin immediately, as 'functioning *in vacuo* could only lead to fumbling and delay, at a moment when delay might be fatal to the prospects of the organization'.[46] Information, arranged both by countries and by subjects, should be prepared forthwith. The recommendation listed one and half pages of topics on which data might be collected.[47] It was soon realised that commission funds and time would limit the studies to samples in a few countries to assist FAO in later planning.[48]

---

44 A copy of Committee A's final draft, approved on 11 August 1943, is in NAA, A989, 43/735/740.
45 NAA, A2937, Post War Food, McDougall to Bruce, 2 August.
46 NAC, RG 25, Vol. 2261, File 5050-B-40C, part 2, C Doc 4, 2 February 1944.
47 Ibid., part 3, 'Draft Report of Committee C to the Interim Commission', n.d.
48 Ibid., part 4, 'Ex. Min 26', 28 April 1944; 'Subcom. on Agric Prod, Min. 4', 15 May.

The Reviewing Panel

McDougall spent much of February and March 1944 in London as part of Australia's delegation to meetings of British Commonwealth officials on international economic collaboration. Pearson cabled him that progress of deliberations was slower than he had wished, but the program of work adopted 'has been more or less as you outlined it'. McDougall's 'panel procedure [that is, the formation of subcommittees to the main panels] should…be applied at once'. By May four such panels—on forestry, fisheries, agricultural production and food management—were established. The secretariat 'would not be in any way responsible for the nature or quality of their work', but there would be a 'Research Reviewing Committee (or some such name) through which these panels would report to Committee C. This will be an extremely important group.' What came to be called the 'Reviewing Panel' was apparently Pearson's idea and he went on to propose that McDougall chair it, adding 'it really would be a key job on the Interim Commission from now on'.[49]

Pearson probably ascribed such importance to the Reviewing Panel because the experts would not be subjected to bureaucratic interference in making their recommendations for the new organisation. The Reviewing Panel, and its chair in particular, was to help them to understand what sort of guidance would be needed by an organisation unlike any that had existed previously.[50] Minutes of meetings between panels and the Reviewing Panel show how the system worked. The latter vetted and amended terms of reference. Should wood pulp come within FAO's ambit? Should there be a statistical study for the years 1945 to 1950? Was there a need for panels on statistics and marketing?[51] It was necessary to explain a broader picture to scientists steeped in the *minutiae* of their own fields. McDougall dealt tactfully with the forestry panel's unanimous desire for a separate deputy director-general for forestry.[52] He asked the fisheries panel to provide 'a general review of the place of fisheries in the world economy', ranging through nutritive values, contributions of fisheries to world food supply, advances in processing and culture, undeveloped fishing areas, problems of organised fishery industries and non-commercial fisheries, and international collaboration in sustaining resources. The report should show how FAO could become 'the central world organization to promote and foster fisheries and fishery industries'.[53]

---

49 Ibid., part 3, Pearson to McDougall, 4 March 1944.
50 See summary by Joint Committee Chairman, Haidari, in ibid., Part 5, Ex. Min. 27, 5 May 1944.
51 Ibid., Part 5, Rev. Pan. Min. 3, 12 May 1944.
52 Ibid., Part 7, Subcom. on Forestry, Min. 2, 26–28 June 1944.
53 Ibid., vol. 2263, part 9, Subcom. on Fisheries, Min. 2, 31 July 1944.

10. 'A New Era in the World's History'

The Reviewing Panel did more than provide guidance for expert committees. Its members mapped out the tasks and the philosophy of the organisation as a whole. In August 1944, they discussed preparation of their report to FAO's founding conference. Although it was assumed that the final report of the Interim Commission would suggest action, McDougall warned against making formal recommendations, thereby inviting a charge of attempting to do the work of the conference, since the conference was to be the supreme policymaking body of the organisation. He approved preparation of a memorandum on the 'economic setting' since, under Hot Springs Resolution XXIV, the general report to FAO 'must be related to the economic setting'. Tolley went further: since purchasing power, an expanding economy and international cooperation were all factors determining food purchases, perhaps FAO should make recommendations to other agencies concerning industrial development, international trade and commodity arrangements. McDougall agreed that industrialisation would be necessary in some countries to achieve the Hot Springs goals, as was international cooperation. 'Should not the FAO Conference, for example, make recommendations to governments concerning this need for industrialization?' There was general agreement that it should.[54]

The panel discussed FAO's role in relation to agricultural commodity arrangements. Wheeler asked: 'If FAO is not to organize and activate commodity agreements, how is it to get its views expressed and acted upon by another body? How is it to see that national plans are properly coordinated?' Tolley argued that Hot Springs resolutions obliged FAO to ensure international action either through another organisation or by organising international commodity agreements itself. McDougall predicted greater government intervention in agricultural production in the future. Without coordination of national programs, 'the trend may become extremely nationalistic'. He proposed a statement explaining the need for coordination and recommending a conference to develop principles governing all commodity agreements. The panel agreed, Tolley emphasising in ensuing discussion the importance of seeing that 'the right kind of agreements are drawn up', particularly in regard to their contribution to agricultural reorientation.[55]

Constitutional Ambiguity

Some Interim Commission members wanted 'a strong food and agriculture organization which could take positive steps to foster economic expansion and help prevent disastrous crises'; others wanted 'a rather narrowly limited

---

54 Ibid., Rev. Pan. Min. 7, 2 August 1944.
55 Ibid., 3 August 1944.

fact-gathering and advisory agency...carefully insulated from positive action'. McDougall took an intermediate view, preferring an advisory body, dependent for influence on 'its own efficiency and competence', on the seriousness with which member nations took the obligation to report, and on arrangements for international investment. He wanted the organisation to have a say in international discussions on issues within FAO's ambit, but believed there should be a separate international advisory body for commodity agreements. He told Twentyman at one point that FAO should not be 'besmirched' by having 'anything to do with the damned merchants'.[56]

In the view of one FAO officer, the constitution devised by the Interim Commission and approved by member states was a compromise between the strong active body and the advisory one. It was 'more on the advisory than the action side but with the way open, constitutionally, to develop in whatever direction the member nations might find most useful'.[57] The relevant paragraphs of Article I listed obligations to 'collect, analyze, interpret, and disseminate information' and to 'promote and...recommend national and international action' in relevant fields. But it was also to 'furnish such technical assistance as governments may request' and to 'organize...such missions as may be needed to assist' governments to fulfil obligations arising from acceptance of the Hot Springs recommendations, and 'to take all necessary and appropriate action to implement the purposes of the Organization'.[58]

This constitutional ambiguity increased British anxiety. Whitehall was alarmed by the committee structures established by Committee C, by its 'ginger group outlook' and by its extension of activity into new fields. Twentyman was instructed to 'remonstrate against the way Committee C is extending its functions to cover postwar economic problems'. Its tendency to tackle 'outlying subjects', albeit lying within its ambit, such as forestry, would stretch its limited resources and 'wreck FAO'. They hoped the chosen Director-General would 'develop the organisation along the lines we envisage, i.e. a fact-finding, educational and advisory body'.[59]

The Search for a Director-General

Members of the Interim Commission wanted the Director-General to be selected before the conference that would formally inaugurate FAO. McDougall's first

---

56 UKNA, FO371/35381, McDougall to Twentyman, 11 August 1943; FO371/35380, Twentyman, 'IC 1, Interim Commission on Food and Agriculture: Commodity Arrangements', 23 July 1943.
57 Hambidge, *The Story of FAO*, p. 53.
58 NAA, A989 44/735/743, text of Australian Bill to approve constitution of FAO, read 19 September, enacted November 1944.
59 UKNA, BT11/2375, FO to Washington, 7 and 14 May 1944.

choice as director-general, approved by Lester Pearson, was Henry Wallace. As it was not yet clear whether Roosevelt would select Wallace as his fourth-term running mate, other options were needed. McDougall thought Bruce would be 'splendid', but preferred him for a bigger job—perhaps chairing the 'Overall Economic and Social Council'. Orr might be possible, although 'Whitehall would be difficult'. Bruce saw that Wallace 'might be regarded by some as a starry-eyed dreamer', an opinion likely to be fostered by 'the reactionary elements in America'. He was dubious about Orr, and agreed that he himself 'might be able to do more useful work in other directions'. McDougall then suggested Bruce take the job 'to give the organisation a flying start', before moving to 'the more central office'. McDougall tossed in several other names without much enthusiasm, but rejected suggestions by some delegates that he be a candidate himself: 'I have warmly repudiated the idea stressing the need for a big man with both a name and abilities.' He would be interested in the deputy job if Wallace were chosen, but if it fell to Bruce, the deputy could not be Australian, and he would settle for a personal assistant position.[60]

The British Government opposed Wallace, carefully avoiding naming him in messages to the embassy in Washington warning against 'the appointment of an eminent US agrarian mystic' or 'a political crusader, especially one connected with US agrarian politics'.[61] They urged the State Department that a director-general should be chosen for administrative and scientific abilities rather than 'general public eminence'.[62] As with the Interim Commission, they preferred a dominion candidate, but remained fearful that Australia would nominate McDougall. The British Embassy in Washington reported that Pearson, not interested himself, suggested McDougall 'ought to get a senior post in the organisation' and could not suggest anyone else outside Britain and the United States.[63]

In fact McDougall's views, particularly on the contentious issue of commodity control, had proved to be moderate and acceptable to British representatives in Washington. Enclosing a letter from McDougall setting out his views on the commission, Twentyman acknowledged him as one of the key members of the commission, and reported his own view that 'whatever may be said about M's general proclivities [he was] fairly sensible and sound…my own present inclination is to temper my natural caution towards him with a goodly measure of appreciation of him as a genuine supporter of our own objectives'.[64] Twentyman told McDougall frankly of British concerns, and that he had warned Appleby

---

60 NAA, M104, 12(4), McDougall to Bruce, 1 and 28 June; Bruce to McDougall, 17 June 1944.
61 UKNA, BT11/92, minute H. J. Habakkuk, 20 May; BT11/2375, FO to Washington, 13 May 1944.
62 Ibid., minute, Habakkuk, 2 May 1944; USNA, 550.AD1, Interim Commission/5, memorandum of conversation, Stinebower and Gore-Booth, 19 May 1944.
63 UKNA, BT11/2375, Snelling, DO, to Coulson, FO, 2 June 1944; Holmes to DO, 17 May 1944.
64 Ibid., FO371/35381, Twentyman to W. J. Hasler, 12 August; McDougall to Twentyman, 11 August 1943.

against making McDougall chair of an important committee, as it 'might cause some U.K. folk to think that the Commission was being dominated by extremists'. McDougall replied that he did not wish to chair a committee.[65] The British did not realise that McDougall was probably as unacceptable to Canberra as he was to Whitehall. Just as senior British representatives in Washington found him more reasonable than did their masters in London, Australian representatives like Coombs formed more favourable views than those at home. J. F. Murphy, then Controller-General of Food, believed that 'for various reasons Mr McDougall would have little influence with his Minister or with the Australian Food Council'.[66]

The New Organisation

Disappointing many expectations, FAO was not established in 1944. A chief reason was the US political process in a presidential election year with an increasingly obstructive Congress. The United States did not formally accept FAO's Constitution until after the election in November. In the meantime, the Bretton Woods Conference in July 1944 had set in train the formation of the International Bank for Reconstruction and Development (IBRD) and the International Monetary Fund (IMF). The war in Europe took a little longer than expected to end, but then the Pacific war ended abruptly only three months later. That ending brought with it devastation previously unimagined. At the Quebec conference inaugurating FAO in October 1945, the US delegate spoke of two scientific facts that made it 'impossible for us to go on taking hunger for granted'. First, the atomic bomb had shown 'that we must not have another war. And that means we must not permit the pangs of hunger to bring about those basic fears and greed which result in war.' Second, science and technology now made hunger unnecessary. Another shadow over the conference warned that wartime cooperation was waning. Members of the USSR delegation participated fully in the conference, as they had done at Hot Springs, and in the Interim Commission. But days passed without their signature to the Constitution; they were finally forced to report that they had been instructed not to sign. The USSR did not become a member.[67] The postwar world was already divided.

Efforts before the conference to find a director-general had failed, though some 17 names had been canvassed.[68] Orr had been persuaded—much against his

---

65 NAA, A2937, Post War Food, McDougall to Bruce, 17 July 1943.
66 NAA/CSIR, series 3, PH/McD/54, Richardson to Rivett, 18 February 1944.
67 HLH, 'Tolley', pp. 619–24.
68 Amy L. S. Staples, *The Birth of Development: How the World Bank, Food and Agriculture Organization, and World Health Organization Changed the World, 1945–1965*, Kent State University Press, Kent, Ohio, 2006, p. 80.

wishes—to go to Quebec as an 'unofficial' member of the British delegation. On the journey across the Atlantic, he told other members of 'the futility of the kind of organisation it was proposed to set up'; he spent much of his time in Quebec as a tourist and renewing scientific friendships. He was persuaded by some delegates to address the conference, and in a stirring speech urged them to reject the recommended constitution. There was no need for more science to show that the world had insufficient food for health. They should seek to give the organisation 'funds and authority to enable [it] to promote the production and better distribution of some of the main foodstuffs as the beginning of a world policy'. Although the proposed constitution was accepted, Orr, reluctant still, agreed to take on the director-generalship:

> I decided to have a shot at putting the organisation on the right lines, and if I failed, as seemed to me likely, to retire with honour satisfied. So I accepted, but only for two years, which seemed long enough to get the organisation established with executive powers if governments could be persuaded to agree, and too long if that should prove impossible.

He began work 'in a gloomy and grim mood'.[69]

Figure 18 FAO Director-General, John Boyd Orr (centre), with Special Advisers F. L. McDougall (left) and S. L. Louwes (right).

Source: E. McDougall.

---

69 Orr, *As I Recall*, pp. 161–5.

Orr immediately appointed two 'special advisers': S. L. Louwes, Netherlands Director-General of Food, and McDougall. Orr's first intention had been to appoint McDougall as Deputy Director-General, but the executive felt that another 'Anglo-Saxon' should not be appointed without two more deputies of other nationalities. It was not, McDougall later wrote, 'very satisfactory from a personal point of view'. It was at first a temporary secondment, but a year later became a five-year appointment as 'Counselor'—effectively 'the second person in the organisation' who would act for the Director-General in his absence. As an international civil servant, McDougall then resigned from all other positions held on behalf of Australia, and even from a personal appointment to the Court of Governors of the London School of Economics.[70] FAO became his life from then on.

FAO had a budget for its first year of US$2.5 million, and $5 million per annum thereafter. Orr adopted what McDougall described as a 'rod and line' approach to staffing, rather than the 'dragnet' employed by the new United Nations Organisation (UN), which by late 1946 had almost 3000 staff, treble that of the League of Nations. FAO, in contrast, had 160 by then, of an expected total of six hundred. For some time they occupied the premises of the Interim Commission, gradually spreading to several other buildings in Washington. FAO's permanent site was expected to be close to UN Headquarters somewhere in the United States.

World Food Crisis

In February 1946 the nucleus organisation faced a crisis. Crops had failed in Europe and in Asia; US production was not increasing to meet the need; machinery, fertilisers and transport were inadequate and world population was rising rapidly. The UN General Assembly, meeting in London, called for immediate action to meet a worldwide shortage of food. Orr responded by offering to take responsibility 'for mobilizing world resources', beginning with a conference on 'Urgent Food Problems' in May.[71] McDougall set out six activities needed for effective action. Three fell within FAO's constitutional ambit: continuous review and assessment of trends; calculation of the amount of basic foods needed by all countries; and agricultural rehabilitation of war-devastated and underdeveloped countries. Constitutional amendments would be needed to allow the remaining three: allocation of supplies on an equitable basis; centralised purchasing to prevent inflationary purchasing and as a reserve; and funding or credit arrangements to allow relief on a business,

---

70   NAA, A1067, ER46/4/8, McDougall to Dunk, 8 July; McDougall to V. C. Duffy, 19 November 1946.
71   Staples, *Birth of Development*, p. 84; NAA, A1067, 46/4/3, UN delegation to EA, 14 February 1946.

rather than a charitable, basis. McDougall thought these should be given to 'an executive body', established under the general supervision of the Economic and Social Council of the United Nations (ECOSOC), which should itself include representatives of relevant agencies.[72]

There were other multinational bodies with responsibility for food supply. The Combined Food Board (CFB) run by the United States, Britain and Canada allocated supplies during the war through a series of commodity committees, but was due to be dissolved in June 1946. The UN Relief and Rehabilitation Agency (UNRRA) and the Food Committee of the Emergency Economic Committee for Europe dealt with the aftermath of war and were expected to be short-lived. In his invitation to the May conference, Orr predicted a need for 'international action on an agreed plan' for four or five years.[73] He feared supply would be slow to return to normal, and that the bogey of unmarketable surplus would hinder efforts to increase production. Means must be found to enable the functions of the temporary organisations to continue while needed. The Conference on Urgent Food Problems agreed that the CFB should remain until December, and an International Emergency Food Council (IEFC)—established, staffed and financed by FAO—should then have power to recommend production increases and direct supplies to areas of most need.

This seemed to solve the problem posed by FAO's ambiguous constitution. But economist J. B. Brigden, then Financial Counsellor at the Australian Legation in Washington, had misgivings. The IEFC was more cumbersome than the CFB and was unlikely to prove any better; a failure would reflect on FAO: 'those nationals who are concerned for the welfare of FAO have been trying…to protect it against such risks, but FAO has chosen to take them.'[74] Brigden also criticised the conference itself. Preparation had been 'negligible' and the conference had to be saved by other organisations. 'I do not expect the FAO Secretariat to cope with all the ambitious objectives which Sir John Orr and our Mr McDougall feel to be within their scope.' Difficulties and the slow pace of recruitment were understandable, but 'FAO is too apt to undertake work which it cannot do'. There had been sufficient goodwill to excuse its failure to foresee the difficulties, but its principals had 'no very clear ideas about the limits of their functions'. He added that FAO was 'only one of a large number of international bodies now trying to get their bearings', and he hoped that the UN Assembly, through ECOSOC, 'would be able to sort out the appropriate functions for each…[but]

---

72 FAO, 3.1, C1, note for Director-General, 'The World Food Policy', 13 May 1946.
73 NAA, A1067, ER46/4/14, Text of invitation, 3 May 1946.
74 W. J. Hudson and Wendy Way, eds, *Documents on Australian Foreign Policy 1937–1949. Volume IX: January–June 1946*, Australian Government Publishing Service, Canberra, 1991, 277, 'The International Emergency Council', Report No. 1, 28 May 1946, pp. 468–70.

the task of the co-ordinating body is extremely onerous'.[75] The IEFC remained in being for some three years, allocating exportable surpluses according to need; Orr for one believed that it had saved millions of lives.[76]

The World Food Board

Encouraged by its establishment of the IEFC, FAO sought to gain the executive powers it lacked for a long-term attack on hunger. Its own figures suggested a rate of population increase in the following 20 years requiring a doubling of world food production. This information was circulated to member governments along with an invitation to a conference in Copenhagen in September 1946 to consider a plan for a World Food Board (WFB). The board would provide long-term credit to food-deficient countries to enable purchases of food and industrial products to modernise their own agriculture. It would have authority to buy and hold stocks of surplus food from exporters, for distribution in time of need and also to help stabilise prices, thereby encouraging greater production in advanced countries. All this would increase world trade and, as Orr later put it, 'be an important step to the evolution of the United Nations Organisation as a World Government without which there is little hope of permanent world peace'.[77]

The plan alarmed the United States and Britain. The United States disapproved in principle of proposals 'inimical' to US international trade policy. Will Clayton, US Under Secretary of State for Economic Affairs, tried to persuade Orr to withdraw it, arguing that the regime of the International Trade Organization (ITO), whose charter and powers were then under discussion at a series of multilateral meetings, would foster an increase in trade, in line with the Department of State's longstanding policy.[78] Orr refused: the ITO would regard trade as an end in itself; food was vital and demanded special treatment.[79] While the Department of Agriculture saw the plan more favourably—it was perhaps an 'ever-normal granary' on an international scale—the department no longer influenced policy and the US delegate, Norris E. Dodd, was instructed to acknowledge the problem and to propose that a committee report on an alternative.[80]

The British Government had 'serious doubts'. The plan was costly and likely to increase import prices, worsening Britain's budget difficulties. As in Washington,

---

75 NAA, A1067, ER46/4/14, 'FAO and IEFC', J. B. Brigden, 28 May 1946.
76 Orr, *As I Recall*, p. 170.
77 Ibid., pp. 170–2. See also Staples, *Birth of Development*, p. 86.
78 NAA, A1067, ER46/4/16, US Legation Canberra to External Affairs, 12 August 1946; Staples, *Birth of Development*, p. 88.
79 Orr, *As I Recall*, p. 173.
80 NAA, A1067, ER46/4/16, Washington Legation to EA, 9 August; US Legation to EA, 12 August 1946.

the Ministry of Agriculture found some merit in proposals to stabilise prices, but failed to persuade Cabinet.[81] There was a danger that the programs of FAO would overlap functions of the ITO; the British delegate would be instructed that solutions to the problem of world food supply went beyond the scope of FAO and should be dealt with by ECOSOC. The problem was serious: Britain had been mobilising imperial resources, holding a conference in Singapore earlier in the year and urging Commonwealth producers like Australia to increase production and withhold domestic consumption. But Britain would support the US proposal to seek alternatives to Orr's plan at Copenhagen.[82]

Australian economic advisers differed in their views on the WFB proposals. Coombs generally favoured its formation. He hoped it would speed up international negotiations on commodity arrangements, which were important to Australia, but acknowledged that the plan presented 'the most contentious and complex problems that FAO has undertaken to solve'. He identified an unstated assumption underlying the plan, that 'western civilization must demonstrate to the world…its capacity for organising in the interests of world needs. Failure is to invite political and economic policies which may be distasteful to some of the Powers.' J. G. Crawford was a member of FAO's independent Standing Committee on Economics and Marketing, which had endorsed the WFB aims of raising productivity in agriculture, increasing food consumption of malnourished peoples and eliminating large price fluctuations in agricultural products. Crawford acknowledged the relevance of the plan to Australian policy, but worried that possible contributions of foodstuffs from 'exporting countries outside commercial markets must be small in relation to the total problem', with a danger that disillusion would follow. He also warned that a suggestion of special prices to meet crisis situations 'raised important issues likely to prove thorny in Washington'. Both Crawford and Coombs approved of the idea that raising health and nutrition standards involved economic policies 'designed to promote higher levels of income and employment, which effectively means industrialization', neatly fitting the Australian positive approach.[83] But Australian views counted for little when delegates met in Copenhagen in September. All 'appeared' to support the WFB, but insisted on considering alternatives, preferably in collaboration with other organisations such as ECOSOC and the preparatory commission for the ITO.[84]

---

81 Staples, *Birth of Development*, p. 89.
82 NAA, A1067, ER46/4/16, Cable D786, 21 August 1946.
83 Ibid., M448, 72[2], Coombs to McDougall, 21 October; 'Proposals for a World Food Board', notes of an address to NSW Branch of Institute of International Affairs, 5 November 1946; J. G. Crawford, 'World Food Board: Preparatory Commission', 4 October; 'Notes on the First Report to the Director-General' by the Standing Advisory Committee on Economics and Marketing, 28 October 1946.
84 Ibid., A1067, ER46/4/16, Bulcock to Dept of Commerce, 5287, 9 September 1946.

The WFB Preparatory Commission

A committee duly recommended formation of a WFB Preparatory Commission to consider Orr's proposal 'and any other relevant proposals'.[85] Australia's delegate to the commission, F. W. Bulcock, warned that in the final analysis the work would be 'largely political'.[86] The task of guiding that work fell to Bruce. As High Commissioner, he had been unable to participate firsthand in the events leading to the formation of FAO. When his London appointment ended in October 1945, he returned to Australia for the southern summer. He may have considered resuming an active part in Australian conservative politics; certainly he was not prepared to predict when he would return to London, and his refusal of a peerage offered in 1945 may have been to allow for that possibility. No such opportunity was presented to him, however, and he was back in London by mid 1946. In 1945 Bruce had probably expected to be offered a significant international appointment, perhaps heading ECOSOC, but now such hopes were fading. Perhaps he was considered too old. He had been in the forefront of Australian, imperial and international politics for nearly three decades; it could well have seemed that he was older than his sixty-three years. Considering possible candidates for UN Secretary-General, Adlai Stevenson and Gladwyn Jebb suggested 'an Australian as good as Bruce but twenty years younger'. Bruce was therefore available to continue his work for social justice as Chair of the Preparatory Commission. Following Bruce's appointment, Attlee again offered him a peerage, arguing that it would strengthen his hand in any international post. This time, Bruce accepted. As Lord Bruce of Melbourne, he was to make good use of the House of Lords as a forum for advocating his international goals.[87]

The announcement of Bruce's appointment to the commission referred to his work in 1935 and on other League of Nations inquiries. Now, in a context of new consensus on the need for expansion of consumption, the three architects of the nutrition initiative were working together in an organisation which they hoped would bring that vision into being. Orr's proposals, argued Bruce, did not exceed the primary tasks of FAO, nor did they mean a new superstructure. They simply showed the 'necessity of greater co-operation between existing bodies' and possible creation of 'some parallel bodies in the future'. Optimism was nevertheless tempered by Bruce's political sense. He feared that it might be impossible to realise proposals for stabilisation and made rough notes of answers to the argument that a drive for greater production would lead to surplus,

---

85   Ibid., FAO Conference, Report of Commission C, 12 September 1946.
86   W. J. Hudson and Wendy Way, eds, *Documents on Australian Foreign Policy 1937–1949. Volume X: July–December 1946*, Australian Government Publishing Service, Canberra, 1993, *116*, Bulcock to Commerce, 13 September 1946.
87   Cumpston, *Lord Bruce*, pp. 246, 249.

thereby renewing the swings from falling prices to reduced production, then to rising prices and surplus again. He concluded it would be safer not to give stabilisation too much prominence: prevention of starvation in times of famine should be at the forefront of arguments in favour of Orr's plan.[88]

Figure 19 Lord Bruce and F. L. McDougall at FAO.

Source: E. McDougall.

After five weeks' work, Bruce stated that his commission was examining three proposals to meet the food crisis: British and American suggestions for programs centred on individual commodities, and Orr's WFB. The commission had been 'struck most forcibly' by the need for 'a co-ordinating authority with respect to all commodity agreements'.[89] He worked hard himself, chairing committees on Development and Food Programs and on Price Stabilization and Commodity Policy, as well as chairing some working groups, because 'he found himself so intensely interested in the subject and desired to get a grasp of all its aspects'.[90] In his statement to the commission's final plenary session, Bruce drew a parallel between the world in 1920 and the world in 1947. He predicted that lack of

---

88 NAA, M104, 14(1), press statement, 8 October; 'Opening Remarks of Rt Hon. S. M. Bruce', 28 October 1946; handwritten notes, 'World Food Board', n.d.
89 Ibid., Address to 60th Annual Meeting of Land-Grant Colleges and Universities, 17 December 1946. Text of full WFB proposals is in NAA, M104, 14(2).
90 FAO, 3.1, A3, McDougall to Philip Noel-Baker, 18 December 1946.

sufficient purchasing power to absorb the output of increased production capacity resulting from war would mean mass unemployment, economic uncertainty and social unrest. Achievement of that purchasing power required international goodwill and cooperation. His commission recommended an annual review to ensure that national agricultural programs would fit into a global food production plan, and a review of national programs; these measures could create in FAO 'a world parliament of food'. The commission also called for development of industry and of the latent resources of less-developed countries, and it urged cooperation with other specialised agencies. Bruce spoke in similar vein a few months later in his maiden speech in the House of Lords: 'Almost every thinking person is now in favour of expansion of production and consumption.'[91] It was a far cry from the nuts and bolts of Orr's WFB.

Bruce had made the best of an impossible situation. He was not, argues David Lee, 'compromising his internationalist principles but rather using his skills as a facilitator to salvage as much as he could of the original plan'.[92] The WFB proposals could not withstand the concerted opposition of the two most influential members of FAO. Orr stuck to his ideals and left FAO a disappointed man in 1948. In the following year, he was created Baron Boyd Orr, made a commander of the *Légion d'Honneur* and awarded the Nobel Peace Prize. Like Bruce, he used the House of Lords to continue his campaign to feed the hungry.

Orr's 'world view, grounded in science and internationalism', had lost to US and British national security policies 'based on alliances, atomic weapons, unilateral international action, large peacetime militaries and a system of managed trade'. But it has been argued that the 'counterhegemonic force', which FAO with its broad membership represented, could not be ignored entirely, and that it was able to 'focus global attention on some of the shortcomings of the emerging world system'.[93] The Cold War could not quite extinguish the flame lit in Geneva in 1935.

FAO after Orr

FAO did change. Its supreme policymaking body was, and remains, its conference, reduced from an annual to a biennial event after the organisation moved to Rome in 1951. In its first years, the organisation was directed between conferences by an independent executive, but in 1948 the conference bowed to pressure and accepted that the World Food Council (WFC), the toothless organisation

---

91 NAA, M104, 14(1), statement by Bruce at Final Plenary Session, 24 January; summary of speech on WFB Proposals, House of Lords, 13 May 1947.
92 Lee, *Stanley Melbourne Bruce*, p. 177.
93 Staples, *Birth of Development*, p. 85.

created instead of the WFB, should also replace the executive. Membership of the council was elected on a national basis, each member country having a power of veto, though its chairman remained independent. Bruce served as its first chairman, from 1947 to 1951 He admitted the WFC lacked power, but argued that it possessed 'powerful moral sanctions': FAO's independent experts would 'confront the world as well as the governments with facts. Governments can't evade these things any more than they can evade the measures to meet these facts.'[94] But the independent experts were now subjected to a further layer of national control.

Figure 20 F. L. McDougall and John Boyd Orr.

Source: E. McDougall.

As FAO moved to take over the functions of the relief organisation IEFC, reports were dire. The Indian harvest had deteriorated and the rice supply was precarious. Europe was 9 million tons short of the 'semi-starvation rations' of grain distributed in the previous winter; the maize crop had failed and drought had deprived European farmers of fodder. The WFC considered a need for legislative powers. 'We can't invent food which doesn't exist but what we can do is to see that the supplies which are available are fairly distributed.' Bruce's long-term aims—never fully realised—remained the maximising of food production by ensuring fair prices all round, and storage of famine reserves.[95]

FAO itself was attracting criticism. Crawford wrote that 'FAO badly needs some leadership…Orr has launched FAO and his task is done. McDougall is a good ideas man but no administrator.'[96] The organisation lacked a good administrator

---

94 Ibid., p. 94; NAA, M104, 14(1), Bruce in interview with Richie Calder, 19 September 1947.
95 Ibid.
96 NAA, M444 [307], Crawford to Coombs, 2 August 1947.

at the top, and coordination between its divisions was inadequate. 'If the new DG is not a capable administrator, he must secure a deputy who is.'[97] Perhaps Orr and McDougall recognised their own failings in this regard. After the 1947 conference, Orr decided to appoint 'one first-class Deputy Director-General', with McDougall designated 'Counselor' at the same level. In this scheme, the Director-General would deal with member governments and large policy questions and attempt to bring together all the specialised agencies to collaborate effectively under ECOSOC. His deputy would be responsible for machinery, personnel and administration, while McDougall would provide advice on policy and relations with international organisations.[98]

Orr undertook to remain at FAO until a successor could be found. The search again proved difficult. There were many jobs for able candidates in the growing international community and such candidates could not easily be spared by nations recovering from war. Many suggestions were made and many feelers put out. McDougall thought Coombs would be excellent, but Coombs recognised his own lack of appropriate expertise and preferred to remain in Australia.[99] Lester Pearson chose Ottawa, and recommended Bruce.[100] Bruce sounded out Casey, who also declined. McDougall did not favour his fellow Special Adviser, Louwes, and was himself thought to be unacceptable to the British and Americans, who might have been prepared to accept Bruce; both would prefer someone younger.[101] Apparently there were no suitable young candidates. At US instigation, a special session of the Conference in April 1948 appointed Norris E. Dodd, then US Under Secretary of Agriculture, with a long background in the Agricultural Adjustment Administration.[102] Dodd, aged sixty-eight, was almost four years older than Bruce, and was to occupy the position until early 1954. W. Noble Clark was seconded from the University of Wisconsin to the position of Deputy Director-General for the first half of 1948, bridging the gap between Orr's departure in April and the arrival of the new Director-General in June. Dodd's deputy was Sir Herbert Broadley, a widely experienced British civil servant who had been Secretary to the IEC for its first two years. Broadley remained at FAO until mid 1958.

Dodd instituted a change of direction, welcomed by the great powers, but disappointing at least some of the staff. One of those disappointed was Howard Tolley, a former head of the US Bureau of Agricultural Economics, who was FAO's Director of Economics and Marketing and Vice-Chairman of FAO's Executive Committee.[103] Tolley respected Dodd as an able administrator, but later recalled

---

97 Ibid., A1068, ER47/4/6, Crawford to Bulcock, 14 August 1947.
98 NAC, MG 26 N1, Vol. 62, McDougall to Pearson, 30 June 1947.
99 NAA, M448 [307], McDougall to Coombs, 15 July; Coombs to Crawford, 26 August 1947.
100 NAC, MG 26 N1, Vol. 62, Pearson to McDougall, 30 June 1947.
101 NAA, A1068, ER47/4/6, Washington Embassy to EA, 10 July 1947.
102 HLH, 'Tolley', pp. 656–60.
103 FAO, 3.1, A3, McDougall to Bruce, 1 October 1946.

that old enmities within the Department of Agriculture resurfaced at FAO. Some staff, including Tolley, were not retained, others were sidelined while yet others abandoned idealism to become 'petty bureaucrats'. Even 'McDougall with all of his zeal, is hanging on today...though he has been shunted clear aside and isn't in position to make any significant contribution whatever'.[104]

There is no evidence that McDougall felt this way. His position, as determined by Orr, remained unchanged until 1955 when he was forced by ill health to retire, and became a part-time 'consultant'. Until then he continued a busy schedule of travel between the United States and Europe, attending meetings of ECOSOC and other bodies under the UN umbrella. By the time FAO moved to Rome in 1951, the postwar food emergency had passed and the Marshall Plan had been implemented in Europe. The focus of FAO's work under Dodd moved to technical assistance to farmers in the undeveloped world, statistical intelligence and 'limited proposals' for disposal of surpluses. The organisation, constrained by a tight budget of $5 million per annum, concentrated on 'small practical steps toward increased agricultural production and freedom from hunger'. Dodd was creating 'an international department of agriculture, overseeing a global agricultural revolution comparable to that in the United States overseen by the US Department of Agriculture'.[105] This was not Orr's grand plan, but it accorded with much of McDougall's thinking. His views on the importance of technical assistance and statistical services, formed long ago in Denmark, persisted. During 1947, before Orr's departure, McDougall had argued that problems of development and of improving the economic and social status of rural populations were 'peculiarly ours', even more than those of food supply and insurance against surplus; they were also important in providing markets for the developed world, as he explained in 'The Challenge to Western Civilization'.[106]

'The Challenge to Western Civilization'

McDougall's last significant memorandum, with the above title, written in March 1947, makes an argument similar to those he had made in the late 1930s, but with a Cold War perspective. 'The faith of the world in Western democracy has been shaken' by the economic failures and restrictions of the 1930s. The West has emerged victorious from the war, but with diminished prestige. A 'rival system [is] challenging our whole concept of a civilization based upon the liberty of the individual'. The 'basic dilemma' is that since 1918 the Western economy works at full capacity only in war: 'the economic problem

---

104 HLH, 'Tolley', pp. 694–5.
105 Staples, *Birth of Development*, pp. 96–7.
106 FAO, 3.1, C1, McDougall to Director-General, 9 June 1947.

of the peace is to find a constructive alternative to the total-war economy of organized waste.' Populations suffering poverty, malnutrition and reduced life expectancy must be convinced that the system of Western democracy is able and willing to help them improve their standards of living. Urgent development must extend beyond improvements in agriculture to industry, transport and education, and must be undertaken by international action. 'International sponsorship' of development is essential. Countries with growing national consciousness are reluctant to depend upon 'any one great nation, since this causes fears of economic domination'. With modern communications, 'poverty-stricken countries know more than ever before about how the well-to-do live… Misery has never been the basis of a stable civilization, but, because it has become so much more conscious, its danger has been increased a hundredfold.' Added to the political argument is an economic one, an international version of McDougall's complementary sheltered markets theory: underdeveloped countries need goods of every description; developed economies need markets.

McDougall was disappointed that the WFB Preparatory Commission had given insufficient emphasis to 'the enormous importance of the development aspects'. He wanted FAO to bridge the hiatus between a developing country's receipt of technical advice and the securing of a major loan from the IBRD by providing small funds to enable preliminary surveys and other preparatory work, but he found that major countries could not accept that such funds were needed. He also wanted coordinated attacks on problems of food, agriculture, transport, health and education. In particular, he stressed the importance of developing industry, disputing a view that it should be the responsibility of the ITO, since that organisation was likely to be largely concerned with trade and 'traders are the very worst people in the world for the development of industrialisation that competes…with existing interests'.[107]

Throughout the memorandum there is a tone of urgency. Western civilisation itself 'depends on how far in the next five to the outside ten years we succeed in convincing the underdeveloped countries that they can obtain the help and assistance they require from countries with our type of institutions'.[108] That sense of urgency was fuelled by the growing political divide. On 12 March 1947, US President Harry S. Truman enunciated his 'Truman Doctrine', signalling the Cold War strategy of containment of Communism. The opportunity for full international cooperation had passed. With some prescience, McDougall wrote that 'in the political battle the development of backward countries will play an increasingly important part'. He went on to suggest development of 'new centres of economic power', which could counteract the division of the world

---

107 NAA, A1068, ER47/4/18, 'The Challenge to Western Civilization', 24 March 1947.
108 FAO, 3.1, A.3, McDougall to Coombs, 28 February 1947.

into opposing blocs.[109] He urged a joint effort from FAO, the IBRD and perhaps the United States in development projects. FAO could provide the technical backing that the bank would need. A development drive might be based on 'individually small but practical projects spread over many underdeveloped countries'. Such a scheme could provide 'a focus for replanning FAO's work as a whole'.[110] Here McDougall seems to have been charting the course to be taken by FAO under Dodd's leadership.

Cooperation with International Agencies

'The one thing which these [international] organisations cannot do is to "ensure".'[111] This statement was particularly true of FAO, limited by its small budget and constitution, and by the influence of the dominant powers. For all the new bodies, demarcation of responsibilities and cooperation with other organisations involved tedious and complicated negotiations over legal minutiae, as well as political savvy and influence. It was probably McDougall's most time-consuming task in the post–Orr era, and it was one for which his aptitude and tolerance had been demonstrated in previous decades. Each agency negotiated, clause by clause, a formal agreement with the United Nations itself, but practical problems of cooperation remained. A belief persisted that specialised agencies were subject to problems of overlapping, a view disputed by McDougall, who declared that all were 'so conscious of the limitations of finance and personnel as to be strictly on guard against duplication'. Outsiders might confuse common fields of interest with common work programs. In rural welfare, for example, there were too few people actually at work and there was 'closest consultation in regard to program preparation'.[112] FAO was happy to see most of the work in many fields undertaken by others, and should concentrate on welfare matters within its own province: nutrition, soil conservation and agricultural extension. But there must be consultation: 'I cannot too strongly emphasise that…some preliminary discussion should occur with the Specialised Agencies concerned before…even an informal commitment', he wrote after the United Nations had organised missions by the specialised agencies to Latin America in which FAO could not afford to participate without extra funding.[113]

A contentious question for FAO in the early years was that of population. McDougall used the old phrase 'intolerable pessimism' for a view that sufficient food could not be produced unless population growth was limited. He objected

---

109   CSIR, 379, 22/16, McDougall to Clunies Ross, 31 March 1947.
110   FAO, 3.1, C1, 'Note for Director General', 14 November 1947.
111   Ibid., 3.1, B3, Leroy D. Stinebower, State Department, to McDougall, 21 June 1948.
112   Ibid., 3.1, A2, McDougall to Alger Hiss, Carnegie Endowment for International Peace, 7 June 1948.
113   Ibid., 3.1, A2, McDougall to Mrs Alva Myrdal, UN Department of Social Affairs, 15 November 1949.

to a proposed UNESCO publication including views of this nature: FAO should have been consulted first; an underlying assumption about the amount of land required was wrong; and 'concentration of attention upon population control would be dangerous and highly unsuitable for an international organization'. And he argued vigorously to ensure that FAO was involved in IBRD consideration of questions relating to food and development.[114] He could not entirely withstand a more general sense of pessimism: 'the deterioration of the political situation [Communists had just seized power in Czechoslovakia, and tensions were mounting over Berlin] is proceeding with quite horrible rapidity and it is difficult to get forebodings out of one's mind. Can mankind find a sufficiently intelligent way of avoiding his own destruction?'[115] But about FAO his optimism prevailed. He noted, after FAO's move to Rome, that while there was still less food per capita than in prewar years, there was greater knowledge about ways to increase productivity. Small things, such as a change from sickles to scythes in Afghanistan, and vaccination of cattle against rinderpest, could achieve much. The most urgent task of all international social and economic problems, perhaps, was to teach farm families, who represented half the world's population. That action could be taken immediately and could achieve 'spectacular results'.[116]

Figure 21 F. L. McDougall, Counselor, FAO.

Source: E. McDougall.

---

114   Ibid., 3.1, A6, McDougall to Dodd, 10 and 18 November 1948.
115   Ibid., 3.1, A1, McDougall to Alan Ritchie, 12 March 1948.
116   Ibid., 3.1, A4, draft article for *Pax et Libertas*, journal of Women's International League for Peace and Freedom, 22 November 1951.

Afterword

At an ECOSOC meeting in Geneva in 1955, McDougall suffered a stroke. After some months of rehabilitation, his secretary wrote of a 'miraculous recovery', though his movements were slower, he tired easily and he 'has lost some of his spark'. He retired officially at the end of that year, aged seventy-one.[1] He remained in Rome, living in a set of rooms in the apartment of a younger colleague, Karl Olsen. He attended his office in FAO regularly until his death from complications of appendicitis early in 1958, at the age of seventy-four. An FAO choir sang at his funeral and FAO later initiated the biennial McDougall Lecture in his honour.[2] Bruce, his senior by one year, remained busy on company boards and as first Chancellor of The Australian National University from 1951 until 1961. He died in London, aged eighty-four, in 1967. Orr retired to his farm in Scotland but travelled constantly in the cause of better nutrition and was lauded by leaders of many countries for his work. The oldest of the three collaborators, he also lived longest: he was ninety when he died in 1971.

Unlike his collaborators, McDougall received little public recognition. The New Year Honours list of 1926 had recorded his elevation as a Companion of the Order of St Michael and St George (CMG), 'in recognition of his services to the Commonwealth', doubtless at the request of Bruce. Some Australians, in his final years, wanted to recognise his later achievements. Creation of an 'Australian' room at FAO had been suggested a few years before McDougall's death. Rivett's successor at CSIRO, Ian Clunies Ross, who had known McDougall well in London, supported the idea with enthusiasm and also recommended a knighthood for McDougall. Others, including Casey, thought Bruce should be included in the tribute. Casey opposed the knighthood as McDougall had not been serving Australia for many years. Clunies Ross countered that he had been at Australia House until 1946 and was 'still regarded as Australian at FAO'.[3] The knighthood proposal seems not to have been taken further, but the room was established, furnished by Australia, and a portrait of McDougall painted and installed. Perhaps the political instinct was to avoid giving McDougall the publicity that a knighthood would bring, but to allow some recognition safely out of the way in Rome.[4]

---

1   CSIR, 379/22/16, McDougall to Clunies Ross, 11 November and 7 December; Betty Steedman to Clunies Ross, 7 December 1955.
2   Notes by sister Margery McDougall, January 1965; these and other recollections supplied by E. McDougall.
3   CSIR, 541, Clunies Ross to Tange, 25 November 1955; 379/22/16, Casey to McEwen, 12 January; Clunies Ross to Tange, 8 February 1956.
4   The Australian Room still exists and was extensively refurbished in 2011. It is considered a prestigious meeting room at FAO headquarters. Information from Department of Agriculture, Forests and Fisheries, Canberra.

Bruce is remembered in Australia chiefly as a prime minister who lost his own seat and, unfairly, as one who identified with Britain rather than Australia. One possible explanation for the neglect of his international work, which gave him high status overseas, is that it took place at times either when Australia lacked its own foreign office and diplomatic service or when both were in their infancy. International relations were distant in fact and in public consciousness, and often seen as the province of Britain, on behalf of the Empire as a whole. The nutrition controversy of the 1930s did not stir public interest in Australia as it did in Britain; wartime measures to limit food consumption existed, but on a far less onerous scale. The activities of the League of Nations were of limited interest. Australians knew little of Bruce's vigorous work in London on their behalf. Bruce held no appeal for his former political colleagues, still less for the Labor Government in power for most of the 1940s; all were content to let his work go unnoticed.

That McDougall's name is unknown in the country with which he is still associated at FAO is understandable. He spent only 10 years of his life in Australia, and those chiefly in a rural settlement in the outlying state of South Australia, far from centres of influence. Relatively few officials in Melbourne and in Canberra knew him firsthand. Those who worked with him overseas were generally impressed by his tireless efforts and his abilities, though there were exceptions, particularly in the 1920s, when his closeness to Bruce, outside the regular establishment, probably caused resentment. In later years, the factors that kept Bruce out of public view did so even more effectively for McDougall, who had avoided prominent positions. Beyond that, there are clues in some of the comments made by Australians in his later years. Coombs's 'virtuous conman' phrase suggests a lack of *gravitas*. Nutritionist and FAO staff member Dr W. A. Aykroyd recalled that despite unfailing kindness, particularly to younger colleagues, McDougall 'did not always suffer fools gladly and his comments on people and things could at times be withering'.[5] Karl Olsen wrote warmly of his generosity and friendship, adding:

> ...his means of shepherding his ideas were wit and hospitality...His dinners were well known. There was no stint of food and drink and they were certainly never dull. If he could not get a man to go to dinner he would buy him a drink. A great deal got done at the bar in between quotes from poets, the Scriptures and hilarious and sometimes naughty stories about his colleagues...Some of his stories were deadly and this is one reason why he had some enemies.[6]

---

5   NLA, MS6890/5/3, undated text of recording by Dr Wallace Aykroyd, Director, FAO Nutrition Division.
6   Ibid., Karl Olsen to John McDougall, 23 February 1965.

In the FAO years, McDougall worked hard to project an Australian image, using his congeniality to emphasise it. Dr A. R. Callaghan recalled

> a very happy party in Rome in 1954 at which every Australian known to F. L. McDougall had been invited to meet the Australian Delegation. The [FAO] 1954 Conference had just finished, the strain was off, and Frank McDougall was ready to celebrate. We sang Australian songs with great gusto, including Waltzing Matilda, mainly, I am sure, because we all felt that Frank McDougall was enjoying his Australian contacts so much.[7]

Coombs wrote that he gave his capacity to consume beer as evidence of his authenticity. 'When I die', Mac would say solemnly, 'I hope they will inscribe on my tombstone the tribute, "He never left his beer unfinished", I have earned it.' Yet Coombs thought that he was not generally considered an Australian and asserted his Australian credentials too much.[8]

Those who knew McDougall best believed the conviviality masked loneliness following the end of his marriage in the late 1920s.[9] The loneliness may also explain his patience with and even enthusiasm for drafting sessions, committee meetings and conferences. Lester Pearson reported to Ottawa from a Wheat Advisory Committee meeting that participants favouring a conference included Australia 'because McDougall loves conferences'.[10] Paul Hasluck, a young diplomat in London when he knew McDougall, later Minister for External Affairs and Governor-General, probably captured the essential McDougall when he wrote: 'He loved being busy in and around the corridors of power. The Heaven which I hope he now enjoys would be an endless succession of conferences, with much plucking of elbows and prompting of delegates and sharing of plans with the archangels'[11]

McDougall was, of course, far more than an eager 'committeeman'. His contribution to the international world formed in the aftermath of war was the result of vision and boundless imagination, driven by the values he had adopted as a young man—pragmatism, independence and above all optimism—and shaped by all that he had learned on his journey of trade, science, economics and of human nature. He had dared to challenge entrenched thinking. Although his ideas had not always been accepted and his plans had sometimes come to nothing, he maintained a determined optimism. And so he became a driving force in the creation of an international organisation aiming to better the lives of millions.

---

7   Ibid., text of broadcast in the 'Countryman's Session', 26 October 1958.
8   Coombs, *Trial Balance*, pp. 41–2; Interview with W. Way, 7 April 1995.
9   Elspeth Huxley, Letter to W. Way, 15 April 1986.
10 NAC, MG 26-N1, Vol. 14, 'Skelton O. D.—Canada—External Affairs', 1939–40.
11 Sir Paul Hasluck to W. J. Hudson, 9 June 1986. Copy in my possession.

Figure 22 F. L. McDougall.

Source: E. McDougall.

Appendix I

F. L. McDougall: Biographical outline

1884	Born 16 April, in Blackheath, London.
1901	Completes education at Blackheath Proprietary School; passes matriculation examination for University of London.
1903–05	Studies German in Godesberg-am-Rhein; prepares to enter Technical University in Darmstadt.
1907–08	Travels to St Helena and Natal and investigates possibility of wattle-growing on behalf of family.
1909	Arrives in Renmark, South Australia, and is assisted by family to purchase fruit block.
1914	Visits England and investigates markets for dried fruits.
1915	Marries Madeleine Joyce Cutlack.
1915–19	Service in AIF.
1918	Birth of son, John.
1919	'Notes on the Dried Fruits Industry in Australia' read at Conference of River Agricultural Bureaux.
1920	Birth of daughter, Elisabeth.
1921–22	Campaign to persuade Australian governments to lobby Britain for tariff preference on dried fruits.
1922	Meets Bruce before departure as member of three-man Fruit Delegation to Britain.
1923	Lobbies in London for fruit sales and tariff preference; assists Bruce at Imperial Conference.
1924	Returns to Melbourne at request of Bruce to prepare legislation for orderly marketing; agreement with Bruce on means of continuing work in London.
1925	Publication of *Sheltered Markets*.
1925–29	Part-time Secretary of the London Agency, Dried Fruits Export Control Board.
1925–44	Economic Adviser, Australia House; represents Australia on Imperial Economic Committee.
1926–30	London Liaison Officer for Development and Migration Commission.
1926–33	Member of Empire Marketing Board.

1926–30s	London Liaison Officer for Council of Scientific and Industrial Research.
1928	Appointed Vice-Chairman, Executive Council of Imperial Agricultural Bureaux.
1928–29	Member of League of Nations Economic Consultative Committee (ECC).
1929	Joins Agricultural Economics Committee of International Institute of Agriculture.
1929–30	Member of ECC Committee of Agricultural Experts.
1930	Adviser to Australian delegation, Imperial Conference, London.
1932	Adviser to Australian delegation, Imperial Conference, Ottawa.
1933	Australian representative at International Wheat Conference.
1933–40s	Member of Wheat Advisory Committee.
1934	Devises 'marriage of health and agriculture'.
1935	Works towards Australia's successful nutrition resolution at League of Nations.
1936	Member of League of Nations 'Mixed Committee' on nutrition.
1936–39	Applies ideas in nutrition initiative to 'economic appeasement' of Germany and to reform of the League of Nations.
1937–39	Member of League Economic Committee
1941–42	Works with US officials to promote the idea of an international agency for food and agriculture, and to prepare for its establishment.
1943	Represents Australia at Hot Springs Conference; appointed member of Interim Commission for Food and Agriculture Organization (FAO).
1945	Formal establishment of FAO; becomes Special Adviser to the Director-General.
1946	Permanent appointment to FAO as 'Counselor'.
1955	Official retirement after suffering stroke.
1958	Dies in Rome.

Appendix 2

Select List of Persons[1]

Acheson, Dean	US Assistant Secretary of State
Amery, L. S.	British Colonial Secretary 1924–29; Dominions Secretary from July 1925; 'imperial visionary'
Appleby, Paul H.	Executive Secretary to US Agriculture Secretary 1933–40; Under Secretary of Agriculture 1940–44; Member of FAO Interim Commission 1943–44
Ashton-Gwatkin, F. T. A.	Head of Economic Relations Section, Western Department, British Foreign Office
Astor, Viscount (Waldorf)	Owner of the London *Observer*; Chairman of the League of Nations Mixed Committee on Nutrition
Aykroyd, W. R.	Nutrition scientist at the League of Nations 1931–35; co-author of *Nutrition and Public Health* (1935); head of Nutrition Division, FAO 1946–60
Baldwin, Stanley	British Prime Minister 1923–24, 1924–29, 1935–37
Barrington-Ward, R. M.	Assistant Editor of *The Times* (London)
Barwell, H. N.	Premier of South Australia 1920–24
Beaverbrook, Lord (Max Aitken)	Canadian-born businessman and owner of the *Daily Express*; President of the Board of Trade 1916; Minister of Information 1918; Minister of Aircraft Production 1940; then Minister of Supply and Lord Privy Seal
Bennett, R. B.	Prime Minister of Canada 1930–35
Berle, Adolf	US Assistant Secretary of State
Boudreau, Frank G.	Deputy Director, League of Nations Health Organisation; Executive Director, Milbank Memorial Fund, New York, from 1937
Brigden, J. B.	Australian economist; member of the committee of inquiry into the Australian tariff 1927–29; Financial Secretary, Australian Legation in Washington 1941–46; Chairman of Finance Committee, FAO

---

1   Positions given are those relevant to this publication.

Bruce, S. M.	Australian Federal Treasurer 1921–23; Prime Minister 1923–29; Resident Minister in London 1932–33; High Commissioner 1933–45; President of the Council of the League of Nations 1936; chaired committees leading to production of the 'Bruce Report' 1937–39; Chairman of World Food Board Preparatory Commission 1946; Chairman of the World Food Council 1947–51; created Lord Bruce of Melbourne 1946
Casey, R. G.	Australian External Affairs Liaison Officer, London 1924–31; Federal Minister from 1933; Federal Treasurer 1935–39; Minister at Australian Legation in Washington 1940–42; British Minister in Cairo 1942–43; Federal President of the Liberal Party 1947–49; Australian Minister for External Affairs 1951–60
Chaffey, W. B.	Canadian engineer and irrigation pioneer; established irrigation settlements at Renmark and Mildura with brother George; remained active in grower concerns at Mildura; member of Fruit Delegation 1922–23
Chamberlain, Joseph	British Colonial Secretary 1895–1903; instituted tariff reform campaign 1903
Chamberlain, Neville	British Minister of Health 1923, 1924–29, 1931; Chairman of the Unionist Party and head of Conservative Research Department 1930–31; Chancellor of the Exchequer 1923–24, 1931–37; Prime Minister 1937–40
Chadwick, Sir David	Secretary of the Imperial Economic Committee from 1927, and of the Imperial Agricultural Bureaux from 1928
Churchill, Winston	British Colonial Secretary 1921–22; Chancellor of the Exchequer 1924–29; Prime Minister 1940–45
Coombs, H. C.	Australian economist; Director of Department for Post-War Reconstruction 1942; leader of Australian delegation at Hot Springs Conference
Cooper, Sir James	London accountant and administrator; Chairman of the London Agencies of the Australian Dried Fruits, Dairy Products and Canned Fruits Export Control Boards

Appendix 2

Crawford, J. G.	Australian economist; founding Director of the Bureau of Agricultural Economics; member of FAO Standing Committee on Economics and Marketing
Curtin, John	Australian Prime Minister 1941–45
Cutlack, F. M.	Brother of Joyce McDougall; lawyer, journalist, writer
Davis, Norman	Financial and foreign affairs adviser to US President Roosevelt; head of American Red Cross
Dawson, Geoffrey	Editor of *The Times* (London)
Deakin, Alfred	Australian Prime Minister 1903–04, 1905–08, 1909–10
Dodd, Norris E.	US Under Secretary of Agriculture; delegate to FAO Conference, Copenhagen 1946; Director-General of FAO 1948–54
Duckham, Sir Arthur	Gas engineer and industrialist; leading member of industrial organisations and inquiries, including the 1928 Business Mission to Australia; cousin of F. L. McDougall
Elliot, Walter	Scottish doctor and Conservative MP; friend of Boyd Orr; Parliamentary Under-Secretary of Health for Scotland 1923, 1924–25; Parliamentary Under-Secretary of State for Scotland 1926–29; Chairman of the Research Grants Committee of the Empire Marketing Board; Minister of Agriculture and Fisheries 1932–36; Secretary of State for Scotland 1936–38; Minister of Health 1938–40
Fadden, A. W.	Australian Prime Minister 29 August – 7 October 1941
Fitzpatrick, Dr A. S.	Technical Assistant to F. L. McDougall 1926–30
Gepp, H. W.	Metallurgist and engineer; Chairman, Australian Development and Migration Commission 1926–30
Grady, Henry	US economist; member of US Tariff Commission from 1937; Assistant Secretary of State for Economic Affairs 1939–41; Member of the League of Nations Economic Committee
Hall, Professor N. F. (Noel)	Head of Department of Political Economy, University College London; Director, National Economic Research Institute, London from late 1937
Hawkins, Harry	Head of Treaties Section, US Department of State
Hopkins, Harry	Adviser to US President Roosevelt

Horne, Sir Robert	British Minister of Labour 1919–20; President of the Board of Trade 1920–21; Chancellor of the Exchequer 1921–22; company director
Hughes, W. M.	Australian Prime Minister (Labor) 1915–17, (Nationalist) 1917–23
Hull, Cordell	US Secretary of State
Huxley, Elspeth, née Grant	Publicity Officer, Empire Marketing Board 1929–32; writer
Jebb, Gladwyn	Civil servant in Economic Relations Section, Western Department, British Foreign Office from 1935; Private Secretary to Sir Robert Vansittart and then to Sir Alexander Cadogan, 1937–40
Julius, George	Australian engineer; Chairman of Council for Scientific and Industrial Research (CSIR)
King, W. L. Mackenzie	Prime Minister of Canada 1921–26, 1926–30, 1935–48
Latham, J. G.	Australian Attorney-General 1925–29, 1931–34; Minister for External Affairs 1931–34
Law, Andrew Bonar	British Prime Minister October 1922 – May 1923
Leith-Ross, Sir Frederick	Chief Economic Adviser to the British Government and member of the League of Nations Economic Committee from 1932; Chairman of the committee 1936–37
Lloyd, E. M. H.	Assistant Secretary, Empire Marketing Board
Louwes S. L.	Netherlands Director-General of Food; Special Adviser, FAO, from 1945
Loveday, Alexander	Director, Financial and Economic Intelligence Section, League of Nations
Lyons, J. A.	Australian Prime Minister 1932–39
MacDonald, Ramsay	British Prime Minister (Labour) 1924, 1929–31; (National Coalition) 1931–35
McDougall, Alexander	Grandfather of F. L. McDougall; amateur chemist; founder of milling and chemical companies
McDougall, Catherine (Kit)	Sister of F. L. McDougall
McDougall, Elisabeth	Daughter of F. L. McDougall
McDougall, John	Son of F. L. McDougall
McDougall, Joyce, née Cutlack	Wife of F. L. McDougall from 1915; separated c. 1929

McDougall, Norman	Half-brother of F. L. McDougall; shared Renmark block
MacLeish, Archibald	Librarian of Congress, Washington; Deputy Director, Office of War Information
Macmillan, Harold	British Conservative MP 1924–29 and from 1931; Prime Minister 1957–63; director of Macmillan & Co. publishers; writer
Menzies, R. G.	Australian Prime Minister 1939–41
Morgenthau, Henry, jr	Head of US Farm Board 1933; Acting Treasury Secretary November 1933; Treasury Secretary 1934–45
Norman, Montagu	Governor of the Bank of England
Ormsby-Gore, William	Under-Secretary at the British Colonial Office 1922–29, except during Labour Government of 1924; member of Empire Marketing Board
Orr, John Boyd	Director, Rowett Institute, Aberdeen; advocate for nutrition; Director-General of FAO 1945–48
Parran, Thomas	US Surgeon General
Pearson, Lester	Counsellor, Canadian Embassy in Washington 1942–45; Ambassador 1945–46; Chairman of FAO Interim Commission 1943–45
Richardson, A. E. V.	Member of Executive Council, CSIR; Director of Waite Agricultural Research Institute, South Australia
Riefler, Winfield	US economist; member of League of Nations Finance Committee; Economic Adviser to US Economic Defense Board, from 1941; Minister in charge of Economic Warfare, US Embassy, London, from 1942
Rivett, A. C. D. (David)	Council Member, CSIR, from June 1926; Chief Executive from 1927
Roosevelt, Eleanor	Wife of President Roosevelt
Roosevelt, Franklin Delano	US President 1933–45
Runciman, Walter	President of the UK Board of Trade 1914–16, 1931–37
Salter, Sir Arthur	Head of the Economic and Financial Section of the League of Nations Secretariat until 1930; writer; Gladstone Professor of Political Theory and Institutions at Oxford University from 1934; Member of Parliament, from 1937
Scullin, J. H.	Prime Minister of Australia October 1929–January 1932

Skelton, O. D.	Deputy Minister, Canadian Department of External Affairs
Smith, A. Stuart	Clerical Assistant to F. L. McDougall 1926–30
Snowden, Philip	British Chancellor of the Exchequer 1924 and 1929–31
Stirling, Alfred	Australian External Affairs Officer, London 1937–45
Stoppani, Paul	Director, Economic Relations Section, League of Nations
Tallents, Sir Stephen	Secretary, Empire Marketing Board
Taylor, H. S.	Editor, *Murray Pioneer*
Thomas, J. H.	Trade unionist; Colonial Secretary 1923; Lord Privy Seal 1929–30; Dominions Secretary 1930–35; member of Empire Marketing Board from February 1927
Tolley, Howard R.	Director, Bureau of Agricultural Economics, Washington; member of FAO Interim Commission; Director of Economics and Marketing, FAO
Twentyman, Edward	British representative on FAO Interim Commission
Van Zeeland, Paul	Prime Minister and Foreign Minister of Belgium; commissioned by Britain and France to study means of reducing obstacles to international trade
Welles, Sumner	US Under Secretary of State
Wickard, Claude	US Secretary of Agriculture
Wilson, Dr M. L.	US Assistant Secretary of Agriculture
Wilson, Senator R. V.	Australian Honorary Minister 1923–25; Minister for Markets and Migration 1925 – June 1926; defeated in November 1925
Winant, John G.	Assistant Director, ILO, Geneva 1935 and 1937–39; Director from 1939; US Ambassador in London from 1941

Bibliography

Archival and Primary Sources

National Archives of Australia (NAA)

A457 correspondence files, Prime Minister's Department, 1915–1923

A458 correspondence files, Prime Minister's Department, 1923–1934

A461 correspondence files, Prime Minister's Department, 1934–1950

A981 correspondence files, External Affairs, 1927–1942

A989 correspondence files, External Affairs, 1942–1945

A1608 correspondence files, Prime Minister's Department, 1939–1947

A1667 international trade relations files, Department of Trade and Customs, 1925–1956

A2937 correspondence files, Liaison Office and High Commission, London, 1924–1945

A9504 selected records of Premiers' Conferences, 1901–

A9816 international trade, Nutrition Approach, 1939–1942

A11804 general correspondence of Governor-General, 1912–1927

A11583 volumes of minutes, notes and papers relating to Imperial Conferences, Imperial Economic Conferences and Imperial Meetings, 1923–1930

CP43/1/1 see A9816, 1943/792

CP103/3 correspondence and other papers relating to Imperial and Economic Conferences, 1921–1932

CP103/12 records of Imperial Conferences, 1897–1933

CP498/1 trade relations files, 1925–1953

Papers of Rt Hon. Viscount Stanley Melbourne Bruce

M100 Monthly War Files, 1939–1945

M103 Supplementary War Files, 1938–1943

M104 Folders of Annual Correspondence, 1926–1964

M111 Correspondence with F. L. McDougall, 1924–1929

Files of Commonwealth Scientific and Industrial Organisation (CSIR)

Series held in NAA (NAA/CSIR)

A10085 correspondence files, 1927–1937

A8520 personal history files

A9778 correspondence files, 1927–1981

A10666 correspondence between A. C. D. Rivett and F. L. McDougall, 1927–1938

Series held in CSIRO (CSIR)

541 papers of Sir Ian Clunies Ross

National Archives of Australia, Victorian Branch (NAAV)

B4242 bound correspondence, cablegrams, reports and papers to and from the London Agency, Dried Fruits Export Control Board, 1925–1954

M1146 Rt Hon. Richard Gardiner Casey: correspondence with organisations, 1915–1978

Department of Foreign Affairs and Trade (DFAT)

Diaries of Alfred Stirling

National Archives of Canada (NAC)

RG 25 C records of the Department of External Affairs: international conferences, committees, commissions and meetings, 1887–1984

MG 26 N1 papers of Lester Pearson

MG 30 E 163 personal correspondence of Norman A. Robertson

National Archives of the United Kingdom (UKNA)

BT 92; **BT** 11/92 records of the Board of Trade

CO 760 records of the Colonial Office

FO 371 records of the Foreign Office

MAF 40/17; MAF 40/144 records of the Ministry of Agriculture and Fisheries

National Archives of the United States (USNA)

RG 25 records of the Department of State, 1940–1944

RG 59 records of Harley A. Notter, 1939–1945

RG 353 records of State Department Committees

Archives of the Food and Agriculture Organization, Rome (FAO)

RG 3.1 Correspondence series of F. L McDougall, A1–A6, B1–B6

Series C1, F. L. McDougall memoranda to directors-general

Series D3, Notes and comments, Mr F. L. McDougall, 1938–1939 [copies of notes by Sean Turnell]

Archives of the League of Nations, Geneva (LN)

1928–1932, 10D, Records of the Economic and Financial Section: Monetary and Economic Conference of 1933

1933–1940, 10A, Records of the Economic and Financial Section: Economic Relations

1933–1946, 8A, Records of the Health and Social Questions Section: Health General

Australian Joint Copying Project (AJCP) on microfilm

DO 35 records of the Dominions Office

National Library of Australia (NLA)

NLA: MS6890, Papers of F. L. McDougall

National Library of Scotland (NLS)*

NLS: Acc. 6721, Papers of Walter Elliot

* An obvious omission from this list is John Boyd Orr. The Scottish Record Office advised in 1989 that Orr had destroyed most of his private correspondence, with some exceptions, principally covering 1948–64. The NLS catalogue lists holdings consisting chiefly of transcripts of talks and articles. There is material in the National Archives in London, and in the Archives of FAO, but in neither case does the listing suggest it to be relevant to this study.

National Library of Wales

Dr Thomas Jones, C. H., Papers

Interviews and Correspondence

Dr H. C. Coombs

Elspeth Huxley

Douglas and Merridy Howie, and David Ruston, Renmark

Elisabeth McDougall

Ian McDougall

Professor J. G. Starke

Contemporary Publications used as Primary Sources

American Journal of International Law, Vol. 38, no. 4, October 1944, 'The Interim Commission on Food and Agriculture and the Food and Agriculture Organization'.

Astor, Viscount, and Seebohm Rowntree, B., *British Agriculture: The Principles of Future Policy*, second revised edn, Pelican, London, 1939.

Boudreau, F. G., 'The International Campaign for Better Nutrition', *Milbank Memorial Fund Quarterly*, Vol. XV, no. 1, 1937, pp. 104–20.

Carr, E. H., *Conditions of Peace*, Macmillan, London, 1942.

Carr, E. H., *The Twenty Years' Crisis 1919–1939: An Introduction to the Study of International Relations*, second edn, Macmillan, London, 1946.

Fisher, Alan J. B., 'World Economic Affairs', in *Survey of International Affairs 1937*, ed. Arnold J. Toynbee, Royal Institute of International Affairs, London, 1938.

Gunther, John, *Inside Europe*, H. Hamilton, London, 1936.

Hancock, W. K., *Australia*, reprint 1961 edn, Jacaranda Press, Brisbane, 1930.

Kruse, H. D., 'Ever Normal Nutrition', *Milbank Memorial Fund Quarterly*, Vol. XVI, no. 1, 1938, pp. 110–18.

McDougall, Ellen, Lady, *The McDougall Brothers and Sisters: The Children of Alexander McDougall of Manchester*, The Blackheath Press, London, 1923.

McDougall, F. L., *Sheltered Markets: A Study of Empire Trade*, John Murray, London, 1925.

McDougall, F. L., 'The International Wheat Situation', *International Affairs*, Vol. 10, no. 4, 1931.

Orr, John Boyd, *Food, Health and Income: A Report on a Survey of Adequacy of Diet in Relation to Income*, second edn, Macmillan, London, 1937.

Phillips, P. D., and Woods, G. L., eds. *The Peopling of Australia*, MUP/Macmillan, Melbourne, 1928.

Official Publications

Commonwealth of Australia

Annual Year Books of the Commonwealth of Australia.

Coleman, P. E., 'Report Upon the Organization of the High Commissioner's Office, London, and the Activities Associated Therewith', London, 1930.

Commonwealth Parliamentary Debates (*CPD*).

Commonwealth Parliamentary Papers (*CPP*).

Commonwealth Bureau of Census and Statistics, *Trade and Customs and Excise Revenue of the Commonwealth of Australia*, 1914–1920.

Development and Migration Commission, *Report on the Dried Fruits Industry of Australia*, Melbourne, 1927.

League of Nations Series of Publications

E: Economic and Financial: *The Problem of Nutrition*:

1: Interim Report of the Mixed Committee on the Problem of Nutrition, 1936 II.B.3.

2: Report on the Physiological Bases of Nutrition, 1936 II.B.4.

3: Nutrition in Various Countries, 1936 II.B.5.

4: Statistics of Food Production, Consumption and Prices, 1936 II.B.6.

'The Development of International Co-operation in Economic and Social Affairs': Report of the Special Committee', LN: A.23. 1939.

League of Nations *Official Journal*.

League of Nations *Monthly Summary*.

United Kingdom

Reports of the Imperial Economic Committee on Marketing and Preparing for Market of Foodstuffs Produced in the Overseas Parts of the Empire: First Report—General, Cmd. 2493.

——*Second Report—Meat*, Cmd. 2499.

——*Third Report—Fruit*, Cmd. 2658.

Empire Marketing Board, *Annual Reports*, 1929, 1930, 1931, 1932, HMSO.

Empire Marketing Board, *The Growing Dependence of British Industry upon Empire Markets*, December 1929, HMSO.

Edited Documents

Hudson, W. J., and North, Jane, eds, *My Dear P. M.: R. G. Casey's Letters to S. M. Bruce 1924–29*, Department of Foreign Affairs, Australian Government Publishing Service, Canberra, 1980.

Hudson, W. J., and Stokes, H. J. W., eds, *Documents on Australian Foreign Policy 1937–1949*. Volume IV: *July 1940 – June 1941*, Department of Foreign Affairs, Australian Government Publishing Service, Canberra, 1980.

Hudson, W. J., and Stokes, H. J. W., eds, *Documents on Australian Foreign Policy 1937–1939*. Volume VI: *July 1942 – December 1943*, Department of Foreign Affairs, Australian Government Publishing Service, Canberra, 1983.

Hudson, W. J., and Way, Wendy, eds, *Letters from 'A Secret Service Agent': F. L. McDougall to S. M. Bruce 1924–1929*, Department of Foreign Affairs, Australian Government Publishing Service, Canberra, 1986.

Hudson, W. J., and Way, Wendy, eds, *Documents on Australian Foreign Policy 1937–1949*. Volume IX: *January–June 1946*, Department of Foreign Affairs and Trade, Australian Government Publishing Service, Canberra, 1991.

Hudson, W. J., and Way, Wendy, eds, *Documents on Australian Foreign Policy 1937–1949*. Volume X: *July–December 1946*, Department of Foreign Affairs and Trade, Australian Government Publishing Service, Canberra, 1993.

Inglis, Alex I., ed., *Documents on Canadian External Relations*. Volume 5: *1931–1935*, Department of External Affairs, Ottawa, 1973.

Neale, R. G., Edwards, P. G., and Kenway, H., eds, *Documents on Australian Foreign Policy 1937–49*. Volume I: *1937–1938*, Department of Foreign Affairs, Australian Government Publishing Service, Canberra, 1975.

Neale, R. G., Edwards, P. G., Kenway, H., and Stokes, H. J. W., eds, *Documents on Australian Foreign Policy 1937–49*. Volume II: *1939*, Department of Foreign Affairs, Australian Government Publishing Service, Canberra, 1976.

Nicholas, H. G., ed., *Washington Despatches 1941–1945: Weekly Political Reports from the British Embassy*, Weidenfeld & Nicolson, London, 1981.

Memoirs and Diaries

Acheson, Dean, *Present at the Creation: My Years in the State Department*, W. W. Norton & Co., New York, 1969.

Appleby, Paul, 'Reminiscences of Paul Appleby', microform, c. 1972, Hollis Library, Harvard, Cambridge, Mass.

Barnes, John, and Nicholson, David, eds, *The Leo Amery Diaries 1896–1929*. Volume I, Hutchinson, London.

Boyd Orr, Lord, *As I Recall*, Doubleday, New York, 1966.

Coombs, H. C., *Trial Balance*, Macmillan, Melbourne, 1981.

Cutlack, F. M., *Renmark: The Early Years*, Nancy B. Basey, Eltham, Vic., 1988.

Huxley, Gervas, *Both Hands: An Autobiography*, Chatto & Windus, London, 1970.

Jebb, Gladwyn, *The Memoirs of Lord Gladwyn*, Weidenfeld & Nicolson, London, 1972.

Macmillan, Harold, *Winds of Change: 1914–1939*, Macmillan, London, 1966.

Pearson, Lester B., *Memoirs 1897–1948: Through Diplomacy to Politics*, Victor Gollancz, London, 1973.

Penrose, E. F., *Economic Planning for the Peace*, Princeton University Press, Princeton, NJ, 1953.

Stirling, Alfred, *On the Fringe of Diplomacy*, The Hawthorn Press, Melbourne, 1973.

Stirling, Alfred, *Lord Bruce, the London Years*, The Hawthorn Press, Melbourne, 1974.

Tolley, Howard, 'Reminiscences of Howard R. Tolley', microform, Hollis Library, Harvard, Cambridge, Mass.

Wallace, Henry Agard, Diaries, University of Iowa, microform copy, Library of Congress, Washington, DC.

Winant, John G., *A Letter from Grosvenor Square: An Account of a Stewardship*, Hodder & Stoughton, New York, 1947.

Principal Newspapers

Murray Pioneer, Renmark, microfilm, Mortlock Library of South Australia.

The Times, London.

Books and Articles

Alter, Peter, *The Reluctant Patron: Science and the State in Britain 1850–1920*, Translated by Angela Davies, Deddington, Oxford, 1987.

Antoniades, Andreas, 'Epistemic Communities, Epistemes and the Construction of (World) Politics', *Global Society*, Vol. 17, no. 1, 2003.

Atkins, Peter J., 'Fattening Children or Fattening Farmers? School Milk in Britain, 1921–1941', *Economic History Review*, Vol. LVIII, no. 1, 2005.

Attard, Bernard, 'The Limits of Influence: The Political Economy of Australian Commercial Policy after the Ottawa Conference', *Australian Historical Studies*, Vol. 29, no. 111, October 1998.

Beer, Samuel H., *Modern British Politics: Parties and Pressure Groups in the Collectivist Age*, W. W. Norton, New York, 1982.

Belich, James, *Paradise Reforged: A History of the New Zealanders from the 1880s to the Year 2000*, Allen Lane, The Penguin Press, London, 2001.

Bendimer, Elmer, *A Time for Angels: The Tragicomic History of the League of Nations*, Alfred A. Knopf, New York, 1975.

Borgwardt, Elizabeth, *A New Deal for the World: America's Vision for Human Rights*, The Belknap Press of Harvard University Press, Cambridge, Mass., 2005.

Bulmer, Martin, 'Mobilising Social Knowledge for Social Welfare: Intermediary Institutions in the Political Systems of the United States and Great Britain between the First and Second World Wars', in *International Health Organisations and Movements 1918–1939*, ed. Paul Weindling, Cambridge University Press, Cambridge, 1995.

Burridge, Trevor, *Clement Attlee: A Political Biography*, Jonathan Cape, London, 1985.

Cain, P. J., and Hopkins, A. G., *British Imperialism: Crisis and Deconstruction, 1914–1990*, Longman, London, 1993.

Coleman, William, Cornish, Selwyn, and Hagger, Alf, *Giblin's Platoon: The Trials and Triumph of the Economist in Australian Public Life*, ANU E Press, Canberra, 2006.

Collingham, Lizzie, *The Taste of War: World War Two and the Battle for Food*, Allen Lane, London, 2011.

Constantine, Stephen, 'Bringing the Empire Alive', in *Imperialism and Popular Culture*, ed. John M. Mackenzie, Manchester University Press, Manchester, 1986.

Constantine, Stephen, *Buy & Build: The Advertising Posters of the Empire Marketing Board*, HMSO, London, 1986.

Constantine, Stephen, 'Anglo–Canadian Relations, the Empire Marketing Board and Canadian National Autonomy between the Wars', *Journal of Imperial and Commonwealth History*, Vol. 21, no. 2, 1993.

Cooper, Andrew Fenton, *British Agricultural Policy 1912–36: A Study in Conservative Politics*, Manchester University Press, Manchester, 1989.

Crozier, Andrew J., *Appeasement and Germany's Last Bid for Colonies*, Macmillan, London, 1988.

Cumpston, I. M., *Lord Bruce of Melbourne*, Longman Cheshire, Melbourne, 1989.

de Waal, Edmund, *The Hare with Amber Eyes: A Hidden Inheritance*, Vintage Books, London, 2011.

Dilks, David, *Neville Chamberlain: Pioneering and Reform 1869–1929*, Cambridge University Press, Cambridge, 1984.

Drummond, Ian M., *British Economic Policy and the Empire 1919–1939*, George Allen & Unwin, London, 1972.

Drummond, Ian M., *Imperial Economic Policy 1917–1939: Studies in Expansion and Protection*, George Allen & Unwin, London, 1972.

Drummond, J. C., and Wilbraham, Anne, *The Englishman's Food: A History of Five Centuries of the English Diet*, Jonathon Cape, London, 1939.

Dubin, Martin David, 'The League of Nations Health Organisation', in *International Health Organisations and Movements 1918–1939*, ed. Paul Weindling, Cambridge University Press, Cambridge, 1995.

Dunlap, Thomas R., *Nature and the English Diaspora: Environment and History in the United States, Canada, Australia and New Zealand*, Cambridge University Press, Cambridge, 1999.

Dunsdorfs, Edgars, *The Australian Wheat Growing Industry 1788–1948*, Melbourne University Press, Melbourne, 1956.

DuPuis, E. Melanie, *Nature's Perfect Food: How Milk Became America's Drink*, New York University Press, New York, 2002.

Eckes, Alfred E., and Zeiler, Thomas W., *Globalization and the American Century*, Cambridge University Press, New York, 2003.

Edwards, Cecil, *Bruce of Melbourne: Man of Two Worlds,* Heinemann, London, 1965.

Fine, Sidney, *Laissez Faire and the General Welfare State: A Study of Conflict in American Thought 1865–1901*, University of Michigan, Michigan, 1956.

Fitzhardinge, L. F., *The Little Digger 1914–1952: William Morris Hughes: A Political Biography, Volume II*, Angus & Robertson, Sydney, 1979.

Foreman-Peck, James, *A History of the World Economy: International Relations since 1850*, Wheatsheaf Books, Brighton, Sussex, 1983.

Gannon, Franklin Reid, *The British Press and Germany 1936–1939*, Clarendon Press, Oxford, 1971.

Gardner, Richard N., *Sterling Dollar Diplomacy in Current Perspective*, Columbia University Press, New York, 1980.

Gillespie, James, 'The "Marriage of Agriculture and Health" in Australia: The Advisory Council on Nutrition and Nutrition Policy in the 1930s', in *Migration to Mining: Proceedings of the Biannual Conference of the Australian Society of the History of Medicine*, ed. Brian Reid, Northern Territory University, Darwin, 1998.

Goodwin, Doris Kearns, *No Ordinary Time: Franklin & Eleanor Roosevelt: The Home Front in World War II*, Simon & Schuster, New York, 1994.

Griggs, Barbara, *The Food Factor: An Account of the Nutrition Revolution*, Penguin, London, 1986.

Grosvenor, G. Arch., *Red Mud to Green Oasis*, Raphael Arts Pty Ltd, Renmark, SA, 1979.

Haas, Peter M., 'Introduction: Epistemic Communities and International Policy Coordination', *International Organization*, Vol. 46, no. 1, *Knowledge, Power and International Policy Coordination*, 1992.

Hambidge, Gove, *The Story of FAO*, D. van Nostrand Company, Princeton, NJ, 1955.

Hamby, Alonzo L., *Liberalism and its Challengers: F. D. R. to Reagan*, Oxford University Press, New York, 1985.

Hammond, R. J., *Food. Volume I: The Growth of Policy*, History of the Second World War: United Kingdom Civil Series, ed. W. K. Hancock, HMSO, London, 1951.

Hancock, W. K., *Survey of British Commonwealth Affairs II: Problems of Economic Policy. Part 1*, Oxford University Press, London, 1940.

Hempton, David, *Methodism: Empire of the Spirit*, Yale University Press, New Haven, Conn., 2005.

Hevesey, Paul de, *World Wheat Planning and Economic Planning in General*, Oxford University Press, London, 1940.

Hill, Ernestine, *Water into Gold*, Robertson & Mullens, Melbourne, 1946.

Hobsbawm, Eric, *The Age of Extremes: The Short Twentieth Century 1914–1991*, Abacus, London, 1994.

Hodge, Joseph M., 'Science, Development and Empire: The Colonial Advisory Council on Agriculture and Animal Health, 1929–43', *Journal of Imperial and Commonwealth History*, Vol. XXX, no. 1, 2002.

Holland, Robert, 'Imperial Collaboration and Great Depression: Britain, Canada and the World Wheat Crisis, 1929–35', in *Theory and Practice in the History of European Expansion Overseas: Essays in Honour of Ronald Robinson*, eds Andrew Porter and Robert Holland, Frank Cass & Co., London, 1988.

Howland, Nina Davis, 'Ambassador John Gilbert Winant: Friend of Embattled Britain, 1941–1946', PhD Dissertation, University of Maryland, 1983.

Hudson, W. J., *Australia and the League of Nations*, Sydney University Press, Sydney, 1980.

Hudson, W. J., *Casey*, Oxford University Press, Melbourne, 1986.

Hudson, W. J., and Sharp, M. P., *Australian Independence: Colony to Reluctant Kingdom*, Melbourne University Press, Melbourne, 1988.

Ignatieff, Michael, *Isaiah Berlin*, Viking, Toronto, 1998.

Kargon, Robert H., *Science in Victorian Manchester: Enterprise and Expertise*, Manchester University Press, Manchester, 1977.

Kenwood, A. G, and Lougheed, A. V., *The Growth of the International Economy 1820–1980, An Introductory Text*, George Allen & Unwin, London, 1983.

Kirby, M. W., *The Decline of British Economic Power since 1870*, Allen & Unwin, London, 1981.

Lake, Marilyn, *The Limits of Hope: Soldier Settlement in Victoria 1915–38*, Oxford University Press, Melbourne, 1987.

Lee, David, *Stanley Melbourne Bruce: Australian Internationalist*, Continuum, London, 2010.

Lewis, W. Arthur, *The Evolution of the International Economic Order*, Princeton University Press, Princeton, NJ, 1978.

Lloyd, T. O., *Empire to Welfare State: English History 1906–1985*, Oxford University Press, Oxford, 1986.

Louis, Wm Roger, *In the Name of God Go!: Leo Amery and the British Empire in the Age of Churchill*, W. W. Norton & Co., New York, 1992.

MacDonald, C. A., 'Economic Appeasement and the German "Moderates", 1937–1939. An Introductory Essay', *Past and Present*, no. 56, August 1972.

McKenzie, Francine, *Redefining the Bonds of Commonwealth, 1939–1948: The Politics of Preference*, Palgrave, New York, 2002.

MacLeod, Roy M., 'Science, Progressivism, and "Practical Idealism": Reflections on Efficient Imperialism and Federal Science in Australia, 1885–1915', in *The 'Creed of Science' in Victorian England*, ed. by Roy M. MacLeod, Variorum, Aldershot, UK, c. 2000.

Malenbaum, Wilfred, *The World Wheat Economy 1885–1939*, Harvard University Press, Cambridge, Mass., 1953.

Marchisio, Sergio, and Di Blasé, Antonietta, *International Organization and the Evolution of World Society. Volume 1: The Food and Agriculture Organization (FAO)*, The Graduate Institute of International Studies, Geneva, Societa Italiana per la Organizzazione Internazionale, Rome, and Martinus Nijhoff, Dordrecht, The Netherlands, 1991.

Markowitz, Norman D., *The Rise and Fall of the People's Century: Henry A. Wallace and American Liberalism, 1941–1948*, Free Press, New York, 1973.

Mayhew, Madeleine, 'The 1930s Nutrition Controversy', *Journal of Contemporary History*, Vol. 23, no. 3, 1988.

Medlicott, William Norton, *Britain and Germany: The Search for Agreement, 1930–1937, Creighton Lecture in History 1968*, University of London, London, 1969.

Middlemas, Keith, and Barnes, John, *Baldwin: A Biography*, Weidenfeld & Nicolson, London, 1969.

Moorhouse, Frank, *Dark Palace*, Vintage, Sydney, 2001.

Morriss, William E., *Chosen Instrument: A History of the Canadian Wheat Board: The McIvor Years*, Reidmore Books/Canadian Wheat Board, Edmonton, 1987.

Nichols, C. S., *Elspeth Huxley: A Biography*, paperback edn, Harper Collins, London, 2003.

O'Brien, John, 'Empire v. National Interests in Australian–British Relations During the 1930s', *Historical Studies* [Australia], Vol. 22, no. 89, 1987, pp. 569–86.

O'Brien, John, 'F. L. McDougall and the Origins of the FAO', *Australian Journal of Politics and History*, Vol. 46, no. 2, 2000.

Offer, Avner, *The First World War: An Agrarian Interpretation*, Clarendon Press, Oxford, 1989.

Offner, Arnold A., 'Appeasement Revisited: The United States, Great Britain, and Germany, 1933–1940', *The Journal of American History*, Vol. 64, no. 2, 1977.

Phillips, Ralph W., *FAO: Its Origins, Formation and Evolution 1945–1981*, Food and Agriculture Organization of the United Nations, Rome, 1981.

Powell, J. M., *An Historical Geography of Modern Australia: The Restive Fringe*, Cambridge University Press, Cambridge, 1988.

Qualter, Terence H., *Advertising and Democracy in the Mass Age*, St Martin's Press, New York, 1991.

Richmond, W. H., 'S. M. Bruce and Australian Economic Policy 1923–29', *Australian Economic History Review*, Vol. 23, 1983.

Rivett, Rohan, *David Rivett: Fighter for Australian Science*, Melbourne, 1972.

Roe, Michael, *Australia, Britain and Migration 1915–1940: A Study of Desperate Hopes*, Cambridge University Press, Cambridge, 1995.

Rooth, Tim, *British Protectionism and the International Economy: Overseas Commercial Policy in the 1930s*, Cambridge University Press, Cambridge, 1992.

Rostow, W. W., *The World Economy: History and Prospect*, University of Texas Press, Austin, 1978.

Rowse, Tim, *Nugget Coombs*, Cambridge University Press, Cambridge, 2002.

Samuels, Walter J., *Henry A. Wallace and American Foreign Policy*, Greenwood Press, Westport, Conn., 1976.

Scammell, W. M., *The International Economy since 1945*, second edn, St Martin's Press, New York, 1983.

Schedvin, C. B., *Shaping Science and Industry: A History of Australia's Council of Scientific and Industrial Research 1926–49*, Allen & Unwin, Sydney, 1987.

Schwarz, Jordan A., *Liberal: Adolf A. Berle and the Vision of an American Era*, Free Press, New York, and Collier Macmillan, London, c. 1987.

Serle, Geoffrey, *John Monash: A Biography*, Melbourne University Press, Melbourne, 1982.

Sharp, Alan, 'Adapting to a New World? Britain's Foreign Office in the 1920s', *Contemporary British History*, Vol. 18, no. 3, Autumn 2004.

Shirer, William L., *The Rise and Fall of the Third Reich: A History of Nazi Germany*, Book Club Associates, London, 1960.

Siriwardana, Mahinda, 'The Economic Impact of Tariffs in the 1930s in Australia: The Brigden Report Re-examined', *Australian Economic Papers*, December 1996.

Skidelsky, Robert, *Oswald Mosley*, second edn, Macmillan Papermac, London, 1981.

Skidelsky, Robert, *John Maynard Keynes. Volume One: Hopes Betrayed 1883–1920*, Macmillan, London, 1983.

Skidelsky, Robert, *John Maynard Keynes. Volume Two: The Economist as Saviour 1920–1937*, Macmillan, London, 1992.

Skidelsky, Robert, *John Maynard Keynes. Volume Three: Fighting for Britain 1937–1946*, Macmillan, London, 2000.

Solberg, Carl E., *The Prairies and the Pampas: Agrarian Policy in Canada and Argentina, 1880–1930*, Stanford University Press, Stanford, Calif., 1987.

Staples, Amy L. S., 'To Win the Peace: The Food and Agriculture Organization, Sir John Boyd Orr, and the World Food Board Proposals', *Peace & Change*, Vol. 28, no. 4, 2003.

Staples, Amy L. S., *The Birth of Development: How the World Bank, Food and Agriculture Organization, and World Health Organization Changed the World 1945–1965*, Kent State University Press, Kent, Ohio, 2006.

Storry, Elizabeth; Bennett, Heather; McIntosh, Robyn; McIntosh, Bill; Taylor, Darnley; and Hersey, Tony; eds, *Pictorial History of Renmark: Celebrating 100 Years: 1887–1987*, Renmark, SA, 1987.

Sullivan, Emmett, 'Revealing a Preference: Imperial Preference and the Australian Tariff, 1901–14', *Journal of Imperial and Commonwealth History*, Vol. 29, no. 1, 2001.

Thomson, David, *England in the Nineteenth Century*, Penguin, London, 1950.

Tsokhas, Kosmas, *Markets, Money and Empire: The Political Economy of the Australian Wool Industry*, Melbourne University Press, Melbourne, 1990.

Tsokhas, Kosmas, 'Protection, Imperial Preference, and Australian Conservative Politics, 1923–39', *Journal of Imperial and Commonwealth History*, Vol. 20, no. 1, 1992.

Tsokhas, Kosmas, *Making a Nation State: Cultural Identity, Economic Nationalism and Sexuality in Australian History*, Melbourne University Press, Melbourne, 2001.

Turnell, Sean, 'Butter for Guns: F. L. McDougall, Nutrition and Economic Appeasement', *Macquarie Economics Research Papers*, no. 14/95, 1995.

Turnell, Sean, *Éminence Grise: The Political Economy of F. L. McDougall*, Macquarie Economics Research Papers, 13 September 1998.

Turnell, Sean, 'F. L. McDougall: Eminence Grise of Australian Economic Diplomacy', *Australian Economic History Review*, Vol. 40, no. 1, 2000.

Wadham, Sir Samuel; Wilson, R. Kent; and Wood, Joyce; *Land Utilization in Australia*, fourth edn, Melbourne University Press, Melbourne, 1964.

Walters, F. P., *A History of the League of Nations*, Royal Institute of International Affairs, London, 1952.

Walters, F. P., 'The League of Nations', in *The Evolution of International Organizations*, ed. Evan Luard, Frederick A. Praeger, New York, 1966.

Weindling, Paul, 'Introduction: Constructing International Health between the Wars', in *International Health Organisations and Movements, 1918–1939*, ed. Paul Weindling, Cambridge University Press, Cambridge, 1995.

Weindling, Paul, 'Social Medicine at the League of Nations Health Organisation and the International Labour Organisation Compared', in *International Health Organisations and Movements, 1918–1939*, ed. Paul Weindling, Cambridge University Press, Cambridge, 1995.

Welles, Benjamin, *Sumner Welles: FDR's Global Strategist: A Biography*, St Martin's Press, New York, 1997.

Wilcox, Walter W., *The Farmer in the Second World War*, Iowa State College, Ames, Iowa, 1947.

Wilson, Derek, *The Astors 1763–1992, Landscape with Millionaires*, Weidenfeld & Nicolson, London, 1993.

Yates, Paul Lamartine, *So Bold an Aim: Ten Years of International Co-operation toward Freedom from Want*, Food and Agriculture Organization of the United Nations, Rome, 1955.

Online Sources

American Dictionary of Biography

'Grady, Henry Francis', Dennis Merrill

'Winant, John Gilbert', Sylvia B. Larson

Australian Dictionary of Biography

'Birks, Rosetta Jane (1856–1911)', Martin Woods

'Bruce, Stanley Melbourne (Viscount Bruce) (1883–1967)', Heather Radi

'Deakin, Alfred (1856–1919)', R. Norris

'McDougall, Frank Lidgett (1884–1958)', Alfred Stirling

'Taylor, Harry Samuel (1873–1932)', Malcolm Saunders

Oxford Dictionary of National Biography

'Elliot, Walter Elliot (1888–1958)', Gordon F. Millar

'Gore, William George Arthur Ormsby, Fourth Baron Harlech (1885–1964)', K. E. Robinson

'Gwatkin, Frank Trelawny Arthur Ashton (1889–1976)', Keith Hamilton

'Hopkins, Sir Frederick Gowland (1861–1947)', H. H. Dale

'Jebb, (Hubert Miles) Gladwyn, first Baron Gladwyn (1900–1996)', Alan Campbell

'Jowitt, William Allen, Earl Jowitt (1885–1957)', Thomas S. Legg and Marie-Louise Legg

'Leeper, Sir Reginald Wildig Allen (1888–1968)', Derek Drinkwater

'Lloyd, Edward Mayow Hastings (1889–1968)', Frank Trentmann

'McGonigle, George Cuthbert Mura (1889–1939)', Susan McLaurin

'Meade, James Edward (1907–1995)', Susan Howson

'Mellanby, Sir Edward (1884–1955)', B. S. Platt, rev. Michael Bevan

'Orr, John Boyd, Baron Boyd Orr (1880–1971)', K. L. Baxter

'Robertson, Sir Dennis Holme (1890–1963)', Gordon Fletcher

'Ross, Sir Frederick William Leith (1887–1968)', Roger Middleton

'Russell, Sir (Edward) John (1872–1965)', N. W. Pirie

'Schuster, Sir George Ernest (1881–1982)', Oliver Van Oss

'Stapledon, Sir (Reginald) George (1882–1960)', Elizabeth Baigent

'Zimmern, Sir Alfred Eckhard (1879–1957)', D. J. Markwell

Miscellaneous

Carpenter, Kenneth J., 'The Work of Wallace Aykroyd: International Nutritionist and Author', American Society for Nutrition, 2007, <http://jn.nutrition.org/cgi/content/full/137/4/873>

Cullather, Nick, 'The Foreign Policy of the Calorie', *The American Historial Review*, Vol. 112, no. 2, 2007, pars 26–7, <http://www.historycooperative.org/journals/ahr/112.2/cullather.html> [accessed 3 July 2008].

Entry for Gunnar Myrdal in <http://nobelprize.org/nobel_prizes/economics/laureates/1974/myrdal-bio.html> [accessed 27 November 2008].

'Four Freedoms', Speech of President Roosevelt in <http://www.fdrlibrary.marist.edu/4free.html> [accessed 8 October 2008].

Fox, Daniel M., 'The Significance of the Milbank Memorial Fund for Policy: An Assessment at its Centennial', *Milbank Quarterly*, Vol. 84, no. 1, 2006, <http://www.milbank.org/quarterly/8401PN.html>

Hill Miller, Helen, in <http://www.cosmos-club.org/web/journals/1995/miller.html> [accessed 19 January 2006].

Proceedings of the Nutrition Society, 1943, 'The Hot Springs Conference', <http://journals.cambridge.org> [published online 28 September 2007; accessed 15 May 2011].

Rutherford, Malcolm, 'Walton Hamilton, Amherst, and the Brookings Graduate School: Institutional Economics and Education' (draft), September 2001, <http://ideas.repec.org/p/vic/vicddp/0104.html> [accessed 2006].

Index

Abbott, E., 98, 109n
Abbott, Miss Grace, 166-7
Abyssinia, 169, 170, 176
Acheson, Dean, 236-7, 245, 247, 248-9, 252, 261, 274, 278
Adelaide, 15, 18
Advisory Committee on Alimentation (US), 156
Advisory Committee on Nutrition (UK), 225, 237
Advisory Council for Agricultural Research (UK), 114
Afghanistan, 302
The Age, 89
Agents-General of Australian States in London, 41, 82
Agricultural Adjustment Acts/ Administration (US), 163, 212
Agricultural Bureaus, South Australia, 24-5, 30-1, 33, 35
Agriculture Department (US), *see* Department of Agriculture (US)
Agricultural Marketing Acts/ Boards (UK), 157-8
agriculture, 1, 159, 164
  credit provision for, 132, 138, 175, 283, 292
  imperial cooperation and rationalisation, 61, 135, 137-9, 159-60
  import duties on, 1, 161, 162, 166
  limits on production, 1, 136, 169, 275, 280
  organisation of, 31, 37, 54, 55-6, 57, 64, 90
  policy for defence, 198
  prices, 1, 11, 107, 135-6, 269-70, 272
    stabilization, 292-5
  reorientation and diversification, 153, 165, 173, 191, 197, 219, 230, 238
  science and increased production, 16, 72, 164, 284, 288
  UK policy, 137-9, 159, 230, 272
  US policy, 272
  see also wheat

AIF Education Service, 25, 30, 82
Amery, Leo, 13, 51, 54, 68, 70, 87-8
Amherst College (US), 212-13, 215
Anderson, Sir Alan, 233
Anderson, Sir John, 233
Angliss, W.C., 109n
Anglo-American coperation, 217, 219-20, 226-30, 232, 233, 234, 236-8, 241-2, 251
  US-British conference and declaration (proposed), 237-8
Angove, T.C., 47
appeasement
  see 'economic appeasement'
Appleby, Paul, 244, 251, 255, 258, 276, 280, 282, 287-8
Argentina, 68, 110-11, 130-1, 136, 137, 141, 143, 145, 148, 150-1, 159, 168, 207, 233
The Argus, 89
Armstrong, P.C., 141
Ashbolt, A.H., 54
Ashton-Gwatkin, F.T.A., 184, 193-4, 203
Astor, Nancy, 184
Astor, Lord Waldorf, 172, 184, 198
Atlantic Charter, 241-2, 252, 272, 275, 277
Attlee, Clement, 225-6, 294
Australia, 64, 142, 162, 167, 168, 183, 207, 222, 237
  loyalties within, 8-9
  industrial development, 10, 12, 43-4, 61, 78, 84, 86, 87-90, 98, 159
  US views of, 263-4
  wheat, 130-1, 136, 140, 145-8, 233
'Australia Unlimited', 34
Australian Advisory Council on Nutrition, 126
Australian Dried Fruits Association, 24, 32, 34-9, 47, 55
  publicity campaign, 32-3
Australian Food Council, 288
Australian High Commission, London, 46, 53, 77, 96-7, 99-100, 206, 217, 221
  External Affairs Liaison Office, 221-3
Australian Industries Protection League, 95

Australian Labor Party, 43
    policy on protection, 93, 95-6, 277-8
Austria, 168, 185, 186
Avenol, J.L.A., 192, 203
Aykroyd, Dr W.R., 155, 166, 304
Bajpai, Sir Girja, 282
Baldwin, Stanley, 37, 50-1, 66, 80
Barrington-Ward, R.M., 102, 187, 188
Barwell, H.N. 36, 37-9
Basey, F.H., 30
Bavin, Thomas, 91
Beaverbrook, Lord, 93, 95, 100-1, 104, 112, 137
Belgian Congo, 183
Belgium, 178, 183, 192, 215
Bennett, R.B., 114, 142, 146
Berle, Adolf, 213-14, 215, 236-7, 243, 247, 249, 251-2, 255, 259-60, 261, 263
Berlin, Isaiah, 245, 251, 264
Biffen, Sir Rowland, 117-18
Bingham, Ambassador Robert, 150, 192
Birks, Charles, 18
Birks, Lucy, 18
Birmingham, 16
Bledisloe, Lord, 71
Board of Economic Warfare (US), 250, 255
Board of Eduction (UK), 156-7
Board of Trade (UK), 46, 67, 87, 107, 185, 189, 193, 240
Boer War, 21
Boudreau, Dr F.G., 214-15, 236-7, 259-60
Bracken, Brendan, 101
Brazil, 136
Bretton Woods, United Nations Economic and Monetary Conference, 277, 288
Brigden, J.B., 89, 277, 291-2
Brigden Report on Australian Tariff, 88-90
British Association for the Advancement of Science, 21, 113
British Empire/Commonwealth, 7, 8, 21-2, 196, 208, 240
    'imperial visonaries', 12-13, 50-1, 119
    proposed organisations to encourage imperial economic cooperation, 101-5, 112-14
British Empire Exhibition, 42, 53

British Medical Association, 155
British Science Guild, 21
British–US cooperation
    see Anglo-American cooperation
Broadley, Sir Herbert, 298
Broken Hill Proprietary Ltd, 87
Brookes, Herbert, 48
Brown, John, 91
Bruce, Ernest, 43-4
Bruce, Ethel, 43-4, 181, 224
Bruce, S. M., xi, 1, 7, 10, 101-2, 119, 129, 134, 159, 171, 179, 181, 189-90, 211, 212, 218, 219, 236, 244, 249-50, 253, 255, 257, 258-9, 264, 267, 276, 281, 295, 303
    childhood, 7, 10-11, 42-3
    correspondence with McDougall, 45-6, 65, 68, 70
    'economic appeasement', 186-7, 192, 195-7, 201
    entry into politics and political aims, 43-5
    FAO, 271-2, 287, 296-7, 298
    Federal Treasurer, 37, 42, 44, 45
    High Commissioner in London, 117, 146-7, 177, 206, 221-3, 294, 304
        influence on Australian policy, 110, 196
    loss of office, 61, 90-1, 93-4
    'men, money and markets' speech', 49-50
    nutrition campaign, 125, 162, 168-73, 177, 201-2, 223-5, 229-33, 239-40, 270, 294
    Ottawa Conference, 107, 109
    peace and war aims incorporating social justice, 223-9, 245-6
    peerage, 294, 296
    personal qualities, 7, 44-5
    President of Council, League of Nations, 175, 177
    Prime Minister, 44, 124, 221
    proposed appointment as League of Nations 'economic superman', 203
    recognition in Australia, 304
    reform of League of Nations, 176-7, 201-2, 204-6

war service, 7, 43, 123
views on agriculture, 37, 54, 75-6, 90
   economy, 58, 75
   increasing imports from Germany, 186-7
   tariff, 88-90
   wheat negotiations, 146-9
   World Food Board Preparatory Commission, 294-6
'Bruce Report', 175, 204-6, 207, 222, 223
Bulcock, F. W., 294
Bulgaria, 131, 148
Bullitt, Ambassador William C., 186
Bureau of Agricultural Economics (US), 238, 251, 298
Bureau of Economic Research (Australia), proposed, 95
Bureau of Economic and Political Research (Australia) proposed, 102-3
Burnet, Dr E., 155
Burnet/Ayckroyd report on nutrition, 155, 169
Butler, H.B., 192, 237, 255
butter, 109-10, 166
Cairns, Andrew, 224, 244, 280, 282
Calder, Ritchie, 158
Callaghan, Dr A.R., 305
Cambridge University, 42, 74
Camrose, Lord, 233
Canada, 8, 12, 22, 28, 62, 109, 168, 221, 233, 237, 280
   and imperial cooperation, 67-8, 114-19
   wheat, 130-1, 136, 140-2, 143-4, 146-7, 148, 149-51, 233
canned fruits, 42, 48, 109
Carlisle, 16-17
Carnegie, Andrew, 211
Carr, E.H., 223, 227
   views on internationalism, 227
Carter, C.B., 109n
Casey, R.G., xi, xiii-iv, 99, 102, 105, 221, 234, 236, 264, 298, 303
cattle, vaccination, 302
Central Statistical Board (US), 212
Chadwick, Sir David, 69-70, 113
Chadwick, Edward, *Report*, [1842], 17

Chaffey, George, 22-3
Chaffey, W.B., 22-3, 32, 39, 41, 47
Chamberlain, Austen, 172
Chamberlain, Joseph, 11-12, 30, 66
Chamberlain, Neville, 66, 101, 105, 113, 137, 138, 185, 188, 197, 198, 199, 226
Chambers of Commerce, 53
Chapman, Sir Sidney, 85
Chick, Dr Henriette, 154
Childs, Marquis, 234
Chile, 168
China, 136, 259
Churchill, Winston, 39, 66, 172, 222, 226, 255, 264
Clark, W.C., 255
Clark, W. Noble, 298
Clayton, Will, 292
Clifford, W.H., 68
closer settlement, 22, 23, 34
Clunies Ross, Ian, 303
Clynes, J.R., 52
coal, 62, 77
Coastal Farmers' Cooperative of NSW, 41
Cold War, 296, 299-301
Cole, G.D.H., 52
Coleman, P.E., 97
Collins, J.R., 97
Colonial Development Fund, 114
Colonial Office (UK), 46, 74, 76
colonial territories, 117, 136, 183, 184
   development of, 72, 228
   'open door' trade policy, 182-3, 191, 197
   proposed international commission for financial and technical assistance, 228
Combined Food Board, 291
Committee of Imperial Defence (UK), 221, 224
Committee of the Privy Council for Scientific and Industrial Research, (UK) 66
commodities 180, 271, 283, 285, 287, 295
Communism, 184, 190, 208, 300, 302
Condliffe, J.B., 179
Conservative Party (UK), 41, 46, 51, 65, 66, 69

Conservative Research Department, 101
    policy on wheat, 137
consumption/consumptionism, 173, 177,
    179-80, 182, 184, 198, 219, 235, 294
Contemporary Review, 18
Cook, Hume, 95-6
Cook, Sir Joseph, 46
Coolidge, US President Calvin, 234
Coombs, H.C., 277, 280, 288, 293, 298,
    304-5
    'positive approach', 277-8, 293
Cooper, Sir James, 54-6, 68
Copland, D.B., 89
Coulson, J.E., 277
Council for Scientific and Industrial
    Research (Aus.), 61, 75-6, 90, 94-5
    animal health, 80
    entomology, 80, 84
    pasture research, 82-4
Country Party (Aus), 43-4
Cranborne, Lord, 203
Crawford, J.G., 293
Cripps, Stafford, 257
Crompton Wood, Brooks, 65
Crowley, Leo, 250
Cummings, H.R., 166
Curtin, John, 244, 261, 264, 267
Curtis, Lionel, 13, 47
Cutlack, F.W., 24, 48
Czechoslovakia, 134, 183, 185, 219, 302
Daily Express, 95-6, 100
Daily Mail, 36, 47
Daily Telegraph, 69
dairy industry
    Australia, 110
    UK, 153, 156-8
    US, 155-6
Dairy Produce Export Control Board
    (Aus), 68
Dale, H.E., 139
Dalton, John, 16
Daly, Senator J.J., 94, 96-7
Darmstadt, 18
Davis, Norman H., 186, 218, 220
Davson, Sir Edward, 106
Dawson, Geoffrey, 12
Deakin, Alfred, 21-2

De Garis, C.J., 32-3
Denmark, 64, 109, 143, 155, 159, 168,
    170, 199, 281
Department of Agriculture (US), 212,
    217, 218-19, 235, 236, 237, 243, 251,
    255, 265, 274, 292, 299
    planning for international
    organisation
        draft plan of action, 258-61, 274
        twelve papers, 258-60, 270
Department of Commerce (Aus), 109n
Department of Health (US), 217, 218, 238
Department of Labor (US), 218
Department of Scientific and Industrial
    Research (UK), 66, 74, 79
Department of State (US), 213, 217, 218,
    231, 238, 243, 253, 257, 258, 259-60,
    263, 274, 276, 287
    dysfunction, 248-9
    postwar planning, 245-7, 249-52, 267,
    272
Department of Trade and Customs (Aus),
    109n
development, international sponsorship
    of, 300
Development and Migration Commission
    (Aus), 61, 75, 90, 94
Devonshire Royal Commission on
    Scientific Instruction (UK), 21
Docker, Dudley, 102
Dodd, Norris E., 272, 292, 298-9, 301
Douglas, Justice William O., 257
Dried Fruits Export Control Board (Aus),
    xiii, 37, 54-7, 96, 99
    London Agency, 55-6
dried fruits industry, 13, 15, 27, 76
    Australia, 27-9, 37-42
    California, 29, 31, 35, 38
    Mediterranean countries, 31
    McDougall memoranda
        'Notes on the Dried Fruits
        Industry in Australia', 30-2
        'A British California on the
        Murray', 33-35
        'A California within the Empire',
        37-8
    marketing, 15, 28-9, 35

organisation, 31, 33, 37, 38, 40, 54-6
prices, 28-9, 35, 38, 40
tariff preference for, 31-6, 37-42, 45-8
  in return for ex-service settlement, 33-6, 37-40
Duckham, Sir Arthur, 102, 105
Duckham, A.N. (Jim), 81
Duffy, M.B., 109n
'dust bowl' drought, North America, 129, 150-1
Dyason, E.C., 89
'economic appeasement', 175, 185-199, 201, 208, 221
  origin of term, 189-90
economic consultation and planning, 101-6
Economic Consultative Committee (League of Nations) [1928-29], 86, 95, 178
  commercial and economic policies, 86
  Committee of Agricultural Experts, 134-7
'economic council of the United Nations' (proposed), 256
Economic Defense Board (US), 212
economic depressions, 76, 188
  prevention and mitigation, 179-80, 200-1
Economic and Social Council of the United Nations, 175, 205, 208, 287, 291-2, 293, 294, 299, 303
economic nationalism, 159, 195
Economist, 171, 240, 245
Eden, Anthony, 185, 186, 192, 233
Edinburgh, 16
education, 16
  as force for change, 125-6, 165, 302
Education Acts (UK), 156-7
Egypt, 159
electrical industry, Australia, 87, 108
Elliot, Walter, 71-3, 80, 81, 101, 104, 105-6, 113, 150-60, 162
  and corporatist structure of economy, 157
  and orderly marketing, 156-8
Emergency Economic Committee for Europe, 291

Empire Free Trade, 95
Empire Marketing Board, 57, 61, 67, 70-5, 93, 156, 217, 228
  abolition, 113-20
  Agricultural Economics Committee, 71, 81-2
  'Buy British' Campaign, 113
  Empire Shopping Weeks, 73, 88
  Film Committee, 71
  finances, 70, 71, 112-14, 118
  proposed remodelling as imperial coordinating body, 112-14
  Publicity Committee, 71, 73
  publications, 72-3, 75
  Research Grants Committee, 71, 73-5, 77-8, 80, 81
Empire settlement
  see migration
Encyclopedia Britannica, 85
Enfield, R., 187
Ephrussi family, 131
Europe, British trade with, 62-3, 183
Evang, Dr K., 275-6
External Affairs Department, Canberra, 195, 220
Ezekiel, Mordecai, 163, 213
Fadden, Arthur, 233, 234
Fairbairn, George, 82
Fascism, 184, 190
Federal Farm Board (US), 146
Federal Reserve Board (US), 212
Federation of British Industries, 46-7, 102, 103-5, 107, 185
Finland, 178, 206
fisheries, 77, 284
Fitzpatrick, Dr A.S., 77, 97
Fleming, J.M., 179
flour milling, 166
Food and Agriculture Organization of the United Nations, xiii, 2, 163, 213, 228, 243, 261, 267, 270, 271-2, 283
  Australia Room, 303
  budget and staffing, 290, 299, 301
  Conference, 296-7
    Quebec [1945], 288-9
    on Urgent Food Problems [May 1946], 290-2

Copenhagen [September 1946], 292-3
commodity arrangements, 285, 286
constitutional ambiguity, 271, 279, 285-6, 289, 290-1
Council, 285
criticism of over-ambition, 291-2
development assistance and loans, 300-1
Directors-General, 271-2, 283, 286-7, 288-90, 296, 297-9
Secretariat, 283
statistical intelligence, 284
technical assistance, 299
views on industrialisation, 285
see also Hot Springs Conference, Interim Commission for Food and Agriculture Organization; World Food Council
food
buffer stocks, 220, 271, 292
conference (proposed) [1942], 259
correlation between productive capacity and requirements, 220, 242, 279, 284
distribution costs, 166, 191, 218, 238, 289, 297
increasing production, 191, 289
'international agricultural and food commission' (proposed) to control trade, 272
liberated areas, committees to assess needs, 269, 290
postwar planning, 247, 258-262
prices, 11, 160, 166, 295-6
relief supplies, 165, 180, 212, 218
shortages in Germany, WW1, 224
'technical commission' (proposed) to report on needs, 272
world crises [1946-47], 290-2, 297
Foreign Economic Administration (US), 250
Foreign Office (UK), 184-5, 186, 187, 193-4, 197, 263
forestry, 33, 77, 284, 286
'Four Freedoms', 217, 229, 256, 262, 268-9

France, 132-3, 134, 136, 154, 155, 159, 168, 170, 181-2, 183, 190, 193-4, 197, 206, 222, 241
fall of France, 217, 223, 226
Frankfurter, Justice Felix, 231, 257
'Freedom from Fear', 229-30, 262, 279
'Freedom from Want', 229-30, 242, 252, 262, 269, 279, 288
free trade, 10, 64
and League of Nations, 86
French, H.L., 139
Fruit Growers, Australian National Conference [1922], 40-1
Fruit Research Station, East Malling, Kent, 80-1
Funk, Casimir, 154
Garran, Sir Robert, 48
General Elections, (UK) [1923, 1924], 50-3, 66
George, Henry, 19
Gepp, H.W., 76-7, 78, 96
Germany, 63, 132-3, 136,176,180-91, 196, 198-9, 208, 219, 220, 222, 241, 244, 246, 254
autarky, 180-1, 197
former colonies, 181, 182-5, 193
industrialisation, 11
McDougall's proposed treaty of cooperation with, 182-3, 190
proposed access to raw materials in Central Africa, 185
rearmament, 180, 181
trade with, 182-5, 187-8, 190-6
UK policy, 184-5
US policy, 185-6
Weimar Republic, 228
Giblin, L.F., 89
Gilmour, Sir John, 139
Goering, Hermann,185, 193
Gordon, T.S., 102
Gordon, W.A., 114
Gore-Booth, P.H., 277
Gospel of Wealth, 211
Gough, A.E., 56
Grady, Dr Henry F., 177, 255

Great Britain
    see United Kingdom
'Great Depression'
    [1870s-90s], 11
    [1930s], 180
Greenwood, Arthur, 233
'Grow more wheat' campaign (Aus), 140
Guiana, 74
Gullett, H.S., 109
Gunther, John, 181
Haberler, Gottfried, 179
Halifax, Lord, 226, 236, 238-9, 273
'Hall Memorandum', 175, 188-9, 193
Hall, Professor Noel, 175, 188, 198, 202-3, 217
    memorandum on inquiry into standards of living, 202-3
Hambidge, Gove, 282
Hamilton, Sir Horace, 87
Hancock, W. K., 9
Hankey, Lord, 224-5
Hansen, Alvin, 248, 255
Hasluck, Paul, 305
Hatton, R.G., 81
Hawkins, Harry, 236, 238, 252, 253, 255, 259-61
'Heidelberg School', 9
Henderson, Sir Nevile, 192
Heysen, Hans, 9
Hilgerdt, Folke, 179
Hirschfeld, H.M., 178
Hoare, Sir Samuel, 225
Holloway, E.J., 91
Hood, John, 221
Hoover, Herbert, 215
Hopkins, F. Gowland, 154, 156
Hopkins, Harry, 231, 235, 252, 267
Horne, Sir Robert, 61-2, 101-2, 113
Hot Springs Conference [1943], 243, 245, 266, 269-70, 271, 272-9, 281, 288
    agenda, 274
    atmosphere, 275-6
    Australian delegation, 271, 277-8
    British delegation, 276-7
    resolutions, 278-9, 285
    US delegation, 276, 277, 278
Howie, H.D., 38

Hudson, R.S., 280
Hughes, W.M., 13, 34, 37-9, 40, 43-4, 91, 124
Hull, Cordell, 192, 220, 231, 248-50, 252, 255, 258, 261
    trade policy, 111-12, 177, 186, 191, 194, 195, 197, 202, 219, 235
Hungary, 131, 148, 206
'Hungry England' inquiry, 155
Huxley, Elspeth, *nee* Grant, 81, 85, 114, 160, 163, 184, 189, 305n
Huxley, Gervas, 81, 85
Hyland, A.E., 56, 96
Ickes, Harold, 231
immigration
    see migration
Imperial Agricultural Bureaux, 81, 113-14, 117-18, 217
Imperial Bureau of Entomology, 74
Imperial Economic Committee, 61, 67-70, 113, 115, 117, 142
imperial economic cooperation, 2, 61, 112-20
    Skelton Committee on, 114-120
Imperial Conference
    [1923], 8, 37, 49, 51, 98
    [1926], 8, 69, 75
    [1930], 8, 98-9, 112, 118, 137, 139, 141, 142
    [1937], 175, 195-6
Imperial Economic Conference
    [1923], 47-50, 66-7
    Ottawa, [1932], 98, 107-12
imperial preference, 42, 89, 187, 193, 272-3, 275
    emotional aspect, 111
    US hostility, 111-12, 263
    see also under dried fruits
imperial trade, 7, 8, 9
    conflicting interests, 10, 22, 109-11, 119
    expanded grouping with tariff protection (proposed), 159
Imperial War Conference [1917], 13, 50
'imperial visionaries'
    see under British Empire
India, 8, 63, 117, 144, 183, 244, 297

individuals, rights and duties of in relationship to the State, 227-8
industrialisation, 10-11, 235, 285, 293, 296
    development of, 300-1
    rationalisation, 61, 87-8, 98, 108, 191
Inside Europe, 181
Institute of Science and Industry (Aus.), 75
Interim Commission for Food and Agriculture Organization, 271, 279, 280-6, 287, 288
    Committees A and B, 282-3
    Committee C, 271, 282-3, 284, 286
    Advisory Panels, 283, 284
    Reviewing Panel, 271, 284-5
international agencies, demarcation of responsibilities, 301-2
International Bank for Reconstruction and Development, 288, 300, 302
International Emergency Food Council, 271, 291-2, 297
International Institute of Agriculture, 1, 134, 160-1, 163, 165, 167, 170-1
    Committee on Agricultural Economics, 134
    relationship with League of Nations, 134
    statistics, 117, 134
International Labour Organization/Office, 153, 165-6, 170-2, 183, 191, 193, 207, 208, 214, 231, 240
International Monetary Fund, 288
International Students' Assembly (US), 252-3
International Trade Organization, 292, 293, 300
International Wheat Council, 243-4, 280
Irish Free State, 8, 68, 178, 244
irrigation, 22-3
Iraq, 159
Italy, 132-3, 134, 136, 168, 170, 176, 182, 183, 184, 206, 223, 244
Jackson, Justice Robert H., 257
Japan, 111, 176, 177, 186, 216, 221, 241, 244

Jebb, Gladwyn, 184-5, 187-8, 189, 193-4, 199, 294
Jones, Jesse, 250
Jones, Thomas, 188
Julius, George, 77-8, 94
Kant, Immanuel, 19
Kelly, W.S., 82
Kennedy, Joseph, 231
Kerr, Philip (Lord Lothian), 13
Keynes, J.M., 178, 180, 213, 271, 273, 276
King, W.L. Mackenzie, 115, 195
Knox, R.W., 109n
de Kock, Dr H.M., 178
Koopmans, Tjalling, 179
Labor Governments (Aus.), 91, 93-9, 277-8, 304
labour conditions, 52-3
Labour Governments (UK)
    [1924], 37, 41, 51-3
    [1929-31], 107, 113, 137, 141
Labour Party, (UK) 51-3, 66, 71, 225, 240
van Langenhove, Ferdinand, 178
Larke, Sir William, 86
Laski, Harold, 225-6
Latham, J.G., 146-8
Latin America, 228, 237, 244, 259, 301
Law, Bonar, 13
Law, Richard, 277
Lawson, H.S.W., 38, 39, 40, 47
The Leader, 34, 35
League of Nations, 1, 2, 43, 95, 103, 123-7, 137, 145, 153, 160, 167, 168, 182, 185, 186, 197, 206, 223, 228, 240, 281, 290, 304
    and Abyssinian crisis, 169, 176
    and agriculture, 134-5
    agent of political change, 127
    agent of social and economic reform, 2, 124, 169, 175, 200, 205, 214
    Assembly [1935], 165
    Assembly [1929], 87
    budget and staff, 207
    building, 176
    Council, 1, 171, 176, 177, 201, 204, 206-7
    Covenant
        Article XIX, 123

Article XXIII, 124
and Versailles Treaty, 176, 182
economic and social agencies, 124-5, 176-80, 201
see also 'Bruce Report'
Final Assembly [1939], 205, 206-7
International Economic Conference [1927], 86, 133
Mixed Committee on Nutrition, 124, 153, 171-3, 200, 204
nutrition debate, 168-171, 231
political weakness, 124-5, 176, 199-200
publications on nutrition, 171-3
restructure of economic and humanitarian work
see 'Bruce Report'
statistics, 86-7, 124, 134-5
tariff policies, 86-7, 124
League of Nations Economic Section and Committee, 175, 177-9, 199-202, 203, 218, 255
League of Nations Financial Section and Committee, 177, 201, 203, 213
League of Nations Health Organization, 125-6, 154, 163, 166, 170, 204
Leeper, R.W.A. (Rex), 187-8
Leith Ross, Sir Frederick, 178, 192, 193, 203, 255
de Lemus, Professor A. Flores, 178
Lend–Lease, 229, 250, 272
Leydon, John, 178
Liberal Party (UK), 51, 53, 66, 71, 107
Lidgett, Dr John Scott, 18
Lindsay, H.A.F., 114
Lister, Joseph, 17
Liverpool, 16
Lloyd, E.M.H., 72, 106, 117
Lloyd George, David, 13
Lloyd, Godfrey, 172
Lloyd-Greame, Sir Philip, 49-50
Lockyer, Norman J., 21
London County Council, 17-18
London School of Economics, 290
Long, Breckenridge, 248
Louwes, S.L., 289, 290, 298
Loveday, Alexander, 192

Lubbock, David, 164
Lubin, Isador, 213
Lyons, Joseph, 9, 93, 99, 146-8
raises nutrition at the Vatican, 167-8
Macarthur, General Douglas, 264
McCall's Magazine, 156
McCarthy, Edwin, 244
McCollum, Elmer, 155-6
McCrosty, Henry, 85
MacDonald, Ramsay, 148, 192
McDougall, Alexander (grandfather of F.L.), 16, 17
McDougall, Alexander (uncle), 17
McDougall, Arthur, 17
McDougall, Catherine (Kit), 18
McDougall, Elisabeth, xi, xiv, 25, 85, 100
McDougall, Ellen, 18
McDougall, F.L., vii, xi, 1, 2, 7, 9, 10, 26, 39, 56, 109n, 161, 168, 170, 175, 176-7, 192, 206-8, 232, 245, 257, 289, 295, 297, 302, 305
and agricultural efficiency/rationalisation, 81, 129, 135-6, 138-9, 143, 153
at FAO, 271-2, 287, 289, 290-2, 297-9, 301-2, 303
at International Institute of Agriculture, 129, 134, 143-4, 160-1, 165
campaign for tariff preference on dried fruits, 37-40, 57
childhood and youth, 7, 10, 11, 18-20
correspondence with Bruce [1920s], 57-9, 85, 89, 94
correspondence with economists, 55-7, 178-9
economic planning and consultation, 101-6
employment and remuneration, 54-5, 65, 77, 93, 95-7, 98-100, 101-3
final years, 303
fruitgrowing, 25-6
Germany and 'economic appeasement', 180-91
Hot Springs conference, 274, 276, 278, 280

liaison for CSIR, 75, 77-84, 94, 95, 96-7, 98-100, 101-3
liaison for DMC, 75-7, 95, 96-7
links poor diets with restrictive agricultural policies, 127, 153, 160-3
marriage and personal life, 25, 84-5, 160, 305
memoranda, 85, 238-9
   see also McDougall, F.L., principal memoranda and published papers
negotiations in preparation for Ottawa Conference, 93, 107-8, 193
nutrition campaign at League of Nations, 161-173, 178
on Economic Consultative Committee, 133-7
on Empire Marketing Board, 70-5, 78, 80, 81-2, 94, 96-7, 112-20, 280
on Expert Panel on Agricultural Problems (ECC), 129, 134-6
on Fruit Delegation to the United Kingdom [1922-23], 39-42, 45-7, 54-5
on Imperial Economic Committee, 68-70, 96, 280
on Interim Commission for Food and Agriculture Organization, 280-6
on League of Nations Economic Committee, 175, 177-80, 199-203
on League of Nations Mixed Committee on Nutrition, 171-2, 178
opposes restriction of production, 159
opposition to in Whitehall, 203, 264, 280-1, 287-8
optimism, 20, 125, 151, 160, 302, 305
partnership with Bruce, 37, 47-50, 52, 53, 57-9, 159, 168
political warfare, 217, 221-3, 240-2
recognition, 303
Secretary, Dried Fruits Export Control Board, 37, 55-6
Sheltered Markets, 61-5, 85, 101
skills of persuasion, 46, 52
spokesman for world beyond Europe, 86-7, 129, 133-4, 137
statistics, 85
views on Anglo-American cooperation, 240

Australian development, 82
Australian tariff, 82, 85-90, 95
consumption, consumptionism, 175
financial economy, 58
reviving British Empire trade with Europe, 183
socialism, 18-20
visits to USA,
   [1938], 211, 213, 217-19
   [1941], 211-12, 217, 233-39
   [1942], 213, 238, 243-67
war service, 7, 25-6, 29-30, 123
wheat negotiations, 133-7, 143-9, 149-51, 159
writing, 46, 64, 84, 85, 238
McDougall, F.L.: principal memoranda and published papers
   'The Agricultural and the Health Problems', 2, 153, 163-6
   'Draft Memorandum for Freedom from Want of Food', 3, 243, 261-2, 266, 267, 269-70
   'Economic Appeasement', 175, 189-94, 200
   'The Dominance of Wheat', 142-3
   The Growing Dependence of British Industry on Empire Markets, 75, 95n, 96
   'The Keynote of the Next Assembly', 199-202, 217
   'Notes on the Re-statement of our Aims', 217, 226-9
   'Progress in the War of Ideas', 217, 240-3, 244, 253-7, 261, 267
   'The Challenge to Western Civilization', 272, 299-301
McDougall, John (son), 25
McDougall, Joyce (*nee* Cutlack), 25, 100
McDougall Memorial Lecture, FAO, 2, 303
McDougall, Norman, 25, 30, 41, 56
McDougall, Sir John, 7, 17-18
McDougall, Sir Robert, 17
McDougall's Powder, 17
McDougall's Self-Raising Flour, 16, 17
The McDougall Memoranda, 2, 261

McGonigle, Dr George, 154, 155, 158, 202
Macgregor, L.R., 98
McKay, S., 109n
Mackinder, Sir Halford, 68-70
MacLeish, Archibald, 252, 258
Macmillan, Harold, 105-6, 158
MacMurray, J.V.A., 150
Makins, R.M., 203
malnutrition, 169, 201, 279
Manchester, 16
Manchester Chamber of Commerce, 86
Manchester Guardian, 114, 240
Manchester Guardian Commercial, 85-6
Manchester Literary and Philosophical Society, 16
Manchester University, 17
Manchuria, 176, 186, 216
margarine, 162
market intelligence, 117
marketing, 62-3, 299
'marriage of health and agriculture', 1, 2, 127, 136, 139, 153, 169, 208
    see also nutrition initiative
Marshall Plan, 299
Maud, J.P.R., 276
May, Sir George, report on economy in government, 112
May, Stacy, 213, 218
Meade, J.E., 178
Means, Gardiner C., 213
Meares, C.E.D., 41, 47, 48
meat, 69, 109-11, 162, 164
Medical Research Council (UK), 74, 155
Mellanby, Edward, 155
Melville, L.G., 109n
Menzies, R.G., 9, 99, 206, 226, 233
Methodism, 16, 18
mercantilism, 11, 75
migration, 197
    within British Empire, 13, 34, 38-9, 49-50, 64, 76
        in exchange for tariff preference, 34-6, 37, 40
Milbank, Albert G., 215
Milbank Memorial Fund, 214-15
Mildura, 45
Mildura Sun, 34, 36

milk, 139, 155-6, 162, 164-6, 225
    importance to UK agricultural economy, 153, 156-8
    school milk, 156-8, 172
    studies of nutritional value, 156-7
Milk Act 1934 (UK) 157
Milk Marketing Board (UK), 156-8
Miller, Francis, 254
Miller, Helen Hill, 254
Milner, Lord, 12, 49, 61
Ministry of Agriculture and Fisheries (UK), 74, 82, 139, 156, 164, 172, 187, 225, 240, 293
Ministry of Food (UK), 172, 276
Ministry of Health (UK), 155, 156, 225
Ministry of Information (UK), 254, 256
Mixed Committee on Nutrition
    see under League of Nations
The Modern Corporation and Private Property, 213
Molony, Parker, 97, 98
Moore, A.C., 109n
Moore, W. Harrison, 9
Morato, O., 177
Morgenthau, Henry, 146, 147
Morning Post, 47
Morrison, Herbert, 233
Morrison, W.S., 225
Mosley, Sir Oswald, 104
Mudaliar, Ramaswami, 178
Munich Agreement, 198-9
Murphy, J.F., 288
Murray Pioneer, 25, 29, 32-4, 36, 40, 41
Murray River, 15, 22, 24, 62
Mussolini, Benito, 133, 184, 199, 220
Mutual Aid Agreement, Article VII, 272, 275, 277
Myrdal, Gunnar, 178
Nash, Walter, 266
Natal, 15, 18
National Government (UK)
    [1931], 107, 113
    [1940-5], 225-6
National Farmers' Union (UK), 82, 156,
National Federation of Iron and Steel Manufacturers (UK), 86

National Institute for Research into Dairying (UK), 74
National Milk Publicity Council (UK), 156
National Policy Committee (US), 254
Nationalist Party, Australia, 43-4
Nature, 21, 171
Nazism, 180, 185, 189, 193, 198-9, 220, 226
Netherlands, 178, 192
Neurath, Baron Konstantin von, 186
New Deal (US), 146, 208, 211-13, 215-16, 234, 263
Newfoundland, 8, 68, 117, 244
Newman, George, 156
New York Times, 255
New Zealand, 8, 28, 63, 64, 68, 109, 117, 136, 162, 167, 168, 222, 237, 266
Nordau, Max, 19
Norman, Sir Montagu, 102, 192
Northcliffe, Lord, 36
Norway, 155, 199
Nurske, Ragnar, 179
nutrition, 162, 198, 214-15, 236, 261, 264
    and 'economic appeasement', 191, 193, 197, 214
    in Asia, 164
    international institute of food (proposed) 217-19, 220
    memorandum of agreement between US and UK on nutrition, draft [1941], 237-8
    national committees, 126-7, 173, 218
    proposed conferences on [1941 and 1942], 237, 239, 269
    relationship to agricultural policies, 160-1, 218
    relationship to national economy, 166, 169-70
    relationship to prices and incomes, 154-5, 160-1
    resolution at League Assembly, 168-71
    scientific knowledge, 153-4, 162, 218
    war aims regarding, 217, 256
'nutrition initiative', 2, 125, 153, 168-71, 176-7, 179

Oakley, R. McK, 48
Observer, 272
Office of Strategic Services (US), 252, 254, 255
Office of War Information (US), 247, 252, 254, 255, 256, 258, 260
Officer, F.K., 195
Olsen, Karl, 303, 304
Ormsby-Gore, William, 71-2, 105
Orr, John Boyd, 2, 61, 72, 80, 82-4, 117, 123, 126, 153, 162, 166, 170, 173, 175, 198, 217, 218, 220, 224-5, 233, 236, 258, 259-61, 276-80, 296, 297, 299
    Director-General of FAO, 271-2, 287, 288-96, 297-8, 299, 303
    Food, Health and Income, 158, 225
    and milk consumption, 156-8
    views on food supply and planning, 261
Opie, Redvers, 255
Orwin, C.S., 82
Osbourne, H.W., 109n
Ottawa Trade Agreements [1932], 2, 93, 157, 159, 160, 182-3, 184-5, 188, 193
    bilateral discussions to prepare (Aus/UK) 107-8
    US hostility, 111-12, 245, 273
    meat quotas, 109-10, 159
Owens College, Manchester, 17
Pacific War, 288
Pacific War Council, 266, 273
Page, Earle, 44-5, 99
Palestine, 74
Parran, Dr Thomas, 218, 236-8, 258, 259-60, 265
pasture improvement
    see Council for Scientific and Industrial Research
Pasvolsky, Leo, 251
Paterson, Laing & Bruce, 42-4
'Peace Ballot' [1935], 181
peace settlement, planning for, 221-2
    and living standards, 222, 223
    and security, 227
    and social justice, 226-9
peasant farmers, 132, 143, 166

Pearl Harbor, 215, 217, 235, 241, 244, 265, 269
Pearson, Lester (Mike), 187, 298, 305
  on Interim Commission, 281, 282, 284, 287
Penrose, E.F., 231, 269, 273-4
Percy, Lord Eustace, 101, 106
Perkins, A.J., 31
Perkins, Milo, 250, 252, 259
Pethick-Lawrence, F.W., 52
Phillips, Sir Frederick, 255
Polak, J., 179
Poland, 131, 168
political warfare, 217, 221-2, 241-2, 249, 257, 260
population sustainability, 301-2
  see also 'Australia Unlimited'
Portugal, 183
postwar planning, 215, 238, 239-40, 244-8, 249, 250-2, 265
poverty, 221-2, 228
Poverty and Health, 202
Premiers' Conference [1921-22], 37-42
price stabilisation, 64, 239, 295
Prime Minister's Department (Aus), 95, 221
Progessivism, 21-22, 72, 211, 265
psychological warfare and weapons, 254-6, 258
Punch, 18
purchasing power, 295-6
ragged schools, 17
Rajchmann, Dr Ludwik, 166
Rank, James V., 141
Ratcliffe, Francis, 80
raw materials, 139, 185, 197, 198, 199, 200, 202, 220, 245
Reading, Claude, 48
Reconstruction Finance Corporation (US), 214, 250
Renmark, 7, 15-16, 18, 20, 22-7, 29, 33, 45, 56, 299
Renmark Fruit Packers Union, 24
Renmark Irrigation Trust, 23, 33
Renmark Pioneer, 25, 33
Renmark Returned Soldiers' Association, 30

Renmark Settlers' Association, 25
Rhodesia, 8, 117
Ribbentrop, Joachim von, 186
Richardson, A.E.V., 109n
Riddle, E.C., 109n
Riefler, Winfield, 212-13, 236, 237, 241, 248, 252, 255
Rivett, A.C.D., 61, 77-9, 81, 94-5, 96, 98-9, 105, 159, 181, 184, 233
  views on EMB, 78, 84, 113, 119-20
Robbins, Lionel, 276
Robert Brookings Graduate School (US), 212-13, 215
Robertson, D.H., 178, 180
Robertson, Norman, 163
Robertson, T. Brailsford, 83-4
Rockefeller Foundation, 125, 176, 207, 213, 218
  International Health Board, 214
Rodgers, A.S., 41-2, 46
Romania, 131, 148
Roosevelt, Mrs Eleanor, 231, 236, 243, 245, 247, 252-3, 261, 266, 267
Roosevelt, President Franklin D., 146, 165, 186, 199, 211, 213, 214, 215, 219, 220, 229-30, 231-2, 233, 234, 235, 243, 247, 248-50, 252, 255, 256, 257, 260-1, 263, 264, 265-6, 287
  inscrutablility, 267-70, 273-4, 275
Roosevelt, President Theodore, 268
Ross, A.E., 36-41
Rothamsted Experimental Station on Soil Science, 81
Round Table, 13, 47
Rowett Research Institute, Aberdeen, 71, 80, 84, 156, 158, 218
Rowntree, Seebohm, 198
Royal Society, 16, 74
Royal Statistical Society, 85
Runciman, Walter, 150, 188, 192, 194
rural populations, economic and social status of, 299
  education, 16, 300, 301, 302
Russell, Sir John, 81
Russia
  see Soviet Union
Ryrie, Sir Granville, 96

Ryti, R., 178
St Helena, 15, 18
Salter, Sir Arthur, 106
Sanderson, John, 48
Scandinavia, 241
Schacht, Dr Hjalmar, 193
Schuster, Sir George, 187, 233
science, 211
    education, 16-17
    imperial cooperation in, 61, 75-81, 114-18
Science Guild, 66
scientists and social reform, 126
Scouler, J.S., 56
Scullin, J.H., 93-5, 98-9, 140
Sheldon, Sir Mark, 68, 69
Sheltered Markets: A Study of Empire Trade, 61-5, 85
Shepherd, M.L., 54
Shudo, Y., 178
Simon, E.D., 104
Sinclair, Archibald, 71
Singapore, 125
Singapore Food Conference [1946], 293
Skelton, O.D., 115, 163
Skelton Committee on Economic Cooperation, 114-19
Smith, Angus, 17
Smuts, Jan, 113, 222
Snowden, Philip, 52, 107
Socialism, 19-20
Social Darwinism, 17, 20-1, 66, 211
social justice, 17-18, 223, 228-30, 241, 268-9
    and security against left and right extremes, 215
South Africa, 8, 15, 117, 136, 178, 233
South Australian Government, 34, 47, 140
Soviet Union, 136, 155, 206-7, 240, 241, 257, 271
    agricultural production, 64
    and FAO, 276, 288
    wheat, 131-2, 136, 142, 143-4, 147, 148
Spain, 178, 206
Spectator, 171

Stalin, Josef, 268
standards of living, 75, 129, 135-6, 179, 190, 197, 200-1, 222, 228, 277, 281, 293
    increase as means to prevent war, 177, 195-6, 232-3, 235, 244, 262, 300
    to revive trade, 190, 230, 277, 300
    proposed inquiry by League of Nations, 201-3
Stanhope, Lord, 187
Stapledon, R.G., 81
State Governments (Aus.), 146-8
The Statist, 171
statistics, provision of and coordination, 116-17
Statute of Westminster, 8
Stevens, B.S., 148n
Stevens, L.E., 109n
Sterling Area, 273
Stevenson, Adlai, 294
Stewart, Sir Frederick, 167
Stimson, Henry, 231
Stirling, Alfred, 186, 195, 206, 221, 222-3, 226
Stoppani, Paul, 172, 192, 202
Strang, William, 193-4
Stuart Smith, A.W., 77, 97, 99
Sumner, William Graham, 20-1
surplus production, fears of, 257, 265, 291, 294, 299
Survey of International Affairs [1937], 189
Swanson, W.W., 141
Sweden, 155, 168
Sydney Morning Herald, 95
Tallents, Sir Stephen, 72, 113, 118
tariffs, 10
    Australian, 12, 124
        criticism, 61, 85-6, 87-90
        Brigden Report, 89-90
        value to UK industry, 49, 62, 86
        imperial preference, 11-13, 31-6, 62, 64, 65
        see also dried fruits industry
    UK, 66, 194
        free trade policy, 10-12, 15
        imperial preference, 240
        'McKenna Duties', 13, 32, 51-3

proposals for broad consulation regarding, 93, 102, 103-5
protection, support for, 107
preference increases [1923], 50-52
proposed modifications to revive triangular trade with Europe, 188, 193
reform campaign [1903], 11-12, 66
revenue tariff, [1931], 107
Tariff Board, proposed, 104-5
US, 111-12, 186
Tariff Board (Aus), 88, 111
Tariff Commission (US), 177
tariff preference, 62, 64, 65
tariffs, consideration at League of Nations, 160-1
at International Institute of Agriculture, 200-2
Taylor, H.S., 25, 29, 32, 33, 35
Theiler, Sir Arnold, 80, 84
Thomas, J.H., 52, 71, 113, 114
Thomas, W.C.F., 56
Tillyard, Robin J., 80
Time Magazine, 213
The Times, 46, 47, 69, 100, 106, 108, 159-60, 162, 171, 189, 202, 240, 266
Times Trade and Engineering Supplement, 64, 86
Tinbergen, Jan, 178
Tizard, Sir Henry, 79
Tolley, H.R., 238, 247-8, 258, 259, 265, 269, 275, 278, 282, 285, 298-9
Toryism and the Twentieth Century, 72
totalitarianism, 208, 227-8
Tout, F.H., 109n
trade and trade policy, 45, 268, 271, 285
Australia, 9, 193-4, 196
'trade diversion' [1930s], 111
Canada, 12
British Empire, 10-13, 62, 182-3, 186, 195
triangular trade with Europe, 180-1, 183, 187, 188, 191, 246
potential postwar US/UK trade war, 223, 229
recession [late 1930s], 185

UK, 10, 62-3, 107, 182-3, 198, 272-3, 276
US, 191, 198, 219, 235, 273
trade unions, 52-3, 64
Trades Union Council (UK), 103
Treasury (UK), 71, 89, 112, 157, 240
Tripartite Declaration/Agreement, 188, 190, 200
Troutbeck, J.M., 194
'Truman Doctrine', 300
Truman, President Harry S., 265-6, 300
Twentyman, Edward, 282, 286, 287-8
unemployment, 11, 36, 50, 62, 76
United Australia Party, 93
United Kingdom, 134, 144, 155, 164-5, 168, 170, 190, 197, 206, 237, 239, 244, 278
and FAO, 271, 292-3, 295
anxiety concerning postwar balance of trade with US, 214, 240
cooperation with US, 217
economy, 107
markets for industry, 62-3, 117
'moral leadership', 181-2
wheat, 137-9, 141-2, 150
see also: Anglo-American cooperation, trade policy; unemployment; tariff reform
'United Nations' (Allies), 256
economic council (proposed), 254, 259
'technical expert commission on food' (proposed), 259, 261-2
US usage of term, 245
United Nations Educational, Scientific and Cultural Organization, 301-2
United Nations Organization, 290, 291, 292, 301
Orr's view as world government, 272, 292
specialised agencies, cooperation between, 301
United Nations Relief and Rehabilitation Agency, 291
United States, 2, 63, 117, 155, 165, 169, 171, 180, 190, 198, 237, 239, 244
and FAO, 271, 288, 292, 295

anti-imperialism, 245, 263-4
changing approaches to international problems, 215-16
Congress, 240, 265, 270, 278, 288
cooperation with UK, 217
farm prices, 269-70
international cooperation, 197
industrialisation, 11
isolationism,182, 197, 215, 135, 245-6, 268
resources for postwar reconstruction, 215, 239-40
trade, 62-3, 292
wheat, 130-1, 141, 144, 146-8, 149, 150-1, 233
Uruguay, 177
US-British Agreement on nutrition, proposed, 217, 237-8
Vanier, Col. G.P., 115-18
Van Zeeland, Paul, 190, 200
Van Zeeland Report [1938], 196, 200
Veblen, Thorstein, 211
Venn, J.A., 82
Versailles, Treaty, 189, 221
    and colonies, 181, 183-4
    calls for revision, 184, 186, 187
    resentment of, 182
vitamins, 154, 155
'voluntary preference', 61, 70-3
Waley, S.D., 188
Wallace, Henry, 163, 211-12, 215, 217, 218, 233, 234, 235-6, 237-9, 243, 244, 246, 247-8, 249-50, 252, 254-7, 258-61, 263, 265, 266, 267, 287
    'Century of the Ccommon Man' speech, 244-5
    'ever-normal granary' idea, 220, 235, 247, 292
Walters, F.P., 203
war aims, 217, 219, 221, 223, 225-9, 269
War Cabinet, (UK), 223, 224, 233, 240
    sub-committee on war aims, 225
War Department (US), 274
Ward, Barbara, 254
Ware, Sir Fabian, 115
Warr, Earl de la, 164, 170, 173
Washington DC, 211, 234-6

angolophile and anti-British factions in, 263-4
Australian representation in, 220, 221, 264, 266, 275, 291
British representation in, 245, 255, 273, 280, 282, 287
Weaving, Mrs Annie, 155
Webb, Sidney, 52
Weekend Review, 155
Welles, Sumner, 199, 215, 236, 237-8, 243, 244, 245-6, 248-9, 250, 251, 255, 258, 260, 261, 263, 265, 268, 269, 276
Welsh Plant Breeding Station, Aberystwyth, 81, 116
Western Australia, 140
Western Civilization, importance of meeting needs of developing world, 293, 299-301
wheat, 129-51, 159, 181, 219, 266
    see also under Argentina, Australia, Canada, United Kingdom, United States
    acreage limitation, 136, 143, 146-7
    conferences, 143-9
        exporters, London [May 1931], 144
        Standing Committee [July 1931], 145
        principal exporters, Geneva, May [1933], 129
        in conjunction with Monetary and Economic Conference, London [1933], 145-9
    European views on production, 132-3, 143
    export quotas, 137, 142, 144, 147-151
    glut, 132, 145-6
    Imperial Conference [1930], 137
    inter-European preference arrangement (proposed), 143-4
    International Wheat Agreement, 159, 160, 233
    prices, 109, 120, 132-4, 135, 136, 140-1, 142, 143, 144, 149-50, 269-70
    millers, 137, 142
    tariffs, 130, 132-3, 136, 142

Wheat Advisory Committee, 129, 149-51, 163, 224, 233, 234, 237
   minimum pricing scheme, 149-50
Wheeler, Leslie, 259
White, Harry Dexter, 273
White House, 246, 252-3, 260, 267
Whiskard, Sir Geoffrey, 113
Wickard, Claude, 237-8, 258
Wickens, C.H., 89
Wilder, Dr Russell, 237
Willits, J.H., 207
Wilson, Sir Horace, 115, 192
Wilson, M.L., 236-8, 259
Wilson, Senator R.V., 48, 54-5
Wilson, Dr Roland, 233
Wilson, US President Woodrow, 215, 254, 268
Winant, Ambassador John, 217, 231-2, 233, 234, 239, 251, 269, 270, 276
wool, 111, 181
Woolton, Lord, 276
World Economic Conference, London, [1933], 119, 248
   see also wheat conference 1933
World Food Board, 271, 292-6
   Preparatory Commission, 271-2, 294-6, 300
World Food Council, 272, 296-7
World Health Organization, 178
World War I, 224, 231, 245
World War II, 173, 206, 234, 239, 263-4, 274, 288
   preparation for, 198-9
The Years of Crisis, 223
Yorkshire Post, 48
Young, Walter, 48
Yugoslavia, 131, 148, 206
Zimmern, Sir Alfred, 188

www.ingramcontent.com/pod-product-compliance
Lightning Source LLC
Chambersburg PA
CBHW060928170426
43192CB00031B/2860